Green Energy and Technology

More information about this series at http://www.springer.com/series/8059

Rocco Papa · Romano Fistola
Editors

Smart Energy in the Smart City

Urban Planning for a Sustainable Future

Springer

Editors
Rocco Papa
Department of Civil, Architectural and Environmental Engineering
University of Naples Federico II
Naples
Italy

Romano Fistola
Department of Engineering
University of Sannio
Benevento
Italy

ISSN 1865-3529 ISSN 1865-3537 (electronic)
Green Energy and Technology
ISBN 978-3-319-31155-5 ISBN 978-3-319-31157-9 (eBook)
DOI 10.1007/978-3-319-31157-9

Library of Congress Control Number: 2016934442

Printed on acid-free paper

This Springer imprint is published by Springer Nature
The registered company is Springer International Publishing AG Switzerland

Preface

This collection of papers is about the energy dimension of a smart city. Its goal is to mark a boundary around the concept of "city smartness", considered with regard to the energy issue and the town planning point of view. From another perspective, the aim of this collection of writings by the main Italian research groups in the field of urban sciences, is to define how the new concept of a smart city can successfully open a new understanding of urban systems and progress towards a new style of management for our metropolis.

If we have been able to see and participate in the smart city debate, one that has been spreading all over the world during the last 2 or 3 years, it is possible to argue that the main factors of this new concept regarding human settlements are energy and technology.

The technological dimension of the smart city movement is inherent to the city itself and represents the engine that moves the urban system in its spatial and temporal development. But today's main issue is energy. Technology without energy is simply useless. It is no exaggeration to say that energy is the main challenge for the future of our cities as well as for human beings. At the same time, cities are the places where this challenge must be played out first, because cities are the main wasters of energy on the earth.

The planning of a smart city will be greatly different from the canonic urban planning of our current cities. Furthermore the energy dimension has to constitute the first issue to be considered in a new planning process. The new urban planning has to consider energy as a starting point and a goal to achieve, at the same time. Technology must be considered as a part of the planning process from the beginning. Technology, in order to know its needs, to understand and to drive urban system towards new, sustainable, and balanced conditions, has to be "adopted" and not merely "added to" the city, as we tend to do today.

Italy is a country particularly exposed to energetic problems for three main reasons:

- The geographic location of the country determines a particular vulnerability to climate change and consequently the need for large amounts of energy;
- The country has no primary energy resources available (Italy imports from abroad more than 80 % of its energy requirement);
- Due to a public referendum, no nuclear plant is available on national territory.

This study has an explicit concern about a city's energy. Again, energy has to be considered inside the urban planning process as well as inoculated within the new idea of a future city. We need new methods, new processes, and new tools to manage the urban system in order to drive it towards a smart dimension. From this concepts arises the idea of this volume which is structured along three main issues: the relationship between energy and city (in its different dimensions), a methodological aspect of energy's contribution to the urban system management, with a special focus about ontological issues, a review of case studies which describes some practices, procedures, and tools of urban planning. At the end this essay could be useful to students of urban planning, town planners, and researchers interested in understanding where the city of the future will go and what the energy contribution to this evolution will be.

The editors wish to express their gratitude to Springer for its professional assistance, and in particular to Mr. Pierpaolo Riva who has supported this publication.

Rocco Papa
Romano Fistola

Contents

About the Editors

Rocco Papa has been Engineer, Full Professor of Land Use Planning at the Engineering Faculty—University of Naples Federico II since 1992. He has been Editor-in-Chief of the Scientific Journal TeMA—Land Use, Mobility and Environment since 2007. He was Director of the Department of Urban and Regional Planning of University of Naples Federico II, from 1999 to 2005. He was Chairman of the Urban Transformation Company Bagnolifutura S.p.A from 2006 to 2010. He was Vice-Mayor of the Municipality of Naples, from 2001 to 2006. He was City Councilor for Livability (appointed to Town Planning and Historical Centre) for the Naples Municipality, from 1997 to 2001. With the research activity carried out continuously since 1974, he has developed according to the following four main lines: the study of the interactions between urban and mobility systems; the management and governance of metropolitan areas; the safeguard of environmental quality in highly urbanized areas; the experimentation of new protocols for urban planning tools connected with the updating of techniques, methods and models of analyses, interpretation, planning and governance of territory. He has been scientific coordinator of many research projects: Research Unit of the RESIS Project 2004–2005 ~ Project of Research and Development for Seismology and Seismic Engineering; European Project 2004–2007 "Applied multi Risk Mapping of Natural Hazards for Impacts Assessments. Contract for "Specific Targeted Research or Innovation Project (STREP)—VI Framework Program"; Research Unit of the CIPE MURST Three-year (2000–2003); Project "The Protection of Seismic Risk: Vulnerability. Analysis and Requalification of Physical and Built Environment by Means of Innovating Techniques"; Research Unit of the National Priority Scientific Research Program 2002–2004 (Projects co-financed by MIUR) "The Safeguard Historical Cultural and Landscape Values in Italian Seismic Areas". He is the author of more than 100 publications.

Romano Fistola has been Architect, Associate Professor at the Department of Engineering of the University of Sannio since 2004. He was permanent researcher at the Institute of Planning and Land Management of the National Research Council of Italy (Naples) from 1996 to 2003 and member of the Scientific Committee. He was also a visiting researcher at the Center for Urban and Regional Development Studies (CURDS) of the University of Newcastle upon Tyne (UK), under the direction of Prof. J.B. Goddard. Since 2004 he has been teaching the courses "Town and Country Planning" and the course of "Management of the Urban Transformations" at the Department of Engineering of the University of Sannio. He was a coordinator and member of research teams in several research projects (national and international) in the field of urban and regional planning. The main scientific topics of his researches are as follows: New technologies and the city, smart cities, complex urban systems, GIS, urban risk, and sustainable cities. He was elected a member of the Board of the Italian Regional Science Association (AISRe) and is currently a member of the Italian National Institute of Urban Planning. Since 2007 he has been the editor of the Scientific Journal TeMA—Land Use, Mobility and Environment. He is author of more than 100 papers published in national and international books and journals.

City SmartNESS: the Energy Dimension of the Urban System

Rosaria Battarra, Romano Fistola and Rosa Anna La Rocca

Abstract This paper proposes a re-thought of the concept of urban smartness, particularly referring to the energy component. Recognizing that the new technologies, which are the most popular aspect of smartness, can play a fundamental role in the new approach, it has been suggested that we consider them in an adoptive way rather than in an adjunctive way, as it is commonly intended in the general sense of a smart city. According to this vision, in the first part of the paper, a new concept of smartness is proposed (SmartNESS: Smart New Energy Saving System). This concept is also related to the possibility of identifying some leading urban functions that can play a strategic role in improving urban smartness. In this sense, in the second part, tourism is considered as a drive function able to make cities more efficient and attractive if it will be integrated inside the urban governance process. The third part of the paper highlights how the rationalization and reduction of energy consumption is one of the essential fields to rely on in order to improve the smartness of a city. This part provides an overview of the most significant initiatives that are being developed on energy efficiency, and investigates some cases particularly innovative addressing the issue with an integrated and non-sectorial approach. Through the analyzed experiences, some possible intervention strategies to integrate the issues of energy efficiency in urban planning are suggested in the conclusive part of the paper.

R. Battarra (✉)
National Research Council, ISSM, Naples, Italy
e-mail: battarra@unina.it

R. Fistola
University of Benevento, DING, Benevento, Italy
e-mail: rfistola@unisannio.it

R.A. La Rocca
University of Naples Federico II, DICEA, Naples, Italy
e-mail: larocca@unina.it

R. Papa and R. Fistola (eds.), *Smart Energy in the Smart City*,
Green Energy and Technology, DOI 10.1007/978-3-319-31157-9_1

1 The City: a Complex, Dynamic, Energy-Intensive System

The debate about the smart city needs a theoretical structure that seems to be difficult to frame (Papa et al. 2013). This approach has to be built in connection with the up-dated theories of urban science and has to be able to describe the new urban phenomena.

Energy and technology are actually the engines of the smart dimension of a city where it is necessary to save energy using new technologies. But first it is necessary to find a way to figure out the parts and interactions among parts inside a city in order to understand how to operate with an effective energy saving.

The first study about a city as a system was made around the Sixties, but actually this paradigm seems to be still today the best way to model urban phenomena.

In 1964, Berry published his famous book: "Cities as a system within systems of cities" (Berry 1964) in which the city is considered as a territorial system. Four years later, Ludwig von Bertalanffy published his seminal volume: "General System Theory" (von Bertalanffy 1968) to whom Berry referred a large amount of its work on this topic.

Considering these first studies, the systemic approach to city interpretation has been deeply developed by a number of urban scientists that have identified sub-systems inside the urban system. Therefore, we can assume that urban subsystems are the best targets for driving the energy saving action of the city. In other words we want to prove the hypothesis regarding the possibility of saving urban energy by acting on the urban sub-systems directly. This is possible in two ways: by reducing the urban entropy (first) and by influencing the social system trying to develop new behaviors. Considering the systemic approach for city analysis and the complex theory as well, it is possible to find an interesting new way to understand the behavior of the urban system and its trend of evolution.

Understanding urban complexity seems to be the most interesting field of research that has emerged in urban disciplines that seem to be slow in developing new paradigms of interpretation (Wallot and Gurr 2014, Mobus and Kalton 2015). In order to discompose the urban system, we can consider many urban subsystems. It is quite difficult to find a rule to understand how to define an urban subsystem. In general, we have to consider that a system has a geometry, composed of elements and relationships among them. However, the rule of geometry may not always be verified. It is surely verified if we consider the generative subsystem of the city: the geo-morphological and the socio-anthropic one. The geo-morphologic subsystem, as a whole, could represent the environmental support to the human settlement and territorial clusters and physical connections among them that compose it; the socio-anthropic subsystem contains the set made by men and human relationships. In order to identify other subsystems, we can consider the systems generated by the main one and define a sort of "dendrogram rule".

From the two main subsystems, it is possible to derive some other urban sub-systems. Starting from a geo-morphologic subsystem, we can identify a physical subsystem, made by the material part of the city; as well as starting from the socio-anthropic subsystem we can identify the functional subsystem, made by human activities. Following this rule, we can identify different subsystems inside the principal ones. For instance, inside the functional subsystem, it is possible to find as many subsystems as the urban activities are: the residential subsystem, the economic subsystem, the education subsystem, the health subsystem and so on.

Urban mobility represents a special subsystem because it is not located in a single site of the urban system but is articulated across the urban space. This subsystem is vital for the city. Some other cross-subsystems act as a sort of connection between two or more subsystems. The psycho-perceptive subsystem, made by the perception that city users have of urban space ("the image of the city" according to the Kevin Linch theory, Lynch 1960) is a bridge between the socio-anthropic and the physical subsystem; also the economic subsystem acts as a connection among many subsystems: socio-anthropic, functional, mobility, etc. All the urban subsystems are elements of the urban system themselves, in a holistic view of the city. At the end, we can imagine the urban system, made by all the subsystems that interact with each other in order to move the system ahead, following its trend of evolution. This moving is supported by energy (coming from natural resources) that the urban system utilizes in order to evolve and to go on. The system evolves throughout space and time by means of energy. The hypothesis is to try to reduce the consumption of energy and its waste byproducts during the development of the system, in other words try to reduce the entropy. So Smartness is related to the reduction of urban entropy.

2 Defining Urban SmartNESS (New Energy Saving System)

Considering the previous formulated hypothesis, it is possible to distinguish two different entropies inside the urban system: an entropy of evolution and an entropy of development. The first one is due to the operation of the system itself and to the interactions among the subsystems. The second is related to the use of energy because of the needs of the system to go ahead along its trend during time. The two entropies are strongly correlated because when the evolving entropy's value is low, this means that the system utilizes the resources in a right way, saving a large part of energy. In this case, the system goes on thanks to this energy that can be used for the development process (Fig. 1). In some way, the two energies are connected to the external and internal complexity described by Jorge Jost (Jost 2004).

In other words, an urban system where the evolving entropy's value is high has no possibility to go ahead along its path of development. So it is possible to say that urban smartness is related to a low level of entropy of evolution and consequently to the possibility of the system going ahead saving energy. Another consideration can be developed about the entropy of evolution: when a system uses energy in a

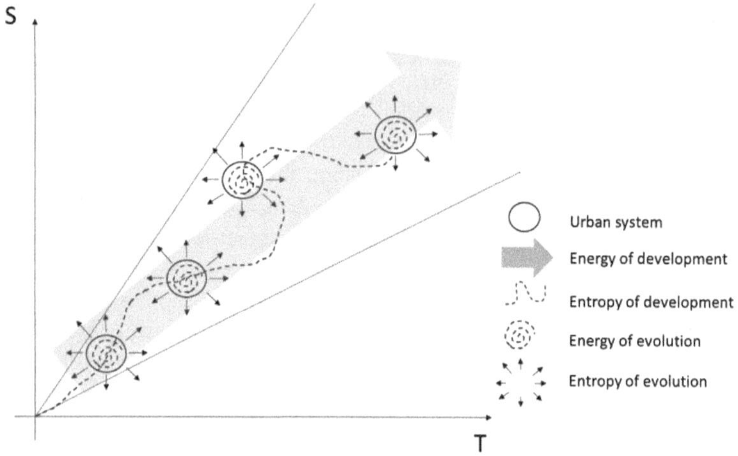

Fig. 1 The trend of the urban system with the energy of evolution conceptualized as a spiral force and the energy of development as a random path moving inside time and space

correct way, avoiding energy waste, this means that the cycle of inside energy of the system is closed. On the contrary, when within the different subsystems, discrepancies occur, these act like points of energy wasting: the cycle breaks down itself and a part of energy is dissipated (Fistola 2012). At the end, it is possible to argue that in an urban system the entropy is inversely related to smartness and moreover that a city with a number of malfunctions, inside the different subsystems (physical, functional, social and so on), cannot be "smart". To bind together these concepts into one, it is possible to say that the real smart city is that one where the urban energy is inside an harmonic-closed circle, correctly used and saved, like in all balanced natural processes (Commoner 1971). Moving from these hypotheses we can find a new syncretic word which is a newly composed acronym: SmartNESS (Smart New Energy Saving System).

Urban SmartNESS is a concept that attempts to link the systemic approach to a need to guide the development of the city towards sustainable configurations characterized by an appropriate and innovative use of energy. The possibility of reducing the entropy of evolution is related to the capability of identifying the causes of malfunction inside different subsystems. In order to improve urban SmartNESS, actions and policies first of all have to identify all the discrepancies existing inside the urban subsystems and try to solve them. In this way, a lot of energy will be saved and the interaction among urban subsystems will transmit a new level of the urban system. The dyscrasias inside the subsystems are the wasting points of energy (resources) and the sources of entropy. This entropy has to be reduced. An example is the socio-anthropic subsystem in which the reduction of entropy is connected to the growing of social capital, an implementation of citizenship (Carta 2014) that produces, as externality, a correct use and a saving of energy within the city.

3 New Technology for Urban SmartNESS: Adoption Versus Adjunction

Urban SmartNESS could be achieved through a new way to use new technology. What it is necessary to clarify is that the innovation, communication technology (ICT) could be a strategic factor in order to activate the process of entropy mitigation inside the urban subsystems. In other words, the ICT have to be considered as an internal element useful to mitigate energy wasting and able to allow a real-time monitoring of the development of an urban system. Moreover, the new technologies, which are the most popular aspect of the smartness, can play a fundamental role in the new approach by considering them in an "adoptive" way and not in an "adjunctive" way, as it is commonly intended in the concept of a smart city.

Nowadays the ICT seems to be an external element (maybe just a showcase element) to classify "smart" a specific human settlement. If the city is equipped with a technological system for traffic control or for air pollution monitoring, it is quickly considered, or worst classified, as a "smart city", even if the buildings are crumbling, the social conflict is high and the queue at the post office is endless.

On the contrary, we want to state that "smartness" is related first to a low level of entropy that can be achieved through a technological adoption rather than a technological adjunction.

Urban smartness (referred to technology) is represented by the capability of the urban system of collecting, transmitting, elaborating and adopting information and data about its states, the active phenomena, the energy flows and physical flows, the intensity of activities and so on. In another words, "smartness" is not just related to the capability of the city to collect and store data but, actually, to the possibility of elaborating and using these data (big data) in order to activate a new organization of the system itself and minimize entropy. This condition (the minimization of entropy) makes a city smarter (Fistola and La Rocca 2013).

In a way, the Tower of Winds in Tokyo by the famous architect Toyo Ito (Fig. 2) could represent an effective example of this capability from an architectural point of view. This building was an element of an old aeration system located inside the district of Yokohama. Ito transformed the old building (located in the historical city) into a new piece of the city, able to catch information from its environment and transform it in light on its skin, with a continuous metamorphosis of its aspect (and role) inside the urban system. Data coming from the urban environment (noises, winds, etc.) are elaborated in order to activate different configurations of the building inside the urban system.

In this sense, it could be possible to say that smartness of a city is strictly connected to its capability of energy saving, in terms of entropy mitigation. In a smart city, the mitigation of entropy has to be a primary target and it can be achieved performing the following actions:

– minimizing entropy through the resolution of dyscrasias existing inside the urban system;

Fig. 2 The tower of winds in Yokohama by Toyo Ito

- minimizing entropy adopting new technologies to monitor and redefining the organization of the urban system;
- minimizing entropy by a saving use of urban energy in a general way (considering natural resources as energy for urban development) as well as in the specific behavior of daily life of inhabitants.
- minimizing entropy by considering a number of driving urban functions that can lead the urban system toward a new state characterized by a high level of energy saving and efficiency.

Furthermore, the new technologies, and in particular cloud computing, enable the creation of repositories of "dynamic knowledge" within which the by-products from the smart city are stored, processed, and reused, even for complex and composite urban systems within the government itself. Some urban functions can be identified to act as the driving function in new urban amenities; for such activities will be necessary to propose a redefinition and a systemic reorganization, including the use of new technologies.

Some recent urban operations seem to be going in this direction and represent interesting examples to analyze.

4 Tourism as a Driving Function

This part explores a specific aspect of analyzing urban tourism as a phenomenon that can affect the competitiveness and the overall well-being of the urban system. The condition to achieve this livability's objective can be identified in the need to integrate tourism development planning within the process of urban transformations' governance.

The key role of tourism for urban economies normally prevails on the impacts that it generates on urban environment, on the general organization of the city and on the socio-anthropic system in terms of conflicts between residents and tourists.

The tourism paradox really consists in the dichotomy of being at the same time a strategic factor for economic income[1]and a generator of negative impacts (overcrowding, pollution, noise, soil and energy consumption) on urban as well as at a larger scale on territorial systems.

The search for equilibrium between these two contrasting aspects must be the main target of urban and regional policies of development to both promote and protect territorial resources. This could be better understood if we referred to tourism as a system composed by its main components: the demand and the supply. Tourism demand concerns the needs expressed by a non-residential population; tourism supply refers to the presence in the city of facilities and structures to satisfy this demand. Town planning can influence the demand requests by intervening on the urban supply system in terms of quantity, distribution and qualities of facilities and structures able to assure efficiency of the city.

The question regards the necessity of defining tools, methods and conditions providing a qualified urban supply (of services, spaces and facilities) that has to be compatible with urban characteristics and resources. In this sense, tourism planning and land use are strictly connected and if properly planned, tourism can be a leading function to drive the urban system towards a "smart" dimension.

The emerging paradigm of the "smart city", in fact, could represent an opportunity to reconsider the current processes of urban planning, but it needs a holistic approach that goes beyond the one applied *per parts* that still seems to prevail in the articulation of its six components (economy, mobility, environment, people, living, governance). The numerous rankings, aimed at "measuring" urban smartness, seem to refer to the prevalence of one component over the others, failing to consider the city as a whole complex and dynamic system.

Urban promotion initiatives seem to concentrate mostly on city branding, rather than on the definition of strategies aimed at making cities able to support an additional urban load expressed by tourism demand; and urban "smartness" seems to be concentrated on the amount of apps available for the tourist use of the city. The application of new technologies, instead, should also strengthen the decisional role in defining adequate policies to manage urban tourism and to optimize availability of urban services and facilities.

The adoption of a systemic logic allows us to propose an innovative approach to the study of the effects of tourism on the organization of the city.

---

[1]The World Tourism Organization highlights that tourism has become one of the largest and fastest-growing economic sectors in the world. About 9 % of the Gross Domestic Product (GDP) comes from tourist activities (direct, indirect and induced) while its contribution to employment is estimated in the order of 7 % of the overall number of jobs worldwide (direct and indirect) and about 1.4 billion of US dollars come from tourism export market (UNWTO 2015).

As pervasive activity[2] where social components have a fundamental role, tourism can influence behaviors and play a driving role in promoting more sustainable use of cities and resources (decrease of waste production, reduction water and energy consumption, etc.).

In this vision, town planning has some responsibilities and needs a general renewal of its tools and procedures to drive the urban system towards compatible states of equilibrium characterized by appropriate and innovative use of resources, and energy in particular, promoting urban SmartNESS (Smart New Energy Saving System).

Referred to tourism, this condition could be represented by the integration between tourism development goals and urban planning targets. This integration would maximize positive aspects of tourism and minimize the impacts of the tourist phenomenon on the city's organization.

5 Urban Tourism from Entropy to Energy

Within urban system, tourism demand concentrates in time (holiday, big events, cultural exhibition, religious celebration) and in spaces that correspond to the areas where factors of attraction are located (monuments, museums, shopping, historical center, cathedrals, etc.). When urban tourism demand exceeds the threshold of the compatibility (with urban social, economic and environmental resources), the urban system collapses and levels of urban livability rapidly decrease. This vision can strictly be connected to the concept of "carring capacity". Referred to tourism carrying capacity can be defined as the ability of the urban system to perform tourist functions without threatening those that are essential for resident population (Thurot 1980; Mathesion and Wall 1982; Grasselli 1989).[3]

---

[2]In spite of global crisis, tourism has an uninterrupted growth over the past six decades. International tourist arrivals have increased from 25 million globally in 1950, to 278 million in 1980, 527 million in 1995, and 1133 million in 2014 (UNWTO 2015). At present, tourism involves all different social levels, being a cross activity, affecting several sectors (mobility, hospitality, leisure, etc.).

[3]Literature is copious about definitions of tourism carrying capacity (Maggi and Fredella 2010). Scholars agree about the complexity of the concept and refers to different components based on three main relationships:

- environmental: refers to the capacity of natural resources that are in de tourist destination and their fruition by tourists.
- cultural: refers to the tourists' satisfaction based on their expectation;
- socio-economic: concerns to the social and economic satisfaction of the residents referred to the presence of tourists in their city.

These relationship can be find into the definition by the UNWTO (1981) "the maximum number of people that may visit a tourist destination at the same time, without causing destruction of the physical, economic and socio-cultural environment and an unacceptable decrease in the quality of visitors' satisfaction".

The levels of crisis (entropy production[4]) arise when the urban system (physical, functional and socio-anthropic) is no more able to support the tourist load (the tourist demand). As we state in this study, there is a threshold (not necessarily numeric but that can be also virtual) that identifies the limit point (not to be crossed) beyond which the system falls in the area of the entropic production.

The balance between the tourist demand (growing and sector-based) and the urban supply corresponds to an "ideal value" (a state) that is difficult to achieve. However, we want to argue that by the planning of adequate actions and by the support of ICT it is possible to lead the system towards compatible states.

Tourism, being characterized by "transversality" and "pervasivity",[5] can be a driving function able to shift the system towards urban smartness conditions that necessarily engages physical, functional and social component of the urban system. In this sense, the "smartness" can identify a condition of possible equilibrium (between tourist demand and supply) where the city achieves widespread urban quality levels for all categories of users: residents, city users, and tourists.

The change that is characterizing the current tourist demand (from tourisms to "smart tourism")[6] denotes an improvement of tourist behaviors and consumptions, and promotes new models of use of the city according to a sustainability paradigm. Although sustainability in tourism is still an object of debate, at present, it refers to a new approach in tourist supply chain (transport, hospitality, entertainment) rather than to a tourist typology. The present tourist demand, on the other side, is more careful about environmental questions making sustainability a principle part of the factor that influences destination choices.

In this sense, the promotion of "sustainable destination" (i.e. zero emissions hotels, management and recycling of waste production; alternative energy applied to lighting of monumental areas and public building as well as to the private building sector, etc.) represents a factor of improving its attractiveness and competitiveness.[7] Energy saving, in particular, is the focus of the recent strategies to promote tourism development but it still lacks a holistic and strategic vision. Actions mainly concern the building scale (in this case the accommodation facilities) and refer to the use of new materials and systems to improve energy performance of the single edifice (lighting, water heating, ventilation). The measures

---

[4]In this study, entropy has been considered as a widespread negative condition of the urban system, which hinders the positive processes to achieve sustainability and tends to move the system towards trajectories totally different from those expected (see also Fistola and La Rocca 2013).

[5]Transversality refers to the multiplicity of sectors (public and private) involved in tourist development. Pervasivity refers both to the constant growth of tourism in the late sixty years and to the trend that characterizes actual demand, at all social levels, impatient of sharing its own experiences rapidly and in real time.

[6]Buhalis and Amaranggana (2014) defines the characteristics of a smart tourism destination referring also to tourists. A smart tourist profile is proposed in La Rocca (2014).

[7]Criteria for destination pointed out by the Global Sustainable Tourism Council propose and establish standards for sustainability in tourist destination recognizing tourism as a potent tool both for preserving resources and reduce poverty (see http://www.gstcouncil.org).

mainly concern the development of procedures to certify the sustainability of existing or new buildings also in accordance with the latest EU legislation[8] that promotes the "zero emission" concept also applied to the accommodation facilities. The hotel sector, after transport, in fact, represents one of the most energy intensive components in the tourist supply chain. Even though few studies exist about the assessment of hotels emissions and consumption, some valuations refer to a range of about 25–285 MJ/guest per night of energy use. In terms of emissions it has been estimated that a range of CO_2 production between <1 kg (in case of renewable energy use) and 125 kg (in case of self-supporting power generation) per guest-night (UNEP and UNWTO 2012). In Italy, the daily energy consumptions for the hotel sector (MJ/presences) are more about the quadruple one than the civil one (MJ/habitants) (ISPRA 2013). Nevertheless, at least as it concerns the Italian situation, there is still a lack of integration among different actors engaged in the promotion of new forms of sustainable supply services for tourism. The procedures for building sustainability licenses (residential, commercial, or tourist) based on voluntary mechanisms that are delegated to the individual initiative of the owners. ECOLABEL for example is the voluntary certification applied to the tourism sector for the acknowledgement of environmental sustainable criteria in the management or designing of tourist buildings. Ecological criteria for allocation of the mark regard the reduction of environmental impacts by the use of high levels of quality in services' supply and in the management of waste production. Even though the mark is considered as a factor of improving attractiveness of a destination, some research (ISPRA 2015), referring to the Italian situation, underline the lack of co-operation between private and public sector in supporting these initiatives in spite of a growing demand of greener structures. More than a third of tourists, in fact, declare it to be favorable to pay between 20 and 40 % more to spend their sojourn in a green accommodation (UNEP and UNWTO 2012). This aspect underlines both the change occurring in the present tourist demand (green tourism demand) and the high potentiality of tourism to affect social behaviors and lifestyles. In this sense, pervasivity refers to the growing trend to share opinion, sentiments and experiences in real time by using social networks (Facebook, Tweeter, Trip advisor, Google+, Istagram, etc.) and could represent a strategic factor to improve more sustainable forms of tourism demand. On the supply side, it contributes at activating new participatory planning processes where all the parts are involved (stakeholders, politicians, residents and tourists).

Tourism, then, is a favorite sector to test the real potentialities of the smart city approach. This regards not only the renewal of supply services by using new technologies, but also it needs co-operation among different actors (public-and private) and different levels (politic and administrative) involved in the implementation of the urban or territorial attractiveness (La Rocca 2014).

[8]The 2010 Energy Performance of Buildings Directive and the 2012 Energy Efficiency Directive.

6 Promoting, Managing and Using Tourism Inside the Smart City

The definition of smart tourism[9] has developed by analogy to that of a smart city and is equally undefined. The recent trends in defining smartness of a city highlight the key role of the social component (Ercole 2013) in the decisional processes and the active contribute of this component to promote more sustainable lifestyles and globally improve the quality of urban life. The relation between tourism and climate change has been the focus of Davos Declaration (2007) where, among others, there has been expressed both the exigency to convert the tourism sector towards a more sustainable growth[10] and the need of integrating tourism into the decisional processes at different scales to implement adaptation and mitigation strategies to face the present urban challenges.

If the role of tourism as a key sector is also able to implement knowledge and stimulate actions, that fact will come out in the framework of international policies; the arising concept of smart tourism seems to be much more connected to technology use and its applications (Buhalis and Amaranggana 2014). Indeed, technology has hardly contributed to the change of "tourism experience" becoming part of it both in the phase of planning and in the way of living and communicating it (Kim et al. 2008).

The term *prosumer* is recently increasingly used to describe the current tourist demand characterized by the ability to interact at any time and at any place as a result of new technologies (Web 2.0, social network, blog, chat, etc.).

Capability of managing, elaborating and sharing these information fluxes becomes strategically important not only to implement the attractiveness of a tourist destination but especially to implement the global planning process of urban supply, both private (tour operator and stakeholder) and public (administrator and politics decisors).

[9]During the first Meeting of the UNWTO Tourism Resilience Committee, in 2009, Smart tourism has been defined as "*clean, green, ethical and quality at all levels of the service chain. A type of tourism able to satisfy the needs for the short-term responses to the economic crisis as well as those one of long term as sustainable development, poverty alleviation and mitigation climate change*".

[10]Referring to this issue in the premises of the document it is stated: the tourism sector must rapidly respond to climate change, within the evolving UN framework and progressively reduce its Greenhouse Gas (GHG) contribution if it is to grow in a sustainable manner. This will require action to: mitigate its GHG emissions, derived especially from transport and accommodation activities; adapt tourism businesses and destinations to changing climate conditions; apply existing and new technology to improve energy efficiency; secure financial resources to help poor regions and countries.

In this context, the smart tourist destination delineates a complex system where new technologies are incorporated in the process of its development and in the planning of tourism attractiveness (Wang et al. 2013; Buhalis and Amaranggana 2014).[11]

Referring to these definition and characteristics, new approaches are needed to facilitate co-creation among actors and co-operation among different institutional levels achieving smartness and competitiveness of a destination.

The priorities for the construction of a Smart Tourism Destinations (STD) refer: (1) to elements and conditions able to enhance tourists' travel experience; (2) to provide more intelligent platforms both to gather and distribute information within destinations; (3) to facilitate efficient allocation of tourism resources; (4) and to integrate tourism suppliers at both micro and macro levels aiming to ensure that benefit from this sector is well distributed to local society (Huang et al. 2012; Rong 2012).

The challenge that seems to be addressed to tourist cities concerns the capability to combine promotion objectives with the need of limiting consumption especially of energy and water, incorporating technologies inside the process of urban development and transformation. As we have stated in this part, tourism can be both a tool to activate new forms of sustainable facilities and feature, at the level of supply (involving private and public sector) and a mean able to affect social behaviors, at the level of the demand (social component). Tourism can be an occasion to promote the use of renewable energies acting as driving functions in the shift of urban system towards a smartness state.

The transition towards urban smartness involves, at least, four different levels/conditions that should interact each other's:

- knowledge sharing,
- integration between public and private sector (Stakeholder engagement),
- research involvement,
- integration between urban planning and tourism development.

As it concerns the first point, potentialities of a smart city approach mainly refers to the large disposability of data produced by technological sensors located inside the city able to measure in real time the efficiency of the physical component of the urban system. On the other side, within the city, the atrophic component, (that is resident population, city users, occasional visitors and tourists) can be considered as "alive sensors" (moving and using the city in different ways) able to share impressions and experiences.

Capability of processing, managing and interacting these flows of information represents the main challenge for the smart city approach (Fistola and La Rocca 2013). Tourism represents a valid area to test potentialities of knowledge-sharing mechanisms that, also by the involvement of users, could improve the urban

[11]Buhalis and Amaranggana (2014, p. 557): *Smart Tourism Destinations can be perceived as places utilizing the available technological tools and techniques to enable demand and supply to co-create value, pleasure, and experiences for the tourist and wealth, profit, and benefits for the organizations and the destination.*

supply systems of services and permit managing the organization of tourist flows inside the city.

Referring to the second point, the promotion of urban smartness as a global condition of urban livability needs a definition of governance processes based on co-operation among economic actors and political levels involved. The tourism sector involves a diverse range of actors and also for this "trasversality" it could act as a driving function to test the effects of actions based on private-public cooperation and improving their fulfillment.

The third point hightlight that the concept of a smart city still suffers from a certain discontinuity in the definition despite its widespread use. In this sense scientific research can contribute to define specific areas where a smart approach could be tested and applied. Mobility, behavioral practices and cycles of sustainable production represent some of the experimental areas where smartness could be better defined going beyond the techno-centric vision that actually still seems to prevail.

As it concerns the last point, we can state that within the present economic framework, tourism is one of the main factors of development that can improve the image of a city and its competitiveness. Tourism also generates impacts on environmental and urban systems affecting their equilibrium. The smart city approach might necessarily consider potentialities committed to urban development and tourist promotion according to the physical, functional and social aspects of the urban system. The impacts generated from a non-governed tourist development affect several aspects (economy, environment, social), they spread through different modes, and through different intensity; this variability can depend on the type of tourist activity, on the resilience of the cities and on the characteristics of the urban supply system composed of services and infrastructure to support tourism. City planning actions (intended as the search for an order according to a plan) should mostly focus on these aspects by a general renewal of tools and procedures of governance of urban transformations.

Some initiatives have been developed in Europe aimed at testing the integration among the four different levels above mentioned and improving urban smartness. The RENFORUS (Renewable Energy Futures for UNESCO Sites) initiative for instance, promotes projects of energy efficiency and the use of renewable energy in a selected number of UNESCO sites proposed as privileged observatories to test models of Sustainable Energy Communities to face global climate change effects. The energy transition from fossil fuel to renewable energy concerns the whole tourism delivery chain (transport, hospitality and leisure activities) creating new opportunities for business between the tourism industry, local communities and developers. Case studies refer to different scales, from buildings (monument) to Island, to cities and their historical center.

The case of a historic site is very significant especially for the restrictions these parts of the city are subjected to. The city of Edinburgh (UK) represents a best practice in Europe according to the national target of eradicating fuel poverty[12] by 2016. The Management Plan for the "Old and New Towns of Edinburgh" includes measures for energy savings through a partnership between Edinburgh World Heritage a not-profit organization for the management of the site, the City of Edinburgh Council and the Governmental Agency of Historic Scotland.

The Swedish isle of Gotland that, since the Nineties, started its energy policies to turn into a self-sustaining community based on the use of local resources and to be greenhouse gas emission neutral within one generation (20–30 years) represents another example of a sustainable community. Methodologies to involve resident population as well as tourists[13] has been adopted by municipalities to exercise a positive influence, disseminate a cultural model of energy consumption, and promote sustainable lifestyle models. Among its strategic objectives, in defining its Energy Plan to reduce the dependence by fossil energies, Municipality supports programs and projects aimed at defining conditions on how the supply of housing, workplaces, services and culture can be designed to reduce the need for car travel and improve the conditions for environmentally-compatible and re-source-economizing means of transport.

These examples and the above-mentioned considerations show how SmartNESS necessarily involves a review of the processes for the governance of the urban system.

Within this dimension, ITC technologies play a primary role that need to be supported, optimized, improved and integrated with city planning and urban government's processes.

The condition for the transition toward smartness refers to the adoption of technology rather than its addition inside the urban evolution process. By the use of ICTs technologies, residents, tourists and city-users can play a dynamic role in monitoring urban functioning permitting us to reduce the lack of efficiency if properly integrated with decisional levels that should be well structured to adopt and elaborate information into action plan.

7 Smart Cities and Energy Efficiency: a Winning Combination

Taking action to reduce the waste of energy, as already described, is essential for the purpose if increasing the SmartNESS of cities. What is more, the subject of energy efficiency, for some time, has been at the center of the strategies and policies

[12]In the UK a household is retained to be in 'fuel poverty' if it spends more than 10 % of its income on heating and power.

[13]The island passes from a population of about 65,000 inhabitants in winter to about 300,000 inhabitants in summer (Municipality of Gotland).

being developed by European cities, also in order to respond to the pressing demands from the European Union in recent years.

In fact the subject of energy, broken down into its various parts, (reducing emissions, using alternative sources, efficiency of distribution networks, and so on) has for some years now been at the centre of the agenda of the EU, which has fixed the strategies related to energy efficiency with a deadline of 2020 (Battarra 2014).

A thorough debate was started in 2014 in order to define the framework of the EU's energy and climate policies up to 2030. New goals were proposed for making the European energy system more competitive and sustainable by, *inter alia*, reducing emissions of greenhouse gas and increasing the use of energy from renewable sources. In particular, in October 2014 the European Council defined the framework for energy and climate policies with a deadline of 2030 and approved four objectives:

- reduction of emissions of greenhouse gas by at least 40 % before the end of 2030, compared to the 1990 levels (mandatory objective);
- consumption of at least 27 % renewable energy in 2030 (mandatory objective);
- improvement of energy efficiency by at least 27 % in 2030 (indicative objective);
- supporting the urgent completion, not later than 2020, of the domestic energy market, achieving the aim of 10 % for the existing electrical interconnections, particularly for the Baltic States and the Iberian peninsula, in order to achieve an aim of 15 % by 2030.

The EU strategy regarding energy is backed by the substantial investments that have also been planned for the 2014–2020 planning period, foreseeing an allocation of more than 17 thousand million euro (European Court of Auditors 2012).

Italy has also followed the policies of the European Commission by adopting a series of documents aimed at setting the aims with regard to energy saving and efficiency and the use of renewable sources. Lastly, the Italian Action Plan for Energy Efficiency (ENEA 2014) approved by the Minister of Economic Development in 2014[14] lays down the actions for achieving the aims fixed by the National Energy Strategy (Economic Development Ministry 2013) for achieving the decarbonisation of Italy in 2050.

For effectively dealing with the challenges in the field of the environment and energy and pursuing the Europe 2020 objectives, many cities are adopting the "smart city" model, which is characterised by the use of ICT and information flows, so as to guarantee careful development of natural resources, and is aimed at guaranteeing high standards of comfort and wellbeing for the community (Papa et al. 2015).

[14]Economic Development Min. Dec. 17 July 2014 Approval of the "Italian Action Plan for Energy Efficiency 2014" (Official Gazette 31 July 2014 no. 176).

Among the many and varied actions being taken by European cities in this field there is no doubt that the "Smart Environment" aspect is the one that attracts the greatest number of projects, research work and experiments in relation to the six characteristics that, according to established literature, define the Smart City (Giffinger et al. 2007). And it is also the one that best characterises the smart city and we can say that the minimum common denominator of smart cities is environmental sustainability broken down into its various parts (The European House 2012).

In the most common breakdown the Smart Environment aspect mainly includes projects aimed at improving energy efficiency on buildings and urban scale but also actions that use ICT for creating smart energy transportation networks and, more generally, for improving urban services, like in the case of innovative public lighting systems, and for reducing atmospheric pollution related to urban transport (European Parliament 2014).

Also in Italy, as shown by a recent experimental research on the subject of implementation of smart cities metropolitan cities,[15] the "environment" aspect is the one with the greatest number of activities. Out of a sample of about 1000 activities surveyed (researches, projects, technologies etc.), 30 % concern matters related to the environment (atmospheric pollution, solid urban waste, greenery etc.) and 25 % of them are aimed at the energy sector in the dual sense of energy saving and efficiency and the use of renewable sources.

Hereunder, to verify how the smart city model can be used for helping to increase SmartNESS, as already defined, we describe some activities being tried out in Europe and in Italy.

8 SmartNESS in Practice: an Overview

For a description of the actions and experiments regarding energy in progress in Europe and in Italy, a breakdown has been made into three macro-categories.

The first includes actions and projects in urban areas. In some cases districts are created *ex novo*, aimed at low energy consumption ("districts with almost 0 energy") and equipped with a complete energy infrastructure (smart networks, alternative and renewable energies, water, and waste management), while in others the projects and actions concern the energy upgrading of existing districts where the buildings, mainly built during the seventies, have serious waste problems and low energy efficiency.

[15]The reference is to the "Smart Energy Master for energy management of territory" project that has been co-financed by the National Operational Programme for Research and Competitiveness 2007—2013 Smart Cities and Communities. The Environment Mobility Workshop of the Department of Civil, Building and Environmental Engineering of Federico II University of Naples drew up the project. For more details: http://smartenergymaster.unina.it/.

There are many actions in progress in Europe for both types of activity and some are well known (just think of the “historic” example of Freiburg), but other examples are Hamburg (HafenCity), Stockholm (Royal Seaport), Vienna (Aspern), etc.

In many cases there are complex activities of upgrading of districts, which not only deal with residential activities but foresee integrated strategies aimed at the development and attraction of new businesses (as in the case of Nordhavnen in Copenhagen), at the provision of services for the community, at the reorganization of the mobility system, and also include actions aimed at the efficient use of energy.

Among the most recent ones we would mention the vast project for upgrading the district of Hackbridge, a mainly residential suburb (with about 8,000 residents) located in the London Borough of Sutton, South West London. The district has various types of building, ranging from late nineteenth century cottages to terrace houses, and the local government has funded an urban restoration activity that includes actions aimed at energy-retrofit of the homes in order to reduce energy consumption and emissions of CO_2 (Deakin et al. 2014).

Many experiments being conducted in Italy are the result of participation by some cities in partnerships that, making use of funding by the European Commission, involve the implementation of pilot projects. Many of these experiments consist of the application of techniques and instruments for improving the energy performance of buildings that were not well built. This is the case of the R2 Cities and Energy Efficiency in Low Income Housing in the Mediterranean project (ELI_Med) being carried out in Genoa, or of European cities serving as Green Urban Gates towards Leadership in sustainable Energy (Eu-Gugle) in Milan.

As far as new projects are concerned, we mention EXPO 2015 in Milan which is intended—as is, of course, the nature of “universal exhibitions”—to be a kind of showcase of the most innovative solutions for creating a district with a low environmental impact. Great emphasis was therefore placed in the project on the adoption of Smart Grid and Energy Management System technological solutions, capable of guaranteeing performances with low energy consumption. In particular, the Energy Management System makes it possible to collect a large quantity of data (electricity consumption and production, presences, temperature, light etc.) in the cloud, process them and display them on an energy map of all the spaces and devices existing in the exhibition area with a view to not wasting energy, but also to “making the most of the unique opportunity offered by the Universal Exhibition, for broad involvement and for visibility, for spreading knowledge and for sharing more sustainable ideas and practices” (Expo Milano 2015 2014).

The second category, called “public services”, includes very dissimilar projects that range from actions on network systems (water, energy, waste, etc.), to the adoption of innovative technologies for creating more efficient public lighting systems, but also to actions on particular types of public building (offices, schools etc.) mostly financed by substantial public funding.

In particular, a field of great activity of experimentation, especially with regard to energy aspects, is that related to the application of ICT to influence systems for management of energy distribution networks and also to make better use of renewable sources of energy.

As defined in the research by the European Parliament (2014), these are «ICT-enabled infrastructures to improve the management of utilities for a city, such as energy, water or electricity, e.g. smart power systems with intelligent management of energy mixes, smart grids, smart metering, heat storage, solar Energy management systems, and surveillance management systems for resources such as clean tap water or wastewater or heating efficiency systems.»

There are very many examples of these types of action in Europe and in Italy, even though in many cases they are still pilot actions applied on a small scale, mostly funded by private companies and firms that operate in the energy sector. By way of example, we mention here the Bremen Environmental Building Management project, Cologne ship-to-grid, Mannheim E Energy, Munich Smart Grid System, Vienna Citizens' Solar Power Plant, etc.

In Italy this category includes many activities aimed at reducing consumption with regard to public lighting systems and public buildings (schools and local government offices). For example, this is the case of the Milan project in Led or the New green NET and WEST (of Turin) Growing in Energy Efficiency Network project, which is aimed at improving the energy efficiency of public property and the public lighting systems of the municipalities belonging to the network (214 public buildings and 23,000 light points) but the project is still being finalised.

Lastly there is the category called "products/devices", which includes actions that foresee great use of technological innovation products for monitoring or controlling various aspects of the city: from traffic to environmental pollution, from electricity consumption to waste management. Generally these are projects developed by firms, which foresee the integration of various types of technology and networks of sensors and use large quantities of data. Cases of large-scale application are still fairly rare whereas there are many cases in Europe of application to pilot areas of the cities. Well known, for example is the Climate Street project in Amsterdam, implemented through a set of technologies that involve the use of: "smart meters", Energy Displays that give a feedback on energy consumption in real time and make it possible to save energy on the basis of the information supplied by the smart meter; an integrated system of low energy consumption street lighting by the use of energy-saving lamps that can be adjusted according to the light, solar panels for lighting up public transport stops, rubbish bins with compactors that make it possible to empty the bins five times more frequently, electric vehicles for rubbish collection etc. This action has been repeated in other European cities, like in the case of Smart Streets of Barcelona suburb Sant Cugat, and of Klima Strasse in Cologne.

In Italy also this category attracts a substantial number of projects, often financed by public funds and of a prototypal kind.

The last two categories of action in the Smart City field, either because of their experimental nature or because of the difficulty of producing them on a large scale, at least as far as Italy is concerned, are those that are less relevant for the purpose of the specific approach of this contribution which, as already mentioned, aims at describing integrated solutions that can influence the organisation of the urban system with a view to increasing the levels of SmartNESS.

9 Planning the SmartNESS: Some Guidelines

From the brief description of the most interesting experiments being carried out in some Italian and European cities we can draw some preliminary conclusions about how, in the light of the adoption of the systematic approach to studying urban phenomena, we can analyse the smart city paradigm.

Another important aspect that calls for careful consideration is if and how implementation of the smart city model can help to define strategies for the development and management of urban systems that increasingly have to deal with complex environmental challenges, the main ones being related to climate change and the need to reduce energy consumption drastically.

The experiments in progress that can be classified as "successful" took integration as their watchword: integration between different types of action but also integration between the different aspects of the city, from social to environmental, from economic to governance.

In particular, as it concerns energy aspects, the most effective actions in this sector refer to cities that are developing network actions, combining various components, bringing together the private sector (which produces innovations) and the public one, (which lays down the action strategies by drawing up plans and programmes), with the social one (for promoting virtuous lifestyles and behavior by the community). In fact, the cities that are leaders in the international classifications (Amsterdam. Copenhagen, Vienna) are the ones that successfully adopted these strategies decades ago.

In other words, a systematic approach to management of the city from a smart point of view calls for the promotion of integrated initiatives that go beyond the approach of sporadic and precise actions and that, learning from the best practices in Europe, succeed in reproducing and scaling actions taking the particular nature of the various urban and social contexts into account.

And in this way, particularly as far as actions in the field of energy are concerned, which are closely linked with specific local characteristics (from the morphology of the settlement to the lifestyles, from climatic factors to the available investments), the critical factor of the scalability of the action can only be dealt with by contextualizing the best practices.

This means establishing for each of the urban subsystems (physical, functional, socio-anthropic and environmental) a specific set of knowledge that also takes their evolutionary trends into account.

Although it is true that ICT can help to perfect sophisticated tools for knowledge about urban systems and that, thanks to the open data policy, the set of data available for characterizing the cities has been greatly expanded, it is also true that, particularly in some countries (like Italy), many data, particularly those related to the energy sector, are fragmentary, hard to collect, or not available on the various territorial scales (urban/metropolitan).

For many towns we are still far from having tools like energy consumption maps, discretized for each building in the urban area, like Amsterdam which within the open data platform carried out the Energy Atlas (City of Amsterdam 2014) project, a series of theme maps relating energy consumption to the types of building, the income of the residents, the value of the housing, etc.[16] But not just data and information related to the various subsystems but also systems for monitoring and controlling the impacts of the various actions. Here it is worth mentioning another very critical matter when we speak of actions and best practices for smart cities or for the setting up of systems of indicators that can provide information about the effectiveness of the actions taken.

Not much has been done about this aspect yet, although the European Community also, particularly in recent planning documents, stresses the need to give more room to the impact evaluation phase of the various activities in order to assess their effectiveness. We still have very different models and assessment indicators whereas it would be better to have a convergence in order to homogenise the information, thus making it possible to learn from more mature experience.

In the energy sector for example this means not only understanding which aspect (or else which subsystem) to give priority to (and therefore the need to have a large quantity of data to compare), but also being able as time goes on to monitor the effects of the actions and compare alternative solutions.

And lastly, adopting once again a systematic approach, if the city is a system made up of several interrelated subsystems it is necessary not only for the various actions, but also the planning tools, to be integrated.

---

[16]The importance of having open data available for finalizing specific strategies is also stressed in the introduction to the Energy Atlas "It would have been impossible to produce the Energy Atlas without the data provided by third parties. The sharing of data and gaining insight into each other's interests and considerations is of huge importance to making energy transition a reality." (City of Amsterdam 2014).

Adoption of the Smart City model needs to be backed up and finalized by a master plan foreseeing integrated strategies that act on the environmental aspects, and on the social and spatial ones in a system approach aimed at increasing the SmartNESS of the city.

10 Conclusions

SmartNESS (New Energy Saving System) as it has been proposed in this paper refers to a condition able to drive the urban systems towards new states characterized by: the sustainability of the evolution process, the mitigation of entropy, and by an energy ethic, which could go beyond that one referred to the simple "energy saving". In this sense, the urban smartness also relates to policies aimed at reducing urban entropy by using the physical system regeneration, the functional system requalification and the mitigation of social conflicts. This can be also achieved by the development of urban driving functions, acting as towing for the system as a whole.

Tourism, considered as a complex urban function and especially referred to the Italian case, can assume this driver role. Tourism development, in fact, involves a number of different sectors that need an integration to assure high level of efficiency in the organization and working of urban system. At the same time, tourism can promote new and more compatible ways of fruition of the city that can affect tourist behavior (green tourism).

Some interesting cases described in this paper can help to strengthen the approach proposal. The case of Copenhagen, for instance, has particularly significance as it connects to two of the main factors of smartNESS: from one end the choice of a strategic set of urban function useful for the regeneration process, on the other end the policies integration oriented to a consciousness use of the energy. Some other cities bring back their smartness to the capability of generating data, also by using some articulated urban grids. As we state, the real capacity of elaborating and making available these data can be identified as one of the main condition for the urban smartness. In other words, the urban smartness is supported by the capability to cluster, analyze and understand urban "big data" in order to predict the future evolution of the urban system. The only disposability of urban big data (that represent a feed-back of the system behavior) is really useless, in order to set up a set of system management actions. These data, therefore, need to be previously ranked and processed by using analytic models of interpretation (also semantic) able to give an effective description of the system condition and its entropy levels of energy dissipation.

The sustainable use of resources and energy savings achieved by acting the management of the urban subsystems and trough the reduction of entropy levels (of the evolutionary process of the city) may be prosecuted to the size "smart" through a new consideration of the technological component. In this dimension, the technology does not replace the man, but it can activate positive processes such as: the conscious use of resources and the development of alternative energy; the increase of the social capital (through an inclusive and modernization social networks); virtualization of urban activities (oriented to the recovery and reuse of the physical spaces); the widespread adoption of ethical sustainable mobility, and so on. The networks are the fundamental infrastructures for the urban system. Smart grids will enable the government of cities, networks increasingly able to automatically steer the supply and demand of energy and the best distribution of resources in the city. Reducing entropy is the action by deployment on urban subsystems and this goal has to be achieved by considering specific urban processes and urban functions.

Acknowledgments The authors have made a joint contribution to the paper's conception and design, background and concluding remarks. Romano Fistola designed the methodological approach to the problem and the definition of urban smartNESS (§§ 1, 2 and 3). Rosa Anna La Rocca analyzed the role of tourism as a driver urban function to shift the city towards smartNESS conditions (§§ 4, 5 and 6). Rosaria Battarra, by the analysis of the most significant and innovative initiatives on energy efficiency, outlined possible guidelines to implement the smart city (§§ 7, 8 and 9).

References

Battarra R (2014) Energy Component and Smart City Planning: Best Practices in Italy. In: Papa R (ed) Towards Smart City a Scientific Approach, Aracne, Rome. ISBN 978-88-548-7024-6

Berry BJL (1964) City as systems within systems of cities. In: Papers and proceedings of the regional science association, vol 13. pp 157–176

Buhalis D, Amaranggana A (2014) Smart tourism destinations. In: Xiang Z, Tussyadiah I (eds) Information and communication technologies in tourism 2014. Springer International Publishing, Switzerland. doi:10.1007/978-3-319-03973-2_40

Carta M (2014) Smart Planning and Intelligent Cities: a New Cambrian Explosion. In: Riva Sanseverino E, Riva Sanseverino R, Vaccaro V, Zizzo G (eds) Smart Rules for Smart Cities. Managing Efficient Cities in Euro-Mediterranean Countries. Springer, ISBN:978-3-319-06421-5

City of Amsterdam (2014) Energy Atlas Amsterdam Southeast. Available at http://www.amsterdam.nl/wonen-leefomgeving/energie/energieatlas/. Last visit May 2015

Commoner B (1971) The closing circle: nature, man and technology, Knopf Inc. (Random House), Munich

Deakin M, Campbell F, Reid A, Orsinger J (2014) The mass-retrofitting of an energy efficient-low carbon zone in the UK 2015. Springer Briefs in Energy. ISBN 9781447166207

Economic Development Ministry (2013) Strategia Energetica Nazionale: per un'energia più competitiva e sostenibile, marzo. Available at: http://www.sviluppoeconomico.gov.it/images/stories/normativa/20130314_Strategia_Energetica_Nazionale.pdf. Last visit May 2015

ENEA (2014) Piano d'azione italiano per l'efficienza energetica. Available at: http://www.mise.gov.it/images/stories/pubblicazioni/PAEE_2014.pdf. Last visit May 2015

Ercole E (2013) Smart tourism: il ruolo dell'informazione social. Annali del turismo vol 2

European Court of Auditors (2012) Cost-effectiveness of cohesion policy investment in energy efficiency, Special Report n. 21. Available at: http://www.eca.europa.eu/Lists/ECADocuments/SR12_21/SR12_21_EN.PDF. Last visit May 2015

European Parliament, Directorate-General for Internal Policies (2014) Mapping smart cities in the EU, IP/A/ITRE/ST/2013–02 January. Available at: http://www.europarl.europa.eu/RegData/etudes/etudes/join/2014/507480/IPOL-ITRE_ET(2014)507480_EN.pdf. Last visit May 2015

Expo Milano 2015 (2014) Rapporto di sostenibilità Expo Milano 2015. Available at: http://www.expo2015.org/it/news/tutte-le-news-/pubblicata-la-seconda-edizione-del-rapporto-di-sostenibilita-di-expo-milano-2015. Last visit May 2015

Fistola R (2012) Urban entropy vs sustainability: a new town planning perspective. In: Pacetti M, Passerini G, Brebbia CA, Latini G (eds) The sustainable city VII, vol 155. WIT press, WIT Transactions on Ecology and The Environment, pp 185–204

Fistola R, La Rocca RA (2013) Smart City Planning: a systemic approach. In: Yiğitcanlar T, Bulu M (eds) Establishing bridges: the 6th knowledge cities world summit, KCWS 2013, Lookus Scientific, ISBN 978-9944-380-11-9

Giffinger R, Fertner C, Kramar H, Kalasek R, Pichler-Milanović N, Meijers E (2007) Smart cities: ranking of European medium-sized cities, final report, centre of regional science (SRF), Vienna University of Technology. Available at: http://www.smart-cities.eu/download/smart_cities_final_report.pdf

Grasselli P (1989) Economia e politica del turismo. FrancoAngeli, Milano

Huang X, Yuan J, Shi M (2012) Condition and key issues analysis on the smartertourism construction in China. In: Multimedia and signal processing. Springer, Berlin

ISPRA (2013) Stato dell'ambiente, http://annuario.isprambiente.it

ISPRA (2015) Indagine sulle strutture ricettive a marchio Ecolabel, UE, ISPRA-Servizio Certificazioni Ambientali

Jost J (2004) External and internal complexity of complex adaptive systems. Theory Biosci 123:69–88

Kim DY, Park J, Morrison AM (2008) A Model of Traveller Acceptance of Mobile Technology. Int J Tourism Res. doi:10.1002/jtr.669

La Rocca RA (2014) The role of tourism in planning the Smart City. TeMA J Land Use, Mobility Environ 7(3):268–282. University of Naples Federico II. doi: 10.6092/1970-9870/2814 e-ISSN 1970-9870, print ISSN 1970-9889

Lynch K (1960) The image of the city. MIT press, Harvard

Maggi E, Fredella LF (2010) The carrying capacity of a tourist destination. In: Conference paper. Available on research gate https://www.researchgate.net

Mathesion A, Wall G (1982) Tourism: economic, physical and social impacts. Longman

Mobus GE, Kalton MC (2015) Principles of systems science. Understanding complex systems. Springer, New York

Papa R, Gargiulo C, Galderisi A (2013) Towards an urban planners' perspective on Smart City. TeMA J Land Use Mobility Plann 6(1)

Papa R, Galderisi A, Vigo Majello MC, Saretta E (2015) Smart and resilient cities. A systemic approach for developing cross-sectoral strategies in the face of climate change. TeMA J Land Use, Mobility Environ 8(1)

Rong A (2012) China economic net. Retrieved July 22, 2013, available at http://en.ce.cn/Insight/201204/12/t20120412_23235803.shtml

The European House—Ambrosetti (2012) Smart Cities in Italy: an opportunity in the spirit of the Reinassance for a new quality of life. Available at: http://www.ambrosetti.eu/it/download/ricerche-e-presentazioni/2012/smart-cities-in-italia. Last visit 15 May 2015

Thurot JM (1980) Capacité de charge et production touristique, Chet, Université d'Aix-Marseille

UNEP and UNWTO (2012) Tourism in the green economy—background report. Printed by the World Tourism Organization, Madrid, Spain

UNWTO (1981) Saturation of tourist destinations. Report of the Secretary General, Madrid

UNWTO (2015) Tourism highlights. Available at http://mkt.unwto.org

von Bertalanffy L (1968) General system theory. George Braziller, New York

Wang D, Li XR, Li Y (2013) China's "smart tourism destination" initiative: a taste of the service-dominant logic. J Destination Marketing Manage 2(2):59–61

Wallot C, Gurr JM (2014) Introducion: towards a transdisciplinary understanding of complex urban systems. In: Wallot C, Gurr JM, Schmidt JA (eds) Understanding complex urban systems: multidisciplinary approaches to modeling. Springer, New York

Smart City, Urban Performance and Energy

Ila Maltese, Ilaria Mariotti and Flavio Boscacci

Abstract Over the last few years there has been a growing demand for more liveable cities, and the notion of "urban smartness" is thus attracting the attention of both policy-makers and academicians. Among many different definitions, the "Smart City" by Giffinger et al. (2007), a functional model with six drivers of urban smartness (economy, environment, governance, living, mobility and people), appears as the most widely recognized. Within this context, increasing attention has been devoted to the "Energy" dimension concerning renewable energy, energy consumption, and energy policy. The present chapter reviews the theoretical and empirical literature on the relation between smartness and energy at the urban scale, specifically focusing on the 103 Italian NUTS3 province capitals. It mainly investigates how and to what extent the province capitals differ in terms of "energy" attributes. The chapter is structured into four sections. The introduction is followed by a literature review of the theoretical and empirical studies defining the concept of Smart City and its characteristics, with a specific focus on the "energy" role in a Smart City. Section three presents data and descriptive statistics exploring the role played by "energy" within the 103 Italian NUTS3 province capitals. The last section focuses on discussion and further research.

I. Maltese (✉) · I. Mariotti · F. Boscacci
Politecnico di Milano, DAStU, Milan, Italy
e-mail: ila.maltese@polimi.it

I. Mariotti
e-mail: ilaria.mariotti@polimi.it

F. Boscacci
e-mail: flavio.boscacci@polimi.it

R. Papa and R. Fistola (eds.), *Smart Energy in the Smart City*,
Green Energy and Technology, DOI 10.1007/978-3-319-31157-9_2

1 Introduction

No doubt that the modern world is totally dependent on electricity, appliances and modern devices it powers; in particular, the supply precariousness, due to growing demand, global warming or peak oil, suggests the shift to renewable energy resources and energy-saving policies and behaviours.

Specifically, with the most of the population and activities concentrated, cities are the largest energy consumer (75 % of the world's energy), and produce 80 % of greenhouse gas emissions (UN 2009); besides, they are also expected to respond to global environmental concerns within tight financial constraints.

Within this context, increasing attention has been devoted to "Energy" concerning renewable energy, energy consumption and Energy Policy at the urban scale.

Actually, high density in the urbanized areas enhances new energetic strategies: (i) new and advanced building technologies to save energy demand and consumption, and to reduce the number of (polluting) power plants; (ii) urbanistic choices (Fistola 2013), which have to take into consideration urban morphology (Salat and Novacki 2010), and compactness (Yamagata and Hajime 2013); (iii) the use of light-emitting diode (LED) and renewable energy like solar and wind power for street lighting, and charging points for Electric Vehicles.

Besides, it is advocated that the environment has to be protected; therefore, new sources of available, diffused, and sustainable renewable energy are required. Furthermore, the traditional relation between the customer and the electricity grid is changing with the new idea of "prosumer", who creates goods for his/her own use and also possibly to sell (McLuhan and Nevitt 1972; Toffler 1980). Changes are also expected for distribution and power networks, which are transforming into an automated Smart Grid, using ICT for improving efficiency, reducing wastefulness, saving energy, and increasing the reliability and sustainability of the electricity's production and distribution (Smart Grid Insights 2014). The Intelligent Energy Europe (IEE) programme, consistent with the Europe 2020 strategy, aims to fulfil the following three main objectives: promoting energy efficiency and encouraging the rational use of energy sources; increasing the use of new and renewable energy sources as well as encouraging energy diversification; stimulating energy efficiency and renewable sources in the field of transport (EU 2012). A general objective (20 % reduction in greenhouse gas emissions, 20 % improvement in energy efficiency and 20 % of renewable in EU energy consumption) has thus been fixed by the European Union in the 2020 European Sustainable Strategy (COM 2010). European cities have thus to increase energy efficiency and decrease total energy consumption, while maintaining secure and continuous access to critical infrastructures like the electric power system (Morvaj et al. 2011). Within this context, urban sustainable development should be based not only on fairer prices, cleaner energy and greater security of supply, but also on different consumer choices (COM 2007). The big challenge to face is thus to combine global competitiveness and local sustainability (Lazaroiu and Roscia 2012). Actually, the cities labelled as *Smart City* (from now on SC) proved to be able to use ICTs to build and integrate critical infrastructures and services (Nam and Pardo 2011). Therefore, it is interesting to explore the urban role of energy, in order to find a possible relationship between energy and smartness.

To this aim, in the present chapter, the 103 Italian NUTS3 province capitals are observed, using the six-dimensions of the "SC" Vienna model by Giffinger et al. (2007), and focusing on the energy indexes. A brief analysis of the evolution of the "SC" concept, which suggests a holistic approach to all the "good" characteristics of a city, introduces the literature on the theoretical models for assessing sustainability and smartness. A review of the empirical studies, considering indexes and indicators adopted for measuring urban success, follows. Since according to the empirical studies the "environment" dimension, which includes "Energy", does not seem to be a driver for urban success, a specific analysis has been carried out, focusing on the role of energy. Specifically, it is investigated how and to what extent the Italian province capitals differ in terms of "energy" attributes, starting from the results provided by Maltese et al. (2013) and Boscacci et al. (2014), which clustered them according to the SC model.

The chapter is structured into four sections. The introduction is followed by a literature review of the theoretical and empirical studies defining the concept and the characteristics of a SC, and the energy variables at the urban scale. Section three investigates differences and commonalities among the 103 Italian NUTS3 province capitals as concerns smartness and energy dimensions. Discussion and further research follow in the last section.

2 Literature Review: Smart City and Energy

2.1 The "Smart City": Theoretical Studies

Among the several definitions of a successful city, that of SC has now got the upper hand, thus attracting the attention of practitioners, policy-makers and academicians (Barca 2009; Barca and McCann 2011; COM 2012; McCann and Ortega-Argilés 2013). Table 1 shows the evolution of the label "successful city" in the last 25 years. Main authors are also indicated, together with the general focus and the different aim, which was supposed to be achieved. As it can be noticed, the original technological dimension has been enriched by quality of life and sustainability (Nam and Pardo 2011; Maltese et al. 2013). Specifically, after the many contributes on the "intelligent city" during the 90s, mainly dealing with ICTs as a key driver, the focus shifted to the "social" aspects of urban development: the higher productivity of a more educated human capital (Shapiro 2006; Winters 2011) and skilled workforces (Berry and Glaeser 2005; Glaeser and Berry 2006), or the triple-featured (tolerance, talent, technology) "creative city" (Florida 2002, 2005; Hall 2000). According to the sustainable approach to growth—considering the three aspects of Economy, Environment and Society—urban studies have flourished first in environmental and social fields (Inoguchi et al. 1999; Hollands 2008; O'Grady

Table 1 Labels for a "successful city"

Period	Authors	Focus	Label	Aim
90s	Mitchell (1995)	ICTs, digital networks	Intelligent, digital, wired	Efficiency
	UN-Habitat (1996), Inoguchi et al. (1999)	Environment	Sustainable	Efficiency
2000s	Shapiro (2006), Berry and Glaeser (2005), Glaeser and Berry (2006)	Human capital: education	Cultural	Efficiency
	Florida (2002, 2005), Hall (2000)	Human capital: creativity	Creative	Quality of life
	Anthopoulos and Fitsilis (2010), Carley et al. (2001)	Social capital: participation and e-governance	Smart	Quality of life
2012	Ratti (2012) (MIT)	Social capital: sensitiveness, adaptability	Senseable	Quality of life/efficiency

Source Authors' elaboration

and O'Hare 2012), and in the last few years—chiefly due to the global crisis—also in the economic one.

Even in the most recent definition of "senseable city" (Ratti 2012), a new path towards urban sustainability is suggested, which entails a deep use of new technologies for improving the quality of life (Ratti 2012; ABB & The European House-Ambrosetti 2012; Legambiente 2012), thus involving not only intelligence and innovation as means, but also inclusion and liveability as goals (Mitchell 1995, 2007; Sassen 2011).

Due to this comprehensive and multifaceted concept, a strict definition of SC is not easy; yet, some working definitions are anyhow available. Among these, the most widely recognized one states that a city is smart "*when investments in human and social capital and traditional (transport) and modern (ICT) communication infrastructure fuel sustainable economic growth, and a high quality of life, with a wise management of natural resources, through participatory governance*" (Caragliu et al. 2011). Indeed, sustainability seems the only common feature to all the feasible definitions of SC (ABB 2012), together with the target of "quality of life" (Legambiente 2012). As a conclusion, according to the majority of the theoretical studies on SCs, whatever the disciplinary approach or the research background (institutional, academic or entrepreneurial), it is possible to state that SC should pursue two main goals: full general sustainability and quality of life, which may be summed up in the concept of "smartness". Last but not least, due to their "wired" nature, which provides a great stock of data, SCs are expected to be easily analysed (Batty 2012). As a consequence, many empirical studies on SC have been carried on, not only by universities and private companies, but also by governmental institutions (COM 2012; MIUR 2012) in order to define the characteristics of a SC, which are presented in the next paragraph.

2.1.1 The Characteristics of a "Smart City"

The most of the empirical studies on successful cities fail to provide a holistic definition of SC and rather end up choosing indicators for the many dimensions; nevertheless, in some cases, useful research tool-frames have also been provided. A first example is represented by the Nijkamp Hexagon, which identifies six typologies of urban capitals for analysing the level of urban sustainability for a United Nations survey (Nijkamp et al. 1993; UN-Habitat 1996). More in depth, the Hexagon model involves six different typologies of capitals which are strongly interconnected: (i) the natural capital (Ecoware) that refers to environment, natural resources, landscape, energy consumption, waste management; (ii) the man-made capital (Hardware) including built environment, technologies, land use and functions, transport system; (iii) the economic and financial capital (Finware) that is sources of capital, both private and public, capability to attract investors, dynamism of economic system; (iv) the institutional capital (Orgware) that concerns policies, governance at local level and partnership or cooperation between public and private sector; (v) the human capital (Software) regarding investments in knowledge, education, training with the aim of promoting innovation and creativity; (vi) the social capital (Civicware) that pertains to intra- and inter-generational equity, stakeholders and community involvement and local quality of life. With reference to the conceptual framework suggested by this multidimensional model, it is clear that the level of sustainability depends on how many dimensions are considered, and on the degree of their mutual relationships, suggesting that effective appraisals of sustainability at the urban scale should be based on an integrated assessment (Ravetz 2000).

Similarly, although in a different business perspective, IBM suggests a research frame for studying urban success, according to which, cities can be seen as complex networks of the following components: citizens, businesses, transport, communications, public utilities (water and energy) and city services. Therefore, Intelligent Cities are those which better manage their stock of instrumentation, interconnection and intelligence (IBM 2009) by combining ICT with its physical infrastructure in order to improve the quality of life and to collect data to make better decisions and deploy resources effectively and efficiently (Kanter and Litow 2009; Hall et al. 2000).

Finally, the most valued description of the SC characteristics has been provided by the Technology Universities of Wien and Delft together with the Ljubljana University (hereinafter "The Vienna model"). According to this functional model, the drivers of urban smartness can be grouped into six "smart" dimensions (Giffinger et al. 2007). Furthermore, in the Vienna model, a working definition of SC is also provided: a SC is "well performing in a forward-looking way in six characteristics, built on the 'smart' combination of endowments and activities of self- decisive, independent and aware citizens" (*ibidem*, p. 11). Focusing on these dimensions, Smart Economy mainly concerns competitiveness; Smart People is about social and human capital; Smart Governance refers to participation; Smart

Table 2 Smart city dimensions in the literature

UN-Habitat (1996)	IBM (2009)	Vienna model (Giffinger 2007)
ECOware: natural capital	**Water/energy**	**SMART environment**
HARDware: hand-made capital	Transport	SMART mobility
FINware: economic and financial capital	Businesses	SMART economy
ORGware: institutional capital	City services	SMART governance
SOFTware: human capital	Citizens	SMART people
CIVICware: social capital	Communications	SMART living

Source Authors' elaboration

Mobility deals with both ICTs and transport; Smart Environment involves natural resources; Smart Living is a synonymous of quality of life.

It is worth noting that, although with small differences, the observed categories in the three models above-mentioned are quite similar, regardless of the field of study (Table 2). Due to this, from here on, the label "smart" will be used for indicating any kind of "successful" city.

In particular, energy management, which is specifically considered by IBM, is also observed in the natural capital (Ecoware) of the UN-habitat survey and in the Smart Environment dimension of the Vienna Model (Table 2, first row in bold).

2.2 *Empirical Studies*

Due to the complexity introduced by the need to make a city "smart", the importance of integrating technologies, systems, infrastructures, services, and capabilities (Jennings 2010) is well reflected in the indicators used for measuring the smartness itself. In particular, as concerns the empirical analysis on SCs, several studies have been conducted at both European and national scales, which can be grouped into two categories: (i) ranking analysis, that classifies cities according to selected indexes/indicators; (ii) more complex analysis (i.e. hedonic prices method, econometric models, and cluster analysis), which try to explain the reasons for success.

2.2.1 The Ranking Studies in Italy

Focusing on Italian cities, the ICity Rate report—where the "I" in the title of the report stands not only for Intelligent, but also for Innovative, Inclusive and Interacting—classifies the Italian NUTS3 province capitals according to the Vienna model dimensions, using about one hundred indicators at the local and provincial scales (Forum 2012) (Table 3). According to this ranking study, the most "intelligent" city is Bologna, which awarded one of the first six positions in all the

Table 3 The ranking studies

Report	Authors	Sample	(stated) Issue	Sources
ICity rate	Forum (2012)	103 province capitals	Intelligent cities	Various sources
Ecosistema Urbano	Legambiente (2012)	104 province capitals	Environmental quality	Istat, municipalities
La Dolce Vita	Colombo et al. (2012)	103 province capitals	Quality of life—real estate price, wages	Various sources

Source Authors' elaboration

dimensions, save for the environmental one (42nd position). Parma follows, with a good placement in four out of the six dimensions; even in this case, environment is less smart than that in the other cities (only 41st), together with the living quality (13th).

Another ranking for the same cities is provided by "La Dolce Vita" (Colombo et al. 2012, 2014), which considers urban quality of life using five framework dimensions (climate, environment, services, society and economy) (Table 3). The best performers show better results in the economy and services dimensions, rather than in those concerning the environment (with climate obviously privileging southern cities) or the society. As far as the geographical pattern is concerned, quality of life is highest in medium-sized towns in the Centre and in the North of Italy: Pisa, Trieste, and Ancona seem to be the best Italian cities to live in, followed by Bologna, Firenze, Pesaro and Venezia.

An additional suggestion can be inferred by the results of the above-mentioned rankings which seem to disregard the environment dimension. Within this context, no surprise, then, if the Intelligent Community Forum (ICF), in selecting 7 out of 21 Smart Communities every year in order to reward them, seeks for the highest scores in the following factors: broadband connectivity, knowledge workforce, digital inclusion, innovation, and marketing and advocacy.

The Environment is, instead, the focus of the Ecosistema Urbano' ranking, yearly edited by Legambiente, focusing on the environmental quality of the Italian NUTS3 province capitals (Table 3). In 2012, for example, the XIX Ecosistema Urbano considers 25 indexes measuring urban environmental performances regarding: air, water, energy and waste management, transport and mobility, green areas, environment, and mobility policies. The best cities present good results in the most of the indexes; among the others: waste management (share of recycled wastes), ciclability—which considers the urban "bike-friendliness" level, and sustainable mobility index—which encompasses broader and different aspects of urban mobility. Verbania proves to be the most "green-friendly" city, together with Belluno, Trento and Bolzano: as expected, the North of Italy shows the highest level of environmental concern.

2.2.2 Econometric and Statistical Studies

Switching to the studies based on statistical or econometric analysis (Table 4), Caragliu and Del Bo (2012) focus on the impact of smart characteristics on urban performance—measured by per capita GDP—and investigate this impact at the local level for a sample of 94 cities in 14 EU countries (1999–2006). They find that urban density is negatively associated to urban performance, while the smartness index is always positive and significant. Besides, cities specialized in industries with high-tech content, with higher amenities, and more attractive as concerns tourist inflows, are better performing.

Colombo et al. (2012, 2014) analyse, on one side, the relationship between quality of life and housing prices; on the other side, the link between quality of life and wages within the Italian province capitals in the 2001–2009 period. Quality of life is defined as the weighted average of a set of local amenities, branched into the five main domains described in Sect. 2.2.1. The result is that housing prices are higher in cities with less pollution, more green areas, located on the coast, with higher teacher-pupil ratio, better transport and cultural infrastructures, higher civic-ness, university enrolment, higher value added per capita and lower unemployment rate. On the other hand, housing prices are lower in those cities with higher crime rates and shares of foreigners.

The recent analysis conducted by Siemens-Anci on the largest 54 Italian province capitals, has grouped the cities by means of a Cluster Analysis on the basis of five synthetic indexes: urban environment (air quality, urban green, water and waste management), real estate stock, energy management, mobility, and health service supply (Siemens-ANCI 2012). Six clusters are found, and five of them are roughly corresponding to the five proposed indexes and present the highest scores for specific features. The left cluster of "becoming cities" (cluster 5) has got scores below average in each measure, but the commonality among the 10 cities of this group, which are small-sized and mainly located in the South, seems to be the growing specialization in one specific sector. The best cluster is the "Ideal Cities" (cluster 3), which is composed by 4 medium sized cities in the North-East of the country (Bergamo, Brescia, Padova and Trento), with the best scores in all the dimensions.

Table 4 Other studies

Authors	Sample	(stated) Issue	Methodology	Sources
Caragliu and Dal Bo (2012)	94 European cities	Smartness	OLS	Urban audit
Colombo et al. (2012)	103 Italian province capitals	Quality of life—real estate prices, wages	Hedonic prices + ranking	Various sources
Siemens-ANCI (2012)	54 Italian major cities	Efficiency	Principal component, cluster	Various sources
Maltese et al. (2013)	103 Italian province capitals	Smartness	OLS, cluster analysis	Various sources

Source Authors' elaboration

It is followed by a small group of 7 big cities belonging to "good living and moving cities" cluster 4, where mobility and real estate stock are excellent if compared to the average values. If Cluster 4 presents low scores in the environmental variable, in the "ideal cities" (cluster 3) environment proves instead to be taken into account. Two other groups (environmental and energy clusters), which account for 25 % of the cities sample, seem concentrated only on the environmental topic, while the wealth cities, a geographically diffused group, show good results referring to the real estate stock and the health service supply as well. In particular, Energy Cluster (6) will be further described in the next Sect. 2.3.

Finally, it is worth mentioning the studies by Maltese et al. (2013) and Boscacci et al. (2014), which investigate the relationship between smartness and urban performance in the 103 Italian NUTS3 province capitals, where the latter is measured by the residential real estate prices in the semi-central area of the city. Besides, they carried out a Cluster Analysis based on the six Smart dimensions of the Vienna model, in order to find differences and commonalities among the cities. Four different clusters have been identified (see Table 7). The most performing one (Cluster 1) is composed by 4 large "competitive" cities with excellent results in Economy, Governance, Mobility and Living. The second best performing group, is composed by 6 large "attractive" cities mainly located in the North, which present the same characteristics of the first cluster, even to a lower level. Cluster 4, concerning 35 "liveable" cities, instead, includes medium-sized cities all over the country with all the variables over the average, except for unemployment in the Economy dimension. Finally, Cluster 2 focuses on the cities located in the South and in some others peripheral areas. These smaller cities have, on average, the worst scores in all the selected variable, but many of them present results above the cluster average for one of the dimension, suggesting a possible future (smart) specialization, which could help in improving the current situation.

Once more, like in the ranking studies, also in these analyses, Environment doesn't seem to be a driver for the city's smartness. Nevertheless, Energy, which is included in the Environment dimension, plays a key role at urban scale, therefore further investigations are acknowledged.

2.3 *The Role of Energy in the Smart City*

As concerns the SC, the European Commission promoted some calls about energy efficiency and buildings, renewable energy and supply networks, and green mobility (public and private) for the largest urban areas. Moreover, secondary goals are identified: involving citizens, affecting their behaviour, improving quality of life, and sharing European best practices in sustainable energy management (EU 2012). In particular, the cities have to elaborate and develop action plans for sustainable energy following the European guidelines. The Italian Ministry for the Economic Development states that a SC has to promote Renewable energy and smart grid, Energy Efficiency, and low carbon technologies (MSE 2012). From the theoretical point of view, and according to the three frameworks presented in the previous Sect. 2.1.1, energy seems to be an important aspect for urban sustainability and smartness.

Actually, in the Nijkamp Hexagon, energy consumption is one of the criteria belonging to the first dimension concerning the natural capital (Ecoware) (Nijkamp et al. 1993). In the IBM report, energy, together with water, represent the infrastructural determinants (IBM 2009), while in the Smart Environment dimension of the Vienna Model the efficient use of electricity (use per GDP) is considered a good indicator for measuring the sustainable management of the resources (Giffinger et al. 2007).

In addition to the three theoretical frameworks, described above, one of the most recent tools for assessing sustainability is the SILENT model (Sustainable Infrastructure, Land-use, Environment and Transport Model): an advanced geographic information system and indicator-based comparative urban sustainability indexing model. This model considers all the major aspects affecting sustainability and highlights the residential resource consumption exploring the energy use levels (Yigitcanlar and Dur 2010).

As described in Sect. 2.2.1, city rankings are another easy tool for identifying the energetic assets of the urban areas, as it has been investigated in recent research reports. Actually, neither ICity Rate (Forum 2012) nor La Dolce Vita (Colombo et al. 2012, 2014) pays close attention to energy indicators, whilst the Ecosistema Urbano provides data on:

- energy consumption *per capita* (annual domestic energy consumption),
- renewable energy supply *per capita* (solar thermal installed and solar electric power on Municipal buildings, district heating volumes)
- energy policies, measured by a composite index (EP) which considers subsidies, rules, open data, the presence of an energy manager and so on (Legambiente 2012, 2013, 2014).

According to the results of the Ecosistema Urbano, only half of the 103 province capitals takes advantage of an energy manager; 73 present solar electric on Public buildings; more than half, instead, have got Solar thermal, while only a third had adopted the heating district measure by 2012. Finally, as concerns the EP index, only Ferrara and Rimini (in the north east of Italy) awarded the maximum score.

In the above mentioned analysis by Siemens-Anci (Sect. 2.2.2), for each considered index, not only the resources stocks, but also the adopted policies have been investigated.

As far as the energy dimension is concerned, they have explored both the energy supply and distribution, focusing on innovative systems (district heating or photovoltaic). Besides, in order to create an index, a principal component analysis has been carried out, starting from the following indicators:

- Solar thermal panels on Municipal buildings (m^2/1,000 inhabitants);
- Solar panels (per 1,000 inhabitants);
- Solar panels power (per 1,000 inhabitants);

- Power of the Solar electric panels on Municipal buildings (kW/1,000 inhabitants);
- Energy domestic consumption (kWh per inhabitant).

The final index is positively correlated to the first three indicators showing a certain propensity for new and renewable energy sources; the fourth one is neutral while the last one, measuring the energy consumption *per capita,* has a negative impact on the index.

According to the general Cluster Analysis conducted on the 54 major Italian cities, six clusters are finally identified, among which, the Energy Cities Cluster presents very good results in the energy index. It is composed by 8 cities in the Centre and the South with a great number of solar panels also on Public buildings. Among these: Foggia, Lecce and Taranto—probably due to the presence of strong Regional regulation—Siracusa, Arezzo, Terni e Latina. As regards the final index, in the North Forlì, Trento e Ravenna seem to pay the closest attention for renewables.

3 Energy, Urban Performance and Smartness: An Empirical Analysis on the Italian Province Capitals

This paragraph is devoted to the analysis, by means of descriptive statistics, of how and to what extent the 103 Italian NUTS3 province capitals differ in terms of "energy" attributes. As presented in Table 5 the considered variables are: (i) solar thermal deployment on public buildings (size); (ii) solar electric deployment on public buildings (power); (iii) presence of Local Energy Plan; (iv) Energy Policy Index by Legambiente. Specifically, the Local Energy Plan collects the main actions for saving energy and for renewables promotion; despite many local differences, it underlines the attention paid by the city to the energy issues, and its

Table 5 Data and sources

Index	Unit of measure	Scale	Year	Source
Solar thermal deployment on Public buildings	m^2/1,000 inhab.	Local	2011	Legambiente
Solar electric deployment on Public buildings	kW/1,000 inhab.	Local	2011	Legambiente
Energy policy (EP) index by Legambiente	Score	Local	2011	Legambiente
Local energy plan	Dummy	Local	2010	Istat
Ecosistema Urbano ranking	Score	Local	2012	Legambiente
ICIty rate	Score	Local	2012	Forum
Checklist (CL)	Score	Local	2012	Authors' elaboration

Source Authors' elaboration

willingness to regulate it. In particular, the Energy Policy Index considers: subsidies, rules, open data, and the presence of an energy manager, etc. (Legambiente 2012, 2013, 2014).

Besides, the following three additional variables, concerning the ranking of the cities, have been considered: (a) the final score of the Ecosistema Urbano-2012 ranking, by Legambiente, focusing on the urban environmental performance in 2012; (b) the final score (ICIty Rate) by the Forum PA, that measures the smartness level of the cities using the six-dimensions' Vienna model by Giffinger et al. (2007); (c) the checklist score (Table 5). The first two variables have been taken into account in order to control for other energy characteristics. The checklist score is calculated starting from the final scores of the three rankings, described in Sect. 2.2.1: La Dolce Vita by Colombo et al. (2012); Ecosistema Urbano by Legambiente; ICity Rate by Forum PA (2012). Following Catalano and Lombardo (1995), every city has been awarded firstly by a score corresponding to the rank in each ranking, and secondly by the sum of these three scores. The best result is represented by the *minimum*. For example, Bologna got the first place in the ICity Rate and the second one in the Ecosistema Urbano, while Colombo et al. (2012, 2014) scored it at firth place. Therefore, following the checklist score, Bologna awarded respectively 1, 2 and 4, whose sum is 7. No other cities did better, so Bologna is the best city according to the checklist method, since it has got the lowest score.

A preliminary analysis at the macro-area level shows that the North-East is best performing in almost all the energy attributes with the exception of solar thermal where the North West performs better (Table 6). Generally, the North West is in the second position for: Ecosistema Urbano, energy policy, local energy plan, while it is superseded by the Centre as concerns the solar electric.

The "energy" attributes are also analysed with reference to the 103 province capitals, and the role played in identifying the urban performance is investigated. To do so, the results of the cluster analysis provided by Maltese et al. (2013) and Boscacci et al. (2014) are taken as starting point. As already stated in Sect. 2.2.2, these studies grouped the province capitals according to the smartness dimensions by Giffinger et al. (2007), but neglecting the "energy" dimension. They identify four clusters (Table 7). Cluster 1, labelled "Competitive Cities" cluster, concerns

Table 6 Variables' average values by macro-area

Variables	Macro area			
	North east	North west	Centre	South and Islands
ICity rate	424.54	388.04	355.12	274.42
Ecosistema Urbano	0.56	0.51	0.48	0.38
Checklist	88.95	139.82	141.12	222.81
Energy policy (EP)	58.86	42.78	32.28	23.36
Local energy plan	0.63	0.43	0.36	0.27
Solar-electric	4.05	1.37	3.58	1.04
Solar-thermal	1.72	2.28	0.84	1.31

Source Authors' elaboration

Table 7 The province capitals belonging to the clusters

Cluster 1	Cluster 2	Cluster 3	Cluster 4
4 competitive cities	58 specializing cities	6 attractive cities	35 liveable cities
Roma, Milano, Firenze, Venezia	Cagliari, Palermo, Mantova, Foggia, Pescara, Potenza, Prato, Pordenone, Trieste, Asti, Campobasso, Pistoia, Cuneo, Novara, Sondrio, Reggio Calabria, Caserta, Grosseto, Terni, Belluno, Alessandria, Biella, Ascoli Piceno, Livorno, Udine, Catania, Latina, Massa Carrara, Matera, Vercelli, Benevento, L'Aquila, Chieti, Rieti, Agrigento, Frosinone, Macerata, Sassari, Siracusa, Taranto, Trapani, Viterbo, Brindisi, Cosenza, Enna, Isernia, Nuoro, Ragusa, Rovigo, Avellino, Catanzaro, Teramo, Lecce, Gorizia, Caltanissetta, Crotone, Oristano, Vibo Valentia	Bologna, Verona, Bolzano, Genova, Torino, Napoli	Rimini, Como, Salerno, Siena, Trento, Brescia, Pesaro e Urbino, Parma, Aosta, Bergamo, Bari, Modena, Padova, Pisa, Piacenza, Reggio Emilia, Treviso, Perugia, Vicenza, Forlì, Lodi, Messina, Ferrara, Ravenna, Imperia, Pavia, Savona, Varese, Ancona, Lecco, Lucca, La Spezia, Arezzo, Cremona, Verbania

Source Maltese et al. (2013)

the best performing cities: Roma, Milano, Venezia and Firenze. These are the largest cities, hosting high added-values activities, a good network of administrations and institutions, and various amenities attracting tourists. On the contrary, despite a good sustainable mobility, pollution in these metropolises is very high; housing market prices are also the highest.

The second best performing group is represented by Cluster 3 ("Attractive cities"), composed by 6 large cities, mainly located in the North, which present the same characteristics of the first cluster, even to a lower extent.

Cluster 4 ("Liveable cities"), instead, includes 35 medium-sized cities all over the country with all the variables above the average, except for unemployment. They do appear good cities to live in.

The tail end is Cluster 2 ("Specializing cities") with the most of the cities located in the South and some other cities in the peripheral areas of the regions they belong to. These smaller cities have, on average, the worst scores in all the selected variables, but many of them show results above the cluster average for one of the dimensions, suggesting a possible future smart specialization, which might help in improving the current situation.

Now, the question is: "are the best performing province capitals even good in saving energy?". To answer to this question a specific analysis of the energy

Table 8 The energy variables and the clusters

Clusters				
	1-Competitive Cities	2-Specializing Cities	3-Attractive Cities	4-Liveable Cities
(8a) Variables				
ICity rate	459.25	309.48	429.33	395.31
Ecosistema Urbano	0.53	0.44	0.53	0.52
Checklist	52.00	198.19	91.50	108.63
Energy policy (EP)	32.00	28.07	62.00	50.44
Energy plan	1.00	0.28	0.50	0.54
Solar electric	0.56	1.08	4.40	4.41
(8b) Dimensions				
Economy	+++	−	++	+
People	−−	+	−	++
Mobility	+++	−	++	+
Environment	−−−	+	−−	−
Governance	+++	−	++	+
Living	+++	−	++	+
Macro area[a]	C	S	NE	−
Size	Metropolis	Small	Large	Medium
Energy	+++	−−−	++	−

Note [a]prevalent macro-area
+/- identify the score being above or below the average score of the Cluster
Source Authors' elaboration

attributes for the four clusters is developed (Table 8a), whose results are compared to those provided by Maltese et al. (2013) and Boscacci et al. (2014), which sintetically identify the clusters according to the six dimensions (Table 8b). It results that all the "Competitive cities" (Cluster 1), with the highest values in the three rankings, have adopted an Energy Plan, while they have not extensively implemented the energy policy, and have not invested in the solar electric, which registers the lowest value. The "Attractive cities" (Cluster 3), instead, show a high value as concerns energy policy, have adopted solar electric, half of them have adopted an energy plan.

The "Liveable cities" (Cluster 4) show the highest value in solar electric, and they are in the second position as concerns: energy policy, and energy plan. Similarly to the results of the previous studies investigating urban performance, the "Specializing cities" (Cluster 2) present the lowest values in the energy characteristics. The energy dimension contributes, therefore, to the success of the cities.

4 Conclusions

Energy plays a key role at urban level: however "smart" may be a city, it surely faces social problems and economic losses if a blackout occurs (Byrd and Matthewman 2014).

Energy is not only dealing with the Environment and Mobility dimensions but also with other Smart dimensions of the Vienna model. First of all, investments in alternative energy can help in creating jobs, thus improving Smart Economy. Smart Governance could also benefit from the smart grid network that could provide: interconnectivity, the development of databases of best practice information for saving energy and time, and the enhancement of e-government and higher levels of collaboration among private and public institutions (regional and local, educational and research institutions, entrepreneurs, and civic organizations). Besides, further benefits are provided by the alternative and renewable sources like the need to reorganize the regional production and the lower dependency from abroad.

As concern Smart Living and People, that is, the quality of life for city-users as well as for citizens, cities have nowadays to cope with new challenges in order to rethink how to manage urban development, and to reduce energy consumption drastically. Lastly, the great opportunity provided by ICT should fill the gap created by the digital divide, eliminating inequalities, and granting a diffused inclusion.

In the present chapter, the importance of energy at urban scale has been investigated, focusing on the major Italian cities. Starting from the analysis of the evolution of the "SC" concept, an extensive review of the theoretical, and empirical studies on urban smartness has been presented. Specifically, the results of the empirical studies have underlined the lack of importance of the environmental dimension in determining the success of the city. Since energy belongs to environment, and represents the focus of the policy agenda, it is worth investigating its role. To this aim, the analysis has focused on how and to what extent the 103 Italian province capitals differ in terms of "energy" attributes. It results that energy matters, nevertheless, it cannot be neglected that not all the several energy characteristics have been considered. Further research might extend this analysis and try to capture the impact of each variable.

References

ABB & The European House-Ambrosetti (2012) Smart cities in Italia: un'opportunità nello spirito del Rinascimento per una nuova qualità della vita. http://www.ambrosetti.eu/it/download/ricerche-e-presentazioni/2012/smart-cities-in-italia. Accessed 10 May 2015

Anthopoulos L, Fitsilis P (2010) Digital cities: towards connected citizens and governance. In: Reddick C (ed) Citizens and e-government. IGI-Global. www.igi-global.com. Accessed 10 May 2015

Barca F (2009) An agenda for a reformed cohesion policy. European Commission. http://www.europarl.europa.eu/meetdocs/2009_2014/documents/regi/dv/barca_report_/barca_report_en.pdf. Accessed 10 May 2015

Barca F, McCann P (2011) Outcome indicators and targets: towards a performance oriented EU Cohesion Policy. European Commission. http://ec.europa.eu/regional_policy/sources/docgener/evaluation/doc/performance/outcome_indicators_en.pdf. Accessed 10 May 2015

Batty M (2012) Smart cities, big data—editorial. Environ Plan B Plan Des 39:191–193

Berry CR, Glaeser EL (2005) The divergence of human capital levels across cities. Pap Reg Sci 84:407–444

Boscacci F, Maltese I, Mariotti I (2014) Smartness and Italian cities. A cluster analysis. TeMA special issue—smart city. Planning for energy, transportation and sustainability of the urban system, June 2014, pp 141–152

Byrd H, Matthewman S (2014) Energy and the city: the technology and sociology of power (failure). J Urban Technol 21(3):85–102

Caragliu A, Del Bo C (2012) Smartness and European urban performance: assessing the local impacts of smart urban attributes. Innovation 25(2):97–113

Caragliu A, Del Bo C, Nijkamp P (2011) Smart cities in Europe. J Urban Technol 18(2):65–82

Carley M, Jenkins P, Small H (2001) Urban development and civil society: the role of communities in sustainable cities. Earthscan, London

Catalano GM, Lombardo S (1995) L' analisi costi-benefici nelle opere pubbliche ed elementi di analisi multicriter. Flaccovio Editore, Palermo

Colombo E, Michelangeli A, Stanca L (2012) Città italiane in cerca di qualità. Università Bocconi, Milano

Colombo E, Michelangeli A, Stanca L (2014) La Dolce Vita: hedonic estimates of quality of life in Italian cities. Reg Stud 48(8):1404–1418

European Commission (2007) Energising Europe: a real market with secure supply. http://ec.europa.eu/energy/electricity/package_2007/index_en.htm. Accessed 10 May 2015

European Commission (2010) A strategy for smart, sustainable and inclusive growth. http://eur-lex.europa.eu/LexUriServ/LexUriServ.do?uri=COM:2010:2020:FIN:EN:PDF. Accessed 10 May 2015

European Commission (2012) Smart cities and communities. www.ec.europa.eu. Accessed 10 May 2015

European Union (2012) Intelligent energy Europe in action. EU Commission, Brussels. http://ec.europa.eu/energy/intelligent/in-action/index_en.htm. Accessed 10 May 2015

Fistola R (2013) Smart city. Riflessioni sull'Intelligenza Urbana. TEMA 1:47–60

Florida R (2002) The rise of the creative class: and how it's transforming work, leisure, community and everyday life. Basic Books, New York

Florida R (2005) Cities and the creative class. Harper Business, New York

Forum PA (2012) ICity rate. La classifica delle città intelligenti italiane. www.saperi.forumpa.it. Accessed 10 May 2015

Giffinger R, Fertner C, Kramar H, Kalasek R, Pichler-Milanovic N, Meijers E (2007) Smart cities—ranking of European medium-sized cities. Centre of Regional Science, Vienna. http://www.smart-cities.eu/. Accessed 10 May 2015

Glaeser EL, Berry CR (2006) Why are smart places getting smarter? Taubman Center Policy Brief 2006-2. John F. Kennedy School of Government, Cambridge

Hall P (2000) Creative cities and economic development. Urban Studies 37(4):633–649

Hall RE, Bowerman B, Braverman J, Taylor J, Todosow H, Von Wimmersperg U (2000) The vision of a smart city. In: Proceedings of the 2nd international life extension technology workshop, Paris, France, 28 Sept. http://www.osti.gov/bridge/servlets/purl/773961-oyxp82/webviewable/773961.pdf. Accessed 10 May 2015

Hollands RG (2008) Will the real smart city please stand up? CITY 12(3):303–320

IBM (2009) How smart is your city? www.ibm.com. Accessed 10 May 2015

Inoguchi T, Newman E, Paoletto G (1999) Cities and the environment: new approaches for ecosocieties. UN University Press, New York

Jennings P (2010) Managing the risks of smarter planet solutions. IBM J Res Dev 54(4):1–9
Kanter RM, Litow S (2009) Informed and interconnected: a Manifesto for smarter cities. Harvard Business School, Working Paper 09-141
Lazaroiu GC, Roscia M (2012) Definition methodology for the smart cities model. Energy 47:326–332
Legambiente (2012) Ecosistema Urbano. XIX rapporto sulla qualità ambientale dei comuni capoluogo di provincia. www.legambiente.it. Accessed 10 May 2015
Legambiente (2013) Ecosistema Urbano. XX rapporto sulla qualità ambientale dei comuni capoluogo di provincia. www.legambiente.it. Accessed 10 May 2015
Legambiente (2014) Ecosistema Urbano. XXI rapporto sulla qualità ambientale dei comuni capoluogo di provincia. www.legambiente.it. Accessed 10 May 2015
Maltese I, Mariotti I, Boscacci F (2013) Smart cities and housing markets: evidence from Italy. In: Proceedings of XXXIV Conferenza AISRE, Palermo, September, 2nd–3rd, 2013
McCann P, Ortega-Argilés R (2013) Smart specialization, regional growth and applications to European Union Cohesion Policy. Regional Studies—Taylor & Francis
McLuhan M, Nevitt B (1972) Take today. Harcourt Brace Jovanovich, New York
Mitchell WJ (1995) City of bits. Cambridge University Press, Cambridge
Mitchell WJ (2007) Intelligent cities. UOC. UOC Papers no. 5. http://www.uoc.edu/uocpapers/5/dt/eng/mitchell.pdf. Accessed 10 May 2015
MIUR (2012) Smart cities and communities and social innovation. hubmiur.pubblicaistruzione.it. Accessed 10 May 2015
Morvaj B, Lugaric L, Krajcar S (2011) Demonstrating smart buildings and smart grid features in a smart energy city. In: Proceedings of the 2011 3rd international youth conference on energetics (IYCE 2011), Leiria, 7–9 July 2011
MSE (2012) Smart cities and communities, Bando PON National Operational Programme. www.ponrec.it/programma/interventi/smartcities. Accessed 10 May 2015
Nam T, Pardo TA (2011) Conceptualizing smart city with dimensions of technology, people and institutions. In: Proceedings of the 12th annual international conference on Digital Government Research, College Park, 12–15 June 2011
Nijkamp P, Oirschot G, Oosterman A (1993) Regional development and engeneering creativity: an instrumental comparison of science parks in a knowledge society. Free University, Research Memoranda, Amsterdam
O'Grady M, O'Hare G (2012) How smart is your city? Science 335:1581–1582
Ratti C (2012) Decalogue for a (smart) senseable city. In: Keynote speech at Smart City Exhibition 2012, Bologna, Italy, Oct 2012
Ravetz J (2000) Integrated assessment for sustainability appraisal in cities and regions. Environ Impact Assess Rev 20:31–64
Salat S, Novacki C (2010) De l'importance de la morphologie dans l'efficience énergétique des villes. Energie et territoires, Liaison Energie-Francophonie 86:141–146
Sassen S (2011) Who needs to become 'smart' in tomorrow's cities. In: Keynote speech at the LIFT conference, Marseille, France, July 2011
Shapiro JM (2006) Smart cities: quality of life, productivity, and the growth effects of human capital. Rev Econ Stat 88(2):324–335
Siemens—ANCI (2012) EfficienCITIES. Città modello per lo sviluppo del Paese. www.swe.siemens.com. Accessed 10 May 2015
Smart Grid Insights (2014) The role of energy in the smart cities, available at http://etsinsights.com/reports/role-energy-smart-cities. Accessed 10 May 2015
Toffler A (1980) The third wave. Bantam Books, New York, USA
UN-Habitat (1996) An urbanizing world: global report on human settlements. http://mirror.unhabitat.org/pmss/listItemDetails.aspx?publicationID=1650. Accessed 10 May 2015
United Nations (2009) World urbanization prospects: the 2009 version, available at http://esa.un.org/unpd/wup/index.htm. Accessed 10 May 2015
Winters JV (2011) Why are smart cities growing? Who moves and who stays. J Reg Sci 51(2):253–270

Yamagata Y, Hajime S (2013) Simulating a future smart city: an integrated land use-energy model. Appl Energy 112:1466–1474

Yigitcanlar T, Dur F (2010) Developing a sustainability assessment model: the sustainable infrastructure, land-use, environment and transport model. Sustainability 2:321–340

Energy and Spatial Planning: A Smart Integrated Approach

Paolo La Greca and Francesco Martinico

Abstract In spite of having a land occupation of 2 % and accommodating 50 % of the world population, cities produce 80 % of GHG emission and consume 80 % of the world's resources. Therefore, spatial planning has a key role in creating urban environments that support less energy-intense lifestyles and communities in order to meet the EU's challenging energy and climate change targets. However, the use of high tech applications is not sufficient: a smarter way of combining tools and approaches that come from the tradition of town and transport planning is required. Spatial planning has a long lasting tradition in defining the shape of urban fabric and the layout of buildings, defining the proper mix of land uses that takes into account the key role of transports. Containing and retrofitting urban sprawl by integrating transport and land use planning will be a considerable part of a new smart approach to city and energy. The new role of land use planning will also be accommodating new forms of distributed energy production in the urban fabric. In addition, planning tools will incorporate incentives aimed at favouring higher energy standards both for new and existing buildings. The technological innovation requires a comprehensive spatial framework, assuming that the energy point of view is a new challenge for innovating spatial planning.

1 New Challenges for Innovative Spatial Planning

There is no certain and unique road to walk through to tackle the challenges unveiled by climate change. In the building industry, both the culture and technics have already produced considerable achievements in the stream of the *green economy,* creating a large amount of energy efficient buildings that features also high architectural quality. Now is the turn of cities as a whole to change the gear to

P. La Greca · F. Martinico (✉)
DICAR, University of Catania, Catania, Italy
e-mail: fmartinico@dau.unict.it

P. La Greca
e-mail: plagreca@dau.unict.it

R. Papa and R. Fistola (eds.), *Smart Energy in the Smart City*,
Green Energy and Technology, DOI 10.1007/978-3-319-31157-9_3

build a new capacity in formulating sounded urban policies and creating adequate new tools for achieving this objective (La Greca 2015). As J. Clos, director of Un-Habitat, as recently pointed out:

> Understanding the contribution of cities to climate change will help us intervene at the local level. With better urban planning and greater citizen participation we can make our hot cities cool again.[1]

It is well known that the rise of interest in the relationship among energy buildings and land use was largely triggered by the 1973 oil crisis, consequent to the Kippur War, that caused an impressive increase in oil prices from 3 $ per barrel to 38 in 1981, a value that was matched again only in 2004. For a long period, the energy issue was no longer boosted by oil prices. However, after the 1970s oil crisis, a new emergency steadily arose so far to become a new paradigm: the climate change.

The Intergovernmental Panel on Climate Change (IPCC) was created in 1988. It was set up by the World Meteorological Organization (WMO) and the United Nations Environment Program (UNEP). Its objective was to devise, based upon the available scientific information, assessments on all aspects of climate change and its impacts, with the aim of formulating realistic response strategies in order to be included in a possible future international convention on Climate.

In 1992, 196 parties signed UNFCCC (United Nation Framework Convention on Climate Change), the parent treaty of the 1997 Kyoto Protocol. The interest on climate change raised considerably in the last 30 years, adding new stimuli to the discussion on the link between planning and energy.

In 2008, the European Parliament approved a resolution entitled *An Action Plan for Energy Efficiency*. This document calls on the Commission to propose a mandatory requirement that all new buildings needing to be heated and/or cooled shall be constructed to passive house or equivalent non-residential standards from 2011 onwards or higher standards (McLeod et al. 2012).

According to this Parliament resolution, the Commission issued two directives in 2010 and 2012. The foundation of these directives is that buildings are responsible for 40 % of energy consumption and 36 % of CO_2 emissions in the EU. While new buildings generally need less than 3–5 L of heating oil per m_2/year, older buildings consume about 25 L on average. Some buildings even require up to 60 L. Currently, about 35 % of the EU's buildings are over 50 years old. By improving the energy efficiency of buildings, we could reduce total EU energy consumption by 5–6 % and lower CO_2 emissions by about 5 %.[2]

The first one is the *Energy Performance in Buildings Directive* (2010/31/EU 2010). Under this directive, all new buildings must be nearly zero energy buildings by 31 December 2020 (public buildings by 31 December 2018). The second one is the *Energy Efficiency Directive* (2012/27/EU). According to the latter, EU countries must draw-up long-term national building renovation strategies which can be included in

[1]N-Habitat report 'Cities: Battle-Ground for Climate Change' 2011.

[2]See http://ec.europa.eu/energy/en/topics/energy-efficiency/buildings.

their National Energy Efficiency Action Plans (NEEAP). To examine in detail the Italian one is particularly interesting to find the limits of the proposed approach.

The Italian NEEAP was approved in 2014. According to 3.2 Paragraph, *Energy efficiency in the building sector*, the following actions are planned: strengthening of the minimum energy standards for constructing new buildings and renovating existing ones, in order to progressively increase the number of nearly zero energy buildings; consolidation of tax deductions; strengthening of incentives for the renovation of government buildings and of targets of the White Certificates scheme. Also in this plan, the focus is on buildings and neither on urban fabric nor on urban structure, intended as city layout and localisation of functions. The only explicit reference to urban planning tools is about the introduction of measures for speeding up the approval of local-level projects and harmonising and simplifying the issuance of building permits. In particular, the plan is reporting new measures approved for simplifying the authorisation process for installing energy efficiency and renewable source systems. By the same token, the paragraph on transport and sustainable mobility focuses only on the vehicles scale, describing the introduction of new electric vehicles charging points, the promotion of the use of bio methane in transport, the incentives for the purchase of low CO_2 emission vehicles and the like. It goes on by saying that at the beginning of 2016 there are about 57 km of underground railways, 20 km of tramways and 20 km of regional railways, for a total cost of about EUR 6.0 billion, already planned. There is also a generic reference to urban mobility plans, introduced by Law No 340/2000 but with limited results, since municipalities adopt them on a voluntary basis. There is also a mention to Urban Mobility Package of European Commission that is supporting the drafting of Sustainable Urban Mobility Plans (SUMP), which take an integrated and cross-sectorial approach to policies and public participation. However, no action is defined for implementing this approach. It is evident that the Italian NEEAP does not include any explicit reference to integration between Energy efficiency and spatial planning.

On the contrary, the EU steadily went on devoting much care to this issue. In May 2011, via a communication to the European Parliament, the Commission has defined the new *Roadmap for moving to a competitive low carbon economy in 2050,* going far beyond the previous *Europe 2020* flagship initiative strategy. It aims at providing member states with a long-term framework for dealing with the issue of sustainability and the cross-border effects of climate change, a phenomenon that cannot be dealt with at the national level alone (Fig. 1).

This roadmap for possible action up to 2050 is considering a key deliverable under the *Resource Efficiency Flagship* in order to reach the target of reducing greenhouse gas emissions by 80–95 % compared to 1990. The document concludes that under current policies the reduction cannot be more than 60 %. However, it argues that it is possible to reach the ambitious target of 80 % if additional policies will be adopted, taking into account technological options available over time.

The quest for these additional policies, and for new and adequate tools to face these challenges, is the reason for the EU to promote the Special project, founded

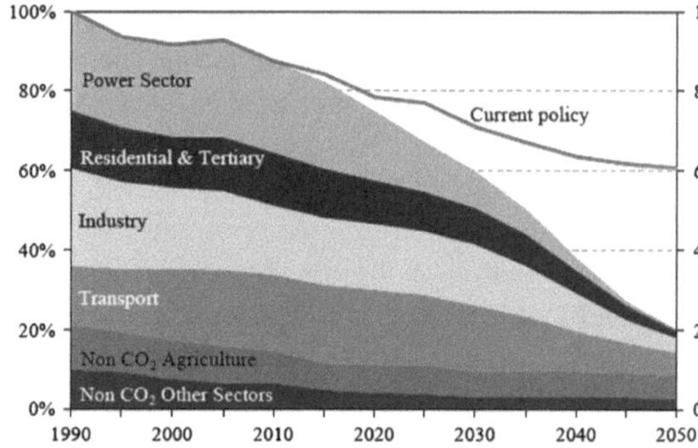

Fig. 1 EU GHG emissions towards an 80 % domestic reduction (100 % = 1990) (Color figure online)

within the Intelligent Energy Europe initiatives.[3] It is a European partnership aimed at enhancing the capability of Town Planning Associations to plan and deliver sustainable energy solutions. Spatial planning has a key role to play in creating urban environments that support less energy-intense lifestyles and communities and planners have a pivotal role in developing energy strategies and actions plans. The project's main objective is to bridge the gap between climate change/energy action planning and spatial and urban planning. Furthermore, it promotes the exchange of best practices and experiences of integrated renewable energy strategies, and building the capacity of the partner planning associations in the planning and delivery of renewable energy solutions. Most importantly, the partners must share that knowledge through their professional networks and maximise the dissemination of their training to others, in a multiplier effect.

This experience is at the basis of the considerations presented in this paper.

2 The Spatial Planners Contribution

2.1 Hints for a Taxonomy

The energy issue in land use planning has been recently boosted by the growing concern about climate change, one of the most discussed topics as far as the environment is concerned.[4] It is also possible to trace this theme back to the history of planning and planning theory (Taylor 1998), even though there are no explicit references to the word 'energy' in seminal books and experiences.

Spatial planning has a long lasting tradition in defining the shape of urban fabric and the layout of buildings taking into account the role of sun and wind. It is also

[3]Special includes planning associations from eight European countries, under the Leadership of Town and Country Planning Association, UK. The authors are in charge for the Special Italian partner: Centro Nazionale di Studi Urbanistici, CeNSU (National Centre for Urban Studies).

[4]Klein (2014) contains a non-neutral account of this raising concern.

concerned with defining the proper mix of land uses considering the key role of transports. Containing and retrofitting urban sprawl by integrating transport and land use planning will be a considerable part of a new smart approach to city and energy in the near future. The new role of land use planning will also be accommodating new forms of distributed sustainable energy production in the urban fabric. In addition, planning tools will incorporate incentives aimed at favouring higher energy standards both for new and existing buildings.

The inclusion of the energy issue in urban planning has to be considered not only as an environmental concern but also as a new urban policy, which has important effects on the public governance approaches and method (Moccia 2011, 2012; Zanon and Verones 2013).

The implementation of the energy approach in spatial plans requires a complex mix of regulations and market stimulating tools. Many studies are dealing with this issue, which is wider than the sole energy consideration (Tietenberg 2007). Impacts of urbanization on urban structures and energy demand have to be considered according to four dimensions: urban production, mobility/transports, infrastructure/urban density and private households (Madlener and Sunak 2011).

The relationship between spatial planning and energy can include the following categories:

- Operating on urban form and localization of urban functions,
- Introducing planning tools that include energy as a way of governing city and regions,
- Including green energy production in spatial plans.

In the following paragraphs we will point out the main issues at stake in each of these three dimensions referring them to relevant good practices that are diffusing worldwide.

2.2 Urban Form and Functions

There is a deeply rooted tradition in town planning on orienting buildings taking into account the role of sun and wind. Among several contributions, the relationship between the shape of buildings and land subdivisions, aimed at assuring solar access to the properties surrounding a given site, has been deeply studied by Knowles (1980) and formalised in the concept of a Solar envelope. Undoubtedly, the approach to orienting buildings according to these principles has been a pillar of modernist planning. Seminal contributions include Le Corbusier's visionary proposals and the more concrete Siedlungen, built in Germany in the 1920–30s. These ideas were strongly influenced by the use of a heliotherapy approach in medicine, which shaped the design of sanatoria where tuberculosis was cured. Health was the centre of the planners' concern, not energy. The quest for energy efficiency became more stringent during the 1970s when *Green Building* practices began to spread, even if their focus was mainly on micro scale. In this sense, a seminal contribution was offered by *PLEA* (Passive and low energy architecture), a network organisation conceived, nurtured and propelled by Arthur Bowen (Kook 2011).[5]

The idea that energy efficiency can be part of the design of the city was included also in Kevin Linch's last book (1981) where he states that energy consumption for heat and transport has to be considered in comparing the cost of settlement that include flats instead of row houses.[6]

Nowadays, the challenge of changing urban development to a more ecological and sustainable model is becoming more and more of a key issue. In considering this, the relationship between transport and land use is becoming central. As Kenworthy (2006) pointed out:

> [...] these dimensions have at their core the issue of urban transport system and their relationship to urban form, and therefore focus primarily on the problems of reducing automobile dependence in cities, building more sustainable urban form and creating more liveable places.

Automobile dependence is highly correlated with settlement shape. Low-density settlements are related to mobility dominated by car usage. One of the most relevant consequences is high-energy consumption and GHG emission. The key concept that relates urban and transport planning is accessibility. A change in the modes and styles of urban transport depends not only on urban and regional settlement patterns but also on socio-economic factors, road networks and answers of public domain to accessibility demands. These factors should not be detected with sectorial approaches for their complex relationships and the variety of aspects involved.

The elements described in up-to-date literature are the following:

- Urban forms (density, building morphologies, amount of urban sprawl),
- Urban functions (spatial distribution and mixed uses),
- People behaviours, with reference to urban conditions (quality and safety), life styles, accessibility not only to working places but also to urban functions in a broad sense.

However, from the review of existing literature it emerges that there is no evidence that compact or a more spread-out urban structure contributes to a higher or lower energy consumption (Næss 2005).

Built environment, travel and vehicle choice, driving behaviour, travel distance and energy consumption intertwine with each other. Single-family home settlements are still very popular not only in the USA in spite of their unsuitability (La Greca et al. 2011). From a study on the metropolitan area in Baltimore (Liu and Shen 2011) it emerges that, "households living in denser areas will choose smaller vehicles that consume less energy. Households in less dense areas might involve more motorized and highway travel, which might cause increases in Vehicle Miles Travelled VMT and energy consumption". Other studies confirm that larger

[5]The initial PLEA conference was held in 1981 in Miami, Florida organized by Arthur Bowen of Coral Gables University, Florida. La Greca, co-author of this paper, had the privilege to be there.

[6]Cpt 12 Efficiency and Justice p. 222.

dwellings are associated not only with higher use of energy for transport but also with higher consumptions of furniture and electronic appliances (Hoyer and Holden 2001).

Therefore, the new emerging paradigm includes the consideration of the role played by the settlement density, but several studies demonstrate that this is not sufficient. More factors are at stake.

However, adopting Smart growth and Transit Oriented Development (TOD) principles is a genuine sustainable perspective for integrating land use and mobility infrastructure (Calthorpe 1993; Kenworthy 2006). Although these approaches cannot be a panacea for all mobility problems, they can induce a change in mobility choices of inhabitants by giving them more opportunities. This will encourage an increase in public transport use and in non-motorised journeys.

The correct way to approach this issue is the one proposed by Handy (1996) who suggests to concentrate research on finding effective ways of providing residents with something they want rather than on finding ways to get them to do what policy makers want. This can be achieved by concentrating housing, retail and services near public transport nodes.

In spite of the success of this approach, which has been validated by several experiences, both in Europe and worldwide in the last 50 years (from Copenhagen to Bilbao, from Stockholm to Freiburg along with Portland and Vancouver), the Italian urban condition shows a quite opposite attitude being almost exclusively car oriented (Campos Venuti 1999).

Several examples can be considered at different scales.

An interesting one is the metropolitan area of Perth in Western Australia. This case demonstrates the possibility of introducing a strong relationship between transport infrastructures and settlements allocation in the region. A wise planning policy is often the result on a long lasting attention to planning. This region has a solid planning history at metropolitan level that begins in the 1950s with the Plan for the Metropolitan Region, Perth and Fremantle (the Stephenson-Hepburn Plan) released by the State Government in 1955. This plan provided the early basis for the city's current spatial form, characterised by urban corridors separated by green wedges of non-urban land uses.[7]

In 2004, the *Network City,* a strategic spatial planning document, was released The plan was developed through a dialogue with the city aimed at improving land use transport integration, in order to achieve more sustainable travel outcomes (Curtis 2008). It introduces three key components: 'Activity corridors' centered on either a main arterial road or suburban railway line; 'Activity centers', developed at intervals along the activity corridor and 'Transport corridors' that provide a higher speed, higher capacity route for inter-urban travel. The issue is a crucial one, considering that in the Perth metropolitan area (with a population of 1.4 million) suburban development averages only 6 dwellings/ha.

---

[7]Western Australian Planning Commission, Directions 2031 and beyond. Metropolitan planning beyond the horizon, August 2011 (www.planning.wa.gov.au).

In 2010, a new strategic document called *Directions 2031 and beyond* was released. The new document envisages the need of providing for new lifestyles, confirming the necessity of integrating infrastructure with land use planning and development. Among the key objectives of the plan, the most energy relevant ones include a settlement pattern defined as the "Connected City". It includes among its main features the following ones: promoting a better balance between greenfield and infill development, planning and developing key public transport corridors, urban corridors and TOD to accommodate increased housing needs and encourage reduced vehicle use.

Accordingly, the 2010 Plan has set a target of 47 % of the required new dwellings as infill development. This will reduce energy dependency and greenhouse gas emissions aiming at a 50 % increase in the current average residential density of 10 dwellings per gross urban zoned hectare (Western Australia Planning Commission 2011).

Back to Europe, the Swiss city of Geneva is facing a relevant urban growth due to its economic attractiveness. 50,000 new dwellings will be built by 2030. In this scenario the energy issue is a key one that requires a close integration with spatial planning. In this direction, the Canton of Geneva has adopted an innovative approach that has produced an unprecedented cooperation between energy and urban planning. Since 1987, this integration is mandatory and in December 2013 the Departments of Town Planning, House Building and Energy have been grouped under the same political direction, the *Département de l'aménagement, du logement et de l'énergie* (DALE). Resource management is included in the new Town and Country Planning 2030 Plan (*Plan Directeur Cantonal 2030*). A territorial Energy Concept (CET) is mandatory for each spatial planning procedure. This approach includes the spatial, temporal and relational dimensions, identifying key participants, co-ordinating them and offering energy supply strategies.

Several relevant projects have been elaborated according to this integration principle. These include a district-wide thermal network using the water from the lake Geneva (GeniLac) and the connection of the industrial zone of Plan-les-Ouates with the housing construction project of Les Cherpins (3000 dwellings and 2500 jobs) aimed at utilising the waste industrial heat in an area with scarce renewable energy resources.[8]

In the Praille-Acacias-Vernets project, the local master plan has been elaborated looking at the energy concept, enabling the establishment of spaces dedicated to deep geothermic. This will allow the use of this source of energy, extending it in the neighbouring areas with lower energy resources. The project is the transformation of the largest industrial area of the city (230 ha, located in its centre) into a mixed-use development.[9]

The excellence of this experience has been recently recognised at European level since the DALE was one of the winners of the 10th European Urban and Regional

[8]http://ge.ch/amenagement/pdq-cherpines.

[9]http://ge.ch/amenagement/praille-acacias-vernets.

Planning Awards 2013–2014, awarded by the European Council of Spatial Planners (ECTP-CEU) as one of the deliverables of Special Project.

Wilhelmsburg, is Europe's largest inhabited river island (835 km^2) located in the heart of the city of Hamburg. From 2006 to 2013 Wilhelmsburg formed the project area for the *Internationale Bauausstellung IBA Hamburg*. The exhibition is based on three themes: *Cosmopolis, Metrozones* and *Cities and Climate change*. The *Cosmopolis* theme includes the social side of the exposition; its goal is to adapt the city to the needs of its residents and to provide them with diverse opportunities for development. This theme is key one in this area, considering that Wilhelmsburg residents come from more than one hundred nations. The *Metrozones* projects are aimed at demonstrating how inner peripheries can be transformed into neighbourhoods worth living in. They include new housing and public buildings in the Inselpark area, bridges, pedestrian and cycling routes. The third theme is the one more directly related to energy. Projects include the production of energy with renewable sources, the retrofitting of existing buildings and high energy standards for the new ones. The objective is the reduction to almost zero CO_2 emission by 2050, in spite of an increase in population form 55,000 to more than 73,000.

According to the proposed energy concept, Wilhelmsburg will be self-sufficient, electricity-wise, by 2030 and more than 80 % heat self-sufficient by 2050 (Internationale Bauausstellung Hamburg 2010).

2.3 *Planning Tools*

One of the main issues for planning discipline nowadays is to find methodologies for designing sustainable districts. The inclusion of sustainability in urban design can be achieved in several ways. Sustainability is highly correlated with energy since it implies a careful use of resources. The simplest way to obtain it is by defining norms that impose sustainability standards higher than the ones that comply with European Union Directives.

However, this implies higher costs for the community with significantly undesirable effects. The main downside to this simple approach is that it risks confining the search for sustainability to wealthy communities, excluding the poor ones (Næss 2011). Several case studies show that in spite of an accurate analysis of the energy flows, in terms of consumption or energy savings and energy production potentialities, it is very difficult to use this knowledge in orienting action. Land use plans have been marginally influenced by the Energy plans and actions are still mainly oriented to management measures like green procurement or upgrading of municipal vehicle fleets. The focus is mainly on building level, both for public or private ones, whereas other key aspects like urban regeneration, densification, brownfield development and TOD are rarely evaluated according to an energy perspective.

This is particularly true in Italy where the approach to energy planning at the municipal level (introduced by Law 10/1991) remains focused on buildings, without addressing the complex relationships between urban patterns and energy consumption. This emerges clearly from the best Italian experiences that include the ones of Bologna, Modena, Venice and Trento (Zanon and Verones 2013). Other more promising experiences are under way in Faenza (Alberti 2013) and Parma.

Roughly, it is possible to classify planning tools into three groups: the first one includes those that are mainly concerned with the measurement of sustainability. The second group includes compensation tools. The third one includes economic incentives mainly intended as tax deductions.

Measurement tools are mainly parameters that allow a control of the level of compliance with sustainability standards. Several parameters have been already introduced in planning practices like the Biotope Area Factor (BAF) in Berlin, the Green Space Factor (Gsg) in Malmo or the *Riduzione Impatto Edilizio* (RIE) in Bolzano. However, these indices do not have an explicit reference to energy, even if their application affects considerably the energy consumption of the city.

A more interesting approach is the one of the second group that include incentives to sustainability that are based on market approach. This can be obtained in several ways that include the provision of higher Floor Area Ratios (FAR) for developments that are based on sustainability standards higher than the mandatory ones.

Some experiences are already explicitly referred to energy. Among these practices it is worth mentioning the UK one called *Carbon Offset Funds*. The zero carbon home policy was defined, in 2006, by the Code for Sustainable Homes (McLeod et al. 2012). In this Code it is required that, at Level 6, all carbon emissions have to be mitigated on site. Although some Code 6 houses have been built, the cost of building to that standard and its impracticability on many sites has reinforced the idea that delivering zero carbon through an entirely 'on-site' strategy was not the right approach for mainstream housing production or the wider commercial sector (Zero Carbon Hub 2012).

Consultation began in response to uncertainty over the existing definition of Zero Carbon Homes and concerns from the construction industry regarding its workability. One of the critiques is that this policy could potentially prejudice smaller urban brownfield developments in favour of larger greenfield sites because larger sites offer greater economies of scale in energy supply technologies. These concerns confirm that energy policies should be strictly integrated with spatial planning.

Following these complaints, the new policy differs from the previous definitions in the inclusion of a third tier, based on cost capped 'allowable solutions'. This term includes any approved carbon-saving measures that would be available to developers from 2016 to allow for the carbon that they would not normally be required to mitigate on site (Zero Carbon Hub 2012).

Also these solutions have been criticised (McLeod et al. 2012) since they can lack in additionality (the carbon offsetting did not create additional carbon savings relatively to what would have happened anyway) and permanence (when a low energy appliance was subsequently sold or replaced with a higher-energy consuming appliance). In contrast, fabric measures are likely to achieve carbon savings that will exceed 60 or even 100 years. Actually, carbon trading and offsetting has been criticised (Muller 2009).

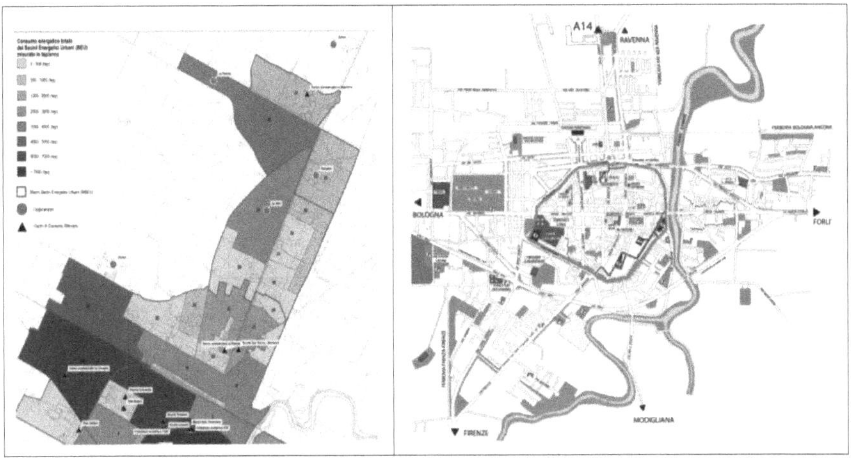

Fig. 2 Faenza' energy town planning scheme and bike sharing locations (*Source* Comune di Faenza: www.comune.faenza.ra.it) (Color figure online)

An interesting case is Faenza, a wealthy medium sized city (about 58,000) in Northern Italy. The city master plan has introduced a set of measures aimed at enhancing the city energy performance. The master plan takes into account the recent trends of the city development that moved from new greenfield housing to retrofitting and urban renewal (Nonni 2013) (Fig. 2).

The new city master plan defines the possibility to increase the building capacity (FAR) when higher standards are met. These measures vary according to the city zoning. For instance, in the built up area, residential buildings can be demolished and rebuilt with a 20 % volumetric increase. In the historic centre, retrofitting is supported by establishing an equivalence between non-emitted CO_2 and increased emission for new developments outside the historic centre. In this way, developers can buy the right to increase the FAR in their project from homeowners in the historic centre. This money will be used for retrofitting the historic building.

The third group is based on tax deductions, aimed at stimulating an increase in the energy efficiency of settlements. The simplest way of applying this is at building level. In Italy, a tax deduction scheme for upgrading buildings has been introduced since 2006.[10] This scheme has produced considerable results with more than 28,000 million euros investment.[11] However, it has no relationship with planning regulation, being a simple tax deduction mechanism, from the personal income or corporate taxes, equal to 50 % of money invested in general building refurbishment, including energy retrofitting for any property owner.

In this case, the relationship with land use planning risks to be weaker than the other two types of tools previously described. A useful development of the tax

[10]Source: Italian Energy Efficiency Action Plan, 2014.

[11]Source: Sole 24 Ore 23/08/2015.

deduction approach could be to link it to the use of tools for energy mapping (Ascione et al. 2013) or for rating energy efficiency at district level. These rating systems should be simple to apply and could also be used to verify the quality of detailed land use plans (Hedman et al. 2014). In this way, it would be possible to graduate the tax deductions according to the priorities that emerge from these classifications, enhancing the positive effects of these policies. A side effect would be also a contribution to the renewal of decaying districts.

2.4 Energy Production

Localization of production facilities, both at urban and regional level, is the third contribution of land use planning toward energy efficiency. Among renewable energy production, the one that has to be carefully considered for its consequences on cities and land use in general is the one of Photovoltaic (PV) power systems. In general, the role of planning is not only related to the introduction of production plants and devices, like combined heat plants or PVs in new or existing buildings but also to the definition of a set of rules for transforming agricultural areas or brownfields. In this case, the introduction of landscape protection norms for reducing negative visual impacts of energy production plants is fundamental.

Planning rules has to be carefully designed both for distributed solar energy systems and utility-scale solar energy plants (larger than 1 MW). Negative environmental impacts tend to be more relevant for the latter. Although solar technology is considered to be much preferable to traditional means of power generation, even considering wildlife and land use impacts (Vahro 2002; Turney and Fthenakis 2011) there are effects that have to be carefully considered. These include land use intensity and visual perception but also changes in the albedo or in groundwater recharge. Visual impact in agricultural and natural areas, more than in built up areas, should be considered. The evaluation of the aesthetic impact of solar power plants is useful for the selection of optimal plant location and most adequate use of panel technology, to minimize this impact (Torres-Sibille et al. 2009). For this reasons the development of large scale plants on degraded lands bear several environmental co-benefits, including the avoidance of additional land-use or land-cover change (Hernandez et al. 2014).

Also for these aspects, the IBA Hamburg experience shows several excellent examples of sustainable energy production.

The *Energieverbund Wilhelmsburg Mitte* (Integrated Energy Network Wilhelmsburg Central) is making a major contribution to energy supply efficiency and therefore climate protection on the Elbe Island. New developments in this part of the island and existing ones will be connected producing considerable synergies. The link of different users with different energy requirements will optimise the reliability of supply and reduce the CO_2 production. This system will allow an higher proportion of solar thermal energy and heath from combined power and heat co-generation, compared with the supply of individual buildings.

Fig. 3 Wilhelmsburg *Energie Bunker* (*Source* Special knowledge pool: www.special.eu)

Heat is generated mainly in central block heating power station, run on biogenic fuels. Energy is also fed into the network by solar thermal systems, located on suitable rooftops. The result is one large 'virtual' power station.

The *Energy Bunker* is a former anti-aircraft bunker that survived WW II and is now a listed monument. The Bunker is a distinctive landmark that has been transformed in a multipurpose facility that not only produces and stores energy but also accommodates a café and an information centre. Visitors can also enjoy a panoramic view of the area from the top of the building. The energy production features of the *Energy Bunker* include a large rooftop photovoltaic system, a solar thermal unit plant on the south facing façade, a biogas and a woodchip CHP plant and a large heat storage that saves surplus heat. In addition, the waste heat from a neighbouring industrial plant will be piped into the heat storage as well. This innovative project is an example of urban resilience, intended as the capacity of an existing infrastructure to be reused on the basis of its own peculiar features (Fig. 3).

The *Energie Colline Georgswerder* is a landfill that was converted into a park and an energy production site. This project is a clear demonstration of how a relevant environmental problem can be turned into a resource for the neighbouring communities. This hazardous landfill was closed in 1979, after decades of activity accumulating household and industrial waste. The energy production features four wind turbines, a 16,000 m^2 photovoltaic plant, a power plant that uses gas produced by the microbiological decomposition of the waste sealed off with plastic barriers. A panoramic aerial path on the top of the hill and an exhibition building contributes to attract visitors to a place that has changed radically from a negative icon of pollution and unsustainability to a positive one (Figs. 4 and 5).[12]

[12]Source: www.iba-hamburg.de.

Fig. 4 Aerial view of Wilhelmsburg *Energie Colline Georgswerder* (*Source* Energy Atlas IBA—Hamburg)

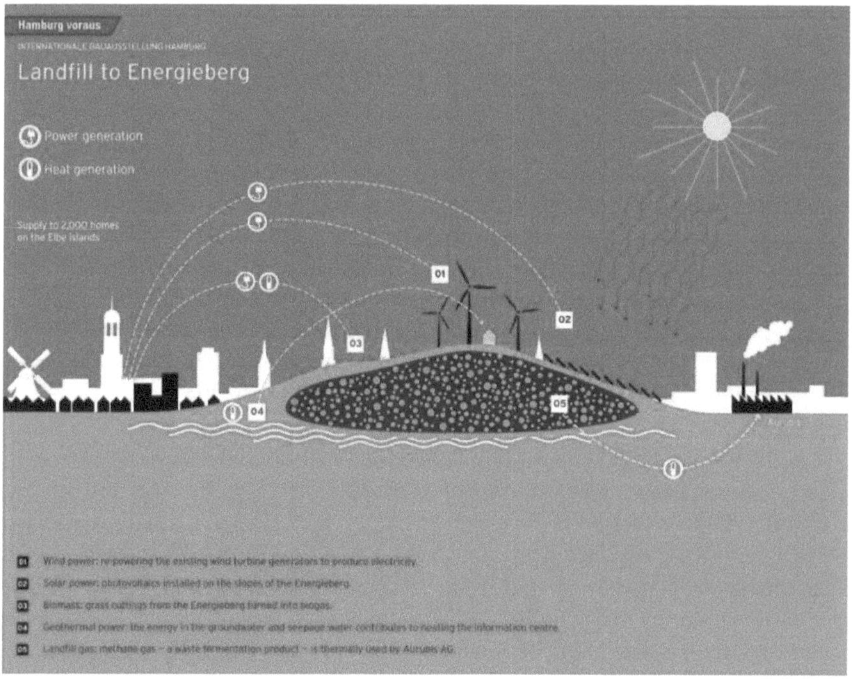

Fig. 5 Cross section of Wilhelmsburg *Energie Colline Georgswerder* (*Source* Energy Atlas IBA—Hamburg)

3 Conclusions

The introduction of energy concepts in spatial planning is a complex and still challenging field of research and practice. The set of good practices and related planning actions, presented in previous paragraphs, require a considerable increase in the level of technical competences, that are among the pillars of planning discipline. Introducing energy efficiency in planning requires also the continuous search of consensus, through participation for involving communities in the decision process, even if this should not prevail on technical competences that sometimes have been lagging behind.

It emerges that spatial planning can be useful to improve the energy efficiency of human settlements in several ways. It is fundamental to distinguish between greenfield settlements and urban renewal. So far, the first category is becoming residual in western countries, whereas the second is going to play a key role. Urban renewal includes both retrofitting of existing residential districts and brownfield settlements, mainly the re-use of defunct industrial areas and, nowadays, also commercial ones.

In planning and designing new sustainable districts, defining high levels of energy performance is straightforward but it can be unfair, since it tends to exclude low-income communities. In this case, good guidance at different scales is required. They span from the district localization in relationship with mobility infrastructure to the building level. A large number of codified procedures to guarantee high energy efficiency are already available.

As far as urban renewal is concerned, the level of complexity is higher, consequently planning tools has to devise smarter strategies. Traditional analytical techniques, popular among spatial planners, have to be updated by using algorithms, models and data commonly used by computational social science. By using these technologies new insights in the knowledge of urban structure and functions can be obtained. This updated level of knowledge is fundamental for fine-tuning some of the planning tools described in the previous paragraphs. In particular, compensation and tax deduction can benefit from this updated knowledge, being essential for the success of energy efficiency actions.

Urban retrofitting is not the only option available. The quality of an existing neighbourhood is multidimensional. It requires a holistic approach where risks evaluation is key. The energy retrofitting of poor quality settlements is undoubtedly a waste of time and money. An unattractive neighbourhood that is vulnerable to seismic and climate change risks, like many social housing districts of the 1960–70s, should be considered as eligible for demolition. On the contrary, devising way for introducing energy retrofitting as a central component of a complete urban renewal process is a new and stimulating field for planning research, since the choice of making sustainability the focus of spatial planning is unavoidable for the future of our planet.

References

Alberti M (2013) The energy enhancement project. In: Nonni E (ed) Biourbanism energy and planning. Valfrido Edizioni, Faenza, pp 181–210
Ascione F, De Masi RF, De Rossi F, Fistola R, Sasso M, Vanoli GP (2013) Analysis and diagnosis of the energy performance of buildings and districts: methodology, validation and development of urban energy maps. Cities 35:270–283
Calthorpe P (1993) The next american metropolis. Princeton Architectural Press, New York
Campos Venuti G (1999) Il trasporto su ferro per trasformare le città: Roma a confronto con le metropoli europee. Urbanistica, 112
Curtis C (2008) Planning for sustainable accessibility: the implementation challenge. Transp Policy 15:104–112
Handy S (1996) Methodologies for exploring the link between urban form and travel behaviour. Transp Res D 1(2):151–165
Hedman Å, Sepponen M, Virtanen M (2014) Energy efficiency rating of districts, case Finland. Energy Policy 65:408–418
Hoyer KG, Holden E (2001) Housing as basis for sustainable consumption. Int J Sustain Dev 4 (1):48–58
Hernandez RR, Easter SB, Murphy-Mariscal ML, Maestre FT, Tavassoli M, Allen EB, Barrows CW, Belnap J, Ochoa-Hueso R, Ravi S, Allen MF (2014) Environmental impacts of utility-scale solar energy. Renew Sustain Energy Rev 29:766–779
Kenworthy JR (2006) The eco-city: ten key transport and planning dimensions for sustainable city development. Environ Urban 18(01):67–85
Klein N (2014) This changes everything. Capitalism Vs the climate. Allen Lane, London
Kook J (2011) Plea yesterday, today and tomorrow. In: Bodart M, Evrard A (eds) Proceedings 27th international conference on passive and low energy architecture. Ucl Presses Universitaire de Louvain, Lovain-la-Neuve
Knowles R (1980) Solar access and urban form. AIA J 69
Internationale Bauausstellung Hamburg (2010) Energy atlas—future concept renewable Wilhelmsburg. Jovis Verlag GmbH and IBA Hamburg GmbH, Hamburg
La Greca P, Barbarossa L, Ignaccolo M, Inturri G, Martinico F (2011) The density dilemma. A proposal for introducing smart growth principles in a sprawling settlement within Catania Metropolitan Area. Cities 28:527–535
La Greca P (2015) Nuove sfide per la pianificazione urbanistica: cambiamenti climatici ed efficienza energetica sostenibile per le città. In: Moccia FD, Sepe M (eds) Una politica per le città italiane. Inu Edizioni, Roma, pp 128–142
Liu C, Shen Q (2011) An empirical analysis of the influence of urban form on household travel and energy consumption. Comput Environ Urban Syst 35:347–357
Lynch K (1981) Good city form. MIT Press, Boston
McLeod RS, Hopfe CJ, Rezgui Y (2012) An investigation into recent proposals for a revised definition of zero carbon homes in the UK. Energy Policy 46:25–35
Næss P (2005) Residential location affects travel behavior—but how and why? The case of Copenhagen metropolitan area. Progress in Planning 63:167–257
Næss P (2011) Economic growth, urban development and environmental sustainability. In: Moccia FD (ed) Abitare la città ecologica/Housing ecocity. Clean, Napoli, pp 49–77
Nonni E (2013) A new environmental vision. In: Nonni E (ed) Biourbanism energy and planning. Valfrido Edizioni, Faenza, pp 13–28
Madlener R, Sunak Y (2011) Impacts of urbanization on urban structures and energy demand: what can we learn for urban energy planning and urbanization management? Sustain Cities Soc 1:45–53
Moccia FD (2011) Disciplinary integration Need. In: Moccia FD (ed) Abitare la città ecologica/Housing ecocity. Clean, Napoli, pp 49–77

Moccia FD (2012) Città senza petrolio. Relazione introduttiva. In: Moccia FD (ed) Città senza petrolio. Edizioni Scientifiche Italiane, Napoli

Muller B (2009) Additionality in the clean development mechanism: why and what? http://www.oxfordenergy.org. Accessed 25 July 2015

Taylor N (1998) Urban planning theory since 1945. Sage, London

Tietenberg T (2007) Tradable permits in principle and practice. In: Freeman J, Kolstad CD (eds) Moving to markets in environmental regulation. Lessons from twenty years of experience. Oxford University Press, Oxford

Turney D, Fthenakis V (2011) Environmental impacts from the installation and operation of large-scale solar power plants. Renew Sustain Energy Rev 15:3261–3270

Torres-Sibille AdC, Cloquell-Ballester VA, Cloquell-Ballester VA, Artacho Ramirez MA (2009) Aesthetic impact assessment of solar power plants: an objective and a subjective approach. Renew Sustain Energy Rev 13:986–999

Vahro V (2002) Environmental impact of photovoltaic electrification in rural areas. Energy Environ 13(1):81–104

Western Australian Planning Commission (2011) Directions 2031 and beyond. Metropolitan planning beyond the horizon. http://www.planning.wa.gov.au/publications/826.asp. Accessed 28 July 2015

Zanon B, Verones S (2013) Climate change, urban energy and planning practices: Italian experiences of innovation in land management tools. Land Use Policy 32:343–355

Zero Carbon Hub (2012) Allowable solutions evaluating opportunities and priorities. www.zerocarbonhub.org. Accessed 28 July 2015

Technological Change and Innovation for Sustainable Cities: A Multiagent-Based Ontological Approach

Dino Borri, Domenico Camarda, Laura Grassini and Mauro Patano

Abstract Debates on sustainable technologies for urban management—be they related to energy, water, transport, or any other major sectors—tend to be restricted to issues about how to support a transition to new and more efficient technologies, as part of a linear path of progress proceeding from old, backward technologies to modern ones. Technologies themselves tend to be treated like black-boxes, somehow exogenously defined by engineering R&D efforts, while societies are mainly conceived of as passive agents, which can, at most, accept or reject them on the basis of their preferences. In our paper we adopt a different perspective, i.e., we start from the acknowledgement of the importance of the work of many distributed agents and micro-learning processes in technological evolutions and innovations as well as from the idea that sustainable solutions are by no means restricted to linear progression from past to future, as they can also encompass returns and recombinations of traditional components and practices together with new ones. After grounding our approach in some of the main contributions from innovation studies as well as in the application of cognitive science to evolutionary studies of technologies, we specifically discuss the possibility to use a multi-agent and multi-indexing ontological system as a device to share and learn technological knowledge across a wide community, thus supporting those micro-learning processes of distributed agents, which—in our perspective—can boost innovation and change in practice. In so doing, we will refer to a case study developed within the EU-funded project ANTINOMOS "A knowledge network for solving real-life water problems in developing countries: bridging contrasts", which has largely dealt with enhancement and management of local/community-based knowledge for water sector management. The major aim is to create an appropriate learning environment for sharing and for actively generating knowledge through multi-actor

The present chapter is the result of a joint research work carried out by the authors. Nevertheless, D. Borri wrote Sect. 3, D. Camarda wrote Sect. 4 and 6, L. Grassini wrote Sect. 1 and 2 and M. Patano wrote Sect. 5.

D. Borri (✉) · D. Camarda · L. Grassini · M. Patano
Technical University of Bari, Bari, Italy
e-mail: dino.borri@poliba.it

R. Papa and R. Fistola (eds.), *Smart Energy in the Smart City*,
Green Energy and Technology, DOI 10.1007/978-3-319-31157-9_4

synergies. In this paper, the above subject is discussed and carried out with a cross-disciplinary, cross-scale, multi-agent approach, considering the different forms of local knowledge and language involved.

Keywords Water resource · Technological evolution · Learning architecture · Semantic navigation · Ontological indexing

1 Introduction

Current debates on technological innovation still tend to be dominated by a linear conception of technological change, which has its roots in mainstream economic thinking and organizational management. Economists have traditionally considered innovation and technological change at the core of economic growth (Schumpeter 1942; Solow 1957) and have long debated about how to foster technical change and about how to define key determinants of innovation and technological change in the attempt to increase the rate of productivity of economic enterprises (Schmookler 1966; Mansfield 1968; Rosenberg 1982; Griliches 1984; Nelson and Winter 1982; Scherer 1986).

In his book "The Theory of Economic Development" Schumpeter argued that development (at that time considered to be equivalent to the concept of growth) was the result of the innovative ability of the entrepreneur and his introduction of new methods of production (Schumpeter 1934). Since then, an heroic conception of technological innovation has been progressing through radical inventions made by a few dominating individuals (Shumpeter 1934, 1942), as well as the idea that innovation proceeds in a linear way from an old, backward technological past towards modernity (Rostow 1960). In accordance to this idea, policies to foster innovation have widely adopted the "technology push" approach, with its core preoccupation on R&D measures and appropriate market policies to sustain wider diffusion of new technologies within society.

Despite the fact that theories and practices still seem to be dominated by enduring convictions about linearity and the breakthrough features of technological changes, real case situations often show opposite evidence. They show, for instance, that technological evolutions may be detached by heroic events, as they may be linked to the work of many distributed agents and be nurtured by several micro-learning processes (Garud and Karnøe 2003; Hendry and Harborne 2011; Grassini 2011). On the other side, they show that trajectories of change do not always follow linear evolutionary patterns, as they may be characterized by possible returns and recombinations of traditional components and practices (for an account in the water sector see also Barbanente et al. 2012; Grassini 2013).

While these observations have constituted the ground for empirical and theoretical research in innovation studies, a cognitive perspective in the analysis of technological change and innovation is still largely to be developed. In our perspective this severely hampers the possibility to understand real determinants and mechanisms of innovation and change in practice. This is the reason why in this paper we try to develop such a perspective. We will specifically reason about technological knowledge and the possibility to develop appropriate tools to share and learn technological knowledge in a multi-agent context. The development of a multi-agent and multi-indexing ontological approach will specifically support micro-learning processes of distributed agents in the definition of feasible options for technological evolution.

Water offers an exceptional stance in the above-mentioned perspective, because of the centrality of water technologies in the ecology of living beings. Moreover, in this field we can observe conflicting dynamics of change, with the strengthening of centralized, high-tech large-scale hydraulics, on one side, and the opposite trend towards the recovery and creative recombination of small and distributed hydraulics produced by local common knowledge, on the other (Borri et al. 2009). Is there any possibility to bridge these antinomies and to define a learning space among these different dynamics of change?

In the attempt to reason about this, we use a case study developed within the framework of the EU funded project ANTINOMOS "A knowledge network for solving real life water problems in developing countries". The ANTINOMOS project rationale lies in the conviction that inadequate knowledge management is at the roots of the present obstacles to solve real-life water problems in most developing countries. One example of such obstacles lies in the existence of enduring divisions and sectoral thinking (Latour 1987; Bijker 1995) across different forms of knowledge and disciplinary approaches to water management, which hampers the capacity to develop more effective and sustainable technological solutions to water problems in most real-life contexts. The enduring conflict between modern technologies and traditional ones is part of a larger opposition between the knowledge systems in which those technologies are embedded.

Starting from this situation, the ANTINOMOS project tries to see if the possibility exists to bridge these antinomies and to define a learning space among them, which may eventually lead to some interesting cases of technological evolutions. In our perspective, the possibility to create these micro-learning processes nurturing technological evolutions depends on the capacity to structure appropriate learning spaces where multiple actors can interact in a meaningful way. This is not easy to do, as their knowledge may be embedded in very different knowledge systems, depending on their disciplinary backgrounds—which entails the use of different taxonomies and knowledge frames—geographical belongings—which entails different knowledge labels and different languages—and different scales—which also means different levels of generalization of concepts. Different actors may indeed belong to different "knowledge communities" (Nonaka 1999), all defining shared codes and jargon for internal communication and rules for sharing information and knowledge. A strong attention to the development of an appropriate cognitive architecture for this interaction thus becomes a core asset. This is what we tried to develop within the ANTINOMOS project, which we describe in the empirical section of the paper. Technologies analyzed are from three countries, namely India, Mexico and South Africa.

The paper is structured as follows: in the next section we discuss major research findings in technological innovation studies with reference to systemic and co-evolutionary approaches to technological innovations and to the role of distributed actors in innovation processes. In the third section some introductory considerations are carried out, relating the multi-agent system approach in dealing with organizations. In the following section, some notes on multi-agent cognition and the role of ontologies in learning system architectures are drawn, as well as key aspects for efficient indexing searching. Finally, Sect. 5 deals with the current aspects, potentials and follow-ups of the learning system architecture, followed by final remarks in Sect. 6.

2 Evolutionary Patterns of Technologies

Innovation and technological change have traditionally been conceived of by economists at the root of economic growth (Schumpeter 1942; Solow 1957). In his book "The Theory of Economic Development" Schumpeter argued that development (at that time considered to be equivalent to growth) was the result of the innovative ability of the entrepreneur and of the introduction of new methods of production (Schumpeter 1934). Technological change was conceived of in three stages, starting with invention as the stage of generation of new ideas, followed by innovation as the locus for the transformation of new ideas into marketable products and processes, and then ending with the diffusion of new products and processes into potential markets.

Since then, economists started debating how to foster technical change and how to identify key determinants of innovation and technological change, in the attempt to increase the rate of productivity of economic enterprises (Schmookler 1966; Mansfield 1968; Rosenberg 1982; Griliches 1984; Nelson and Winter 1982; Scherer 1986). Directly connected to the Schumpeterian concept of innovation is the "technology push" approach to technological change, which basically conceives of innovation as proceeding from the engineering application of scientific discovery through manufacturing to the marketplace. In this perspective, R&D efforts are the real boosts of innovation while users are mainly reduced to passive consumers of technological products. The "demand pull" model is but another version of this linear model of innovation, whose only difference from the "technology push" model is the attribution of a more active role to users through the possibility they have to articulate a market demand, which somehow contributes to give a direction to R&D efforts (Schmookler 1966).

Nevertheless, both models pay little attention to the socio-cultural environment where innovations are supposed to take place. For them local communities, and users in general, just constitute a sort of "selection environment" for exogenously driven technologies (Breschi and Malerba 1997). In those approaches, R&D efforts are confined to specialist knowledge and technical expertise, outside local communities. Technological change itself is a sort of "black box" (Rosenberg 1982), exogenously driven, proceeding with breakthrough events made by a few talented innovators. The success and failures of innovation activities are attributed primarily to specific individuals (Gartner 1988), as heroic individuals who have the ability to discover, create and exploit opportunities that lie beyond the reach of most.

Like science, technological change and economic growth are conceived of as proceeding through a linear pattern from a backward past to modernity, with “take off” being deeply related to the capacity of countries to unload the traditions of the past and adopt modern technologies and models of production (Rostow 1960). This strongly supported the transfer of technology as a key approach to spur economic growth within lagging behind economies. This quick-fix approach to the use of new technologies to solve complex development problems of developing economies glibly dismisses the achievements of the past (Agarwal and Narain 1997) and underestimates and minimizes the many difficulties some of the new technologies have brought in their wake.

This is particularly evident in relation to technological innovations for environmental management, where increasing critiques emerged, highlighting the environmental as well as the social and cultural side effects of application of the technology-push and the transfer of technology approaches (for the water sector see also Escobar 1996; McCully 1996; Postel 2000; Shiva 2001; Grassini 2012).

At the same time, this conception neglected the differences of local contexts in terms of factor endowments and assets, including natural resources, knowledge and social and cultural capital. In particular, the productivity of technological innovation along imitation curves runs the risk of being significantly lower than the one obtained from innovation efforts rooted in local factor endowments (Hicks 1932). This is particularly evident in the field of natural resources management, where traditional technologies—which developed in close tune with the capacity of local population to exploit the potential of their endowments—tend to be substituted by modern technologies ubiquitously transferred across the globe (Ahmad 1966; Binswanger and Ruttan 1978). One particular case of this tendency, and a critique to its effect in wider terms, has been recently discussed in the field of water resources management (Kubursi et al. 2011).

Initial attempts to open the black box of technology and to show the more intricate and complex dynamics leading to innovation and technological changes were made within evolutionary economics, with its claim for a non-deterministic description of technology development, the role of limited rationality of involved actors and the co-evolutionary linkages between technologies and organizational settings (Nelson and Winter 1977, 1982; van den Bergh and Gowdy 2000).

The need to adopt a more systemic approach to technological innovations through a co-evolutionary framework of analysis encompassing technological artifacts as well as social, institutional and policy environment has been reinforced by contributions within the field of history and sociology of technology, in particular with the research stream known as Social Construction of Technological System (Bijker et al. 1987; Hughes 1987). More recently, several approaches were developed in the attempt to encompass the multi-dimensional dynamics, which affect at the same time technology, user practices, policies and institutional structures through complex multi-actor processes. One of these is the multi-level perspective on socio-technical systems (Rip and Kemp 1998; Geels 2002), with its focus on larger “socio-technical systems” (Geels 2004; Geels and Schot 2010)—encompassing at the same time material artifacts and techniques, but also knowledge related to them (Raven and Geels 2010), networks of actors, and institutions (socio-cultural norms, technical standards, regulations) (Voβ et al. 2009; Farla et al. 2012). It develops a theoretical framework to explain the non-linearity of technological transitions by the interplay of dynamics at three levels: niches, regimes and landscapes (Geels 2002). The field of technological innovation systems is also interesting (Hekkert et al. 2007; Bergek et al. 2008), in that it applies methodologies coming from the analysis of National Systems of Innovation or Sectoral Innovation Systems (Carlsson and Stankiewicz 1991; Archibugi and Lundvall 2001) with its attempt to overcome the narrow concept of market failure as the only reason responsible for poor diffusion of innovations and to single out a broader set of system failures, which encompass institutional, infrastructure, organizational aspects (Negro and Hekkert 2008).

These frameworks have been successfully employed in the environmental domain, where they contributed to contrasting the simplistic and overenthusiastic faith in the "technology fix" approach for solving environmental problems, which could only provide partial and temporary solutions due to rebound effects or other unintended consequences (Farla et al. 2012). In so doing, they have supported the birth of a newly formed field of research dealing with environmental innovation and "sustainability transitions" (Elzen et al. 2004; Geels et al. 2008; Smith et al. 2010), which have specifically developed both detailed accounts of the formation of new socio-technical configurations, as well as frameworks for analyzing determinants of radically new modes of production and consumption.

As a result of application of the above-mentioned analytical frameworks to the issue of innovation and technological change, several steps have been made in that direction to contrast long lasting convictions about innovations and technological changes. One of them is certainly the idea that technological innovation only proceeds through heroic and breakthrough events, which is increasingly challenged by a growing body of research dealing with those cases where innovations are not based on any new dramatic and breakthrough inventions or scientific discoveries, but rather on the steady accretion of inputs from many actors (Garud and Karnøe 2003; Kamp et al. 2004; Hendry and Harborne 2011; Grassini 2011). The case of the wind turbine development pattern in Denmark is one of such cases. It shows how technological change was linked to the interplay of different actors: designers and producers, on one side, which tried to incorporate inputs of several actors into low-tech design; users offering continual feedbacks; policy makers modulating the emergence of the market to keep the technological path alive. The emergence and success of this incremental pattern of technological change in Denmark happened at the time when a technology-push approach to wind turbine development was failing in the US and in other parts of the world, despite huge investments in R&D efforts and large subsidies to jump-start new technologies (Kamp et al. 2004; Hendry and Harborne 2011).

Acknowledgement of the importance of micro-learning processes in technological evolution and a critique of the breakthrough models of technological innovations are being attempted in several other sectors, including the water sector, where negative effects induced by an heroic technology-push approach to technological change have been analysed in comparison to a series of micro-adaptive changes happening at the grassroots level (Grassini 2012).

These types of cases shed light on different issues.

On one side, they highlight the importance of a multiplicity of learning modalities—learning by doing, learning by using, learning by interacting—besides the much celebrated learning by search (Kamp et al. 2004). This in turn leads to the acknowledgement of inherent fragility of high-tech breakthrough development patterns deriving from the technology-push approach as far as they tend to overcome and to stifle multiple micro-learning processes from distributed agencies (Garud and Karnøe 2003).

On the other side, these cases highlight the existence of a different pattern of technological change, beginning with a low-tech design but ramping up progressively through multiple inputs and micro-learning processes. Furthermore, it is the very capacity of the emergent pattern to learn from multiple inputs from distributed actors that makes it prevail even over a high-tech breakthrough development pattern deriving from the technology-push approach, which instead ends up stifling multiple micro-learning processes from distributed agencies (Garud and Karnøe 2003). This happened in the case of the wind turbine development in Denmark, but is increasingly evident in several other sectors.

The embeddedness of these innovation patterns on micro-innovation processes made by distributed actors and their reliance on the capacity of those actors to bring into the process their technological and practical knowledge lead to other interesting observations, pointing to the capacity of those actors to use, re-interpret and innovate technological components and knowledge deriving from ancient and traditional technologies. Starting from the Schumpeterian definition of innovation as a process of search and recombination of existing components, which Schumpeter defined as "factors" of innovation (Schumpeter 1934), a whole range of research developed about the direction of innovation efforts within firms, being somehow dependent upon the fundamental tension between exploration of untried possibilities and exploitation of previous successes (March 1991). Although potential components and their combination are infinite, recombination usually occurs between components that are salient, proximal and available for the inventor (Fleming 2001).

In this respect, literature on firm and organizational studies have long debated the opportunity of firms to adopt local versus distant search for components, thus addressing the divide between searching within existing competencies in a path dependent way and exploring new components outside their traditional expertise (Nelson and Winter 1973; March 1991; Stuart and Podolny 1996), and the level of depth of search within each direction (Katila and Ahuja 2002). Costs and benefits of those opposite tendencies have been widely discussed, with often mixed evidence (Ahuja and Lampert 2001; Fleming 2007; Cohen and Levinthal 1990; Fleming 2001; Messeni Petruzzelli and Albino 2012).

As a particular type of search for local components, in the last years increasing attention is being given to the age of components leading to innovation within firms. Despite the mainstream conviction that "old" components are "obsolete" as they do not fulfill current needs and expectations (Sørensen and Stuart 2000), some research is attempting to demonstrate how firms may successfully innovate by re-interpreting re-employing past competencies and solutions, due to their reliability and/or unexploited potential and/or renewed interest of the consumers towards the rediscovery of the "past" (Katila 2002; Nerkar 2003; Brown et al. 2003; Messeni Petruzzelli et al. 2012; Messeni Petruzzelli and Albino 2012).

The interest in innovation patterns, rooted in re-interpretation of traditional components and knowledge, seems particularly strong in relation to technological innovations for environmental management, partly due to increasing critiques of the complex negative effects of exogenously driven, high-tech technologies and the acknowledgement of potentials of hybridization pattern of technological change, which are able to re-interpret pieces and rational of traditional technologies in innovation dynamics (for an account in the water sector see also Barbanente et al. 2012; Grassini 2013).

But if and how traditional/old components and knowledge can become available within a complex innovation pattern is still an under-debated issue. How can this knowledge and practice be shared among different actors involved in the process? How can a real learning environment be created that can lead to learning about the past and finding root innovation in it? This is the core concern we try to address with this paper.

3 Multi-agent Systems and Technological Organizations

Multiagent modeling methods (MMMs), in both theory and practice, have rarely dealt with modeling knowledge when this is highly distributed among agents (societal knowledge, at its different from local to general levels) and mostly has non-expert features (common knowledge, end users knowledge) (Ferber 1999).

Due to the prevalent operational orientation of MMMs, knowledge elicitation mostly selects well structured sources (oriented to specific tasks, clearly localized in space and time, having accountable expertise in the concerned knowledge domain, etc.), neglecting the ill structured ones (Buchanan and Shortliffe 1977).

But research reflection on Technological Change and its cognition-in-action aporias needs the adoption of special adaptive analytical lenses which fit the observation of both explicit and non-explicit (latent, unclear), single and multiple (community, not task-oriented) agents: when spatial-temporal location of agents is problematic, participant observation is impossible and different ways of knowledge elicitation (conjectural, compilation-based) have to be adopted which are usually cumbersome and do not fit the heuristic shortcut typical of knowledge elicitation for automatic reasoning (Johnson-Laird 1988).

Adopting adaptive analytical lenses for cognitive change, which is embodied in the technological change, means adopting a multiscale tool (frame) and a much wider system inasmuch as agents are a distributed multiplicity and changes come from knowledge interaction at the level of the 'social' system and not merely from the knowledge individual generation (Minsky 1987). A difference in scale and a difference in dimension, both evoke a difference in knowledge structure and/or knowledge organization (for the relevance of this couple in systems analysis see Zeleny 2005) and make MMMs for distributed and unclearly structured (not task-oriented societal organizations) multiple agents that are much more problematic than MMMs for polarized (concentrated) and clearly structured (task-oriented organizations). The monotonic (at the level of the single agent or of the multiple agents operating in a task-oriented organization) reasoning architecture typical of MMMs for polarized knowledge have to be changed into a non-monotonic reasoning architecture because of the continuous adjustments and microchanges that feed the TC when this is analyzed at the societal level (Collins 1990; Minsky 1988) (Figs. 1 and 2).

Rather than being made by monolithic production processes and problem solving abilities, technologies are made by technological chains that in turn are made by huge decentralized systems of production processes, by technological chains in which decisions and actions are taken by multiple agents. As usual in the expert domain of TC, for instance, some basic ideas of the current policies for continuous innovation technology in Europe is that the agents' behaviors must be changed according to the innovation lines provided by the experts, that the technological policies must be coherent, integrated, and multisectoral, that the behavioral economists must be among the main leaders of the effort, and that the "European technology" must become the standard international reference for the sake of present and future generations (see Eu Sti Report for 2030: Giovannini 2016, p. 9). Nothing of new, of course. But the fact is that the idea of centralized and global models of technology is increasingly challenged by both the recurrent

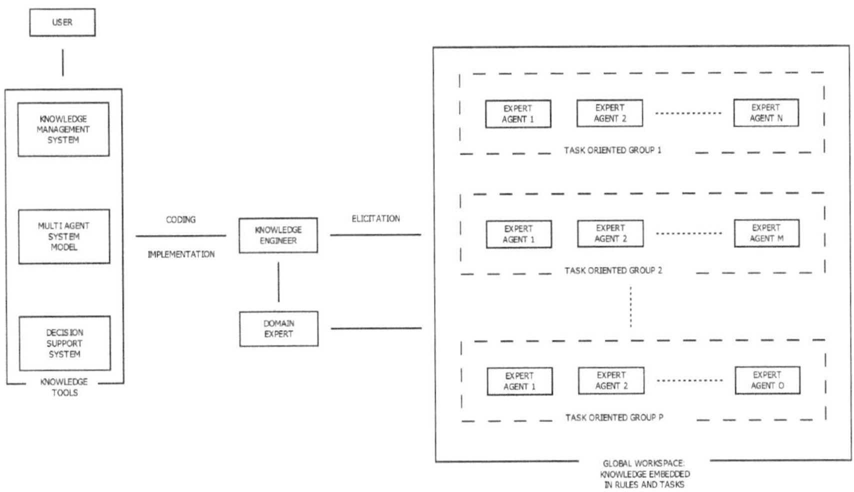

Fig. 1 Conventional model of knowledge elicitation for expert, task-oriented, non-distributed agent

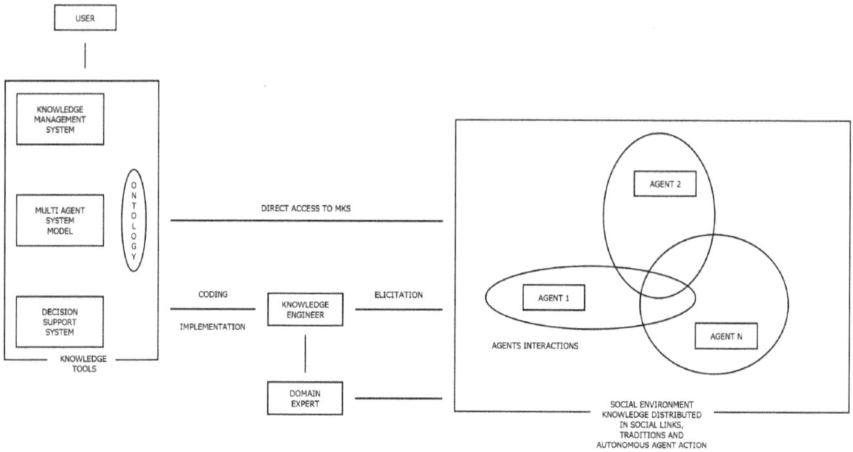

Fig. 2 Alternative model of knowledge elicitation for commonsense, non-task-oriented, distributed agents multiplicity

disasters that are created in many cases by the introduction of new technologies and models of 'industrial' production (see the appearance of a new generation of man-induced micro-seismic activity with the diffusion of the 'shale' techniques for the extraction of gas from soils) and the continuous rediscovery of the good performances of old technologies abandoned (see the coming back in the building sector for thermal isolation to materials and solutions of the past, even of a 'tribal' and 'pre-civilization' past, as in the case of reintroducing non artificial, vegetal, and bio-components) face to the bad ones of novelties in the field; see the coming back to pedestrian movements in cities; see the rediscovery of cannabis as a powerful and beneficial pain killer for human health) (Giovannini 2016, p. 9).

Future technology depicts scenarios in which human routine problem solving is substituted with intelligent machines while only creative problem solving, manual and/or intellectual, remains human prerogative because of its complex and unforeseeable features, a thesis which continues to support the idea of a Promethean, heroic, individual, non-societal character of any intelligent organization: an advanced Technologic Change will be the virtuous seedbed of this intelligent selective knowledge-in-action evolution; in this vein, the intelligent and self-reflexive evolution of technology (for Aeschylus, Griffith 1983; Severino 1979, 1998) will easily outpace society and its traditional non-competitive rhythms (Comelli 2016, p. 10).

But, for better understanding the nonlinear and not always positive evolution of technology, the fact that in many cases a given technical solution to a problem remains strictly localized without any exportation, confined in a sort of local life (in the ancient world, the disappearance of the traditional technique used by Roman builders for covering wide spaces by light circular domes brought to generalized spatial-temporal absence for one thousand years of this architectural performance; more recently, many sustainable solutions to problems of production of energy, water, food, health, etc. come back to old traditional solutions), having attention to the peculiar spatial-temporal and social organization of any relevant technological system, for instance with its typical input-output layout of distributed agents and related processes, is a basic condition (Zhang 1998).

If attention is paid to multiagent spatial-temporal organization of any relevant technology (a technological system which creates and comes from mass production processes, presents complex needs of maintenance, whose total or partial failure creates mass problems to the involved agents and users) the routine and molecular structure of many technological processes and systems and their socially distributed multiagent roots and features become evident, technological memory (Borri 2011) integrates technological creative and heroic disruption, and spatial-temporal knowledge modeling becomes much more complex if we compare it to the simple single (which for us means also a coherent but in general not massive group of agents) task-oriented agent-based organization which is involved and observable in action without hard spatial-temporal granularity and complexity. In parallel, also the conventional structure of knowledge elicitation models (in primis its dyadic observation attitude) is changed by the appearance of highly distributed in space-time multiagent knowledge sources into an alternative structure based on a multiplicity of reflexive and selfreflexive agents participate in the technological effort in complex spatial-temporal dimensions and scales and can continuously adapt their theories-in-action at both single agent and multiagent levels. In this vein, knowledge elicitation for Technological Change simulation models goes closer to the anthropological architecture of a unsupervised self-reflexive neural nets which learn from a trial-and-error massive exploration of existing and known technological resources with their performances of use (Winograd and Flores 1987; Dosi 1988; Mignolo 2000; Brodt 2001; Howells 2002; Bathelt et al. 2004; Rinkus et al. 2005; van de Kerkhof and Wieczorek 2005; Keeney et al. 2007; Barbanente et al. 2012; Silver et al. 2016).

4 Ontological Indexing in Multi-agent Knowledge Environments

Organizing and managing social services to support local communities is more and more dependent on multiscale levels of infrastructure technologies. From extensive projects oriented to serve larger settlements to small installations for villages and neighbourhoods, infrastructuring efforts seem to be typically focused on diverse and frequently unconnected scale levels. Yet, while extensive projects are usually able to bring about significant—perhaps just formally—knowledge deposits for technology dissemination, smaller initiatives are often unshared, confined to a local and narrow use, even if they may prove to be successfully applied in daily life. Further, because of manifold—geological, physical, socioeconomic, organizational, political—problems in delivering large-scale infrastructures to all communities, smaller groups basically continue to be dependent on local-based technological initiatives.

However, particularly developing Countries show typical contexts of small and scarcely connected communities (in terms of cognitive, organizational interactions and information) where technologies can hardly spread temporally and spatially. This situation frequently leads to short-lived, fragmented, non-replicable and often unsuccessful experiences. In case-studies of water delivery and sanitation, technological knowledge gaps seem to lay behind that circumstance, rooted on how to build, manage and maintain technology infrastructures (Unver 1997; Torregrosa Armentia et al. 2006). Yet the awareness of that complex issue is able to stimulate peculiar research actions aimed at narrowing and bridging gaps by fostering learning environments for knowledge raising and exchanging.

The support to knowledge interaction processes based on information technology has increasingly shaped local development initiatives over the last few decades. Positive outcomes have shown the potentials of such approaches, yet mixed up with an evidence of critical aspects (Borri et al. 2005, 2009). The diffusion of information technologies in Developing Countries, in particular, is often hampered by subtle political constraints (well beyond commonplaces of culture and literacy). In turn, this results in asymmetrical information systems, beneficial to a bunch of corporative agents rather than to the population as such (Greenwald and Stiglitz 1986, 1990). However, a customized IT-based approach involving real-time processes, multiple agent interactions, multiple source knowledge management can set up articulated knowledge networks and learning environments potentially powerful and effective.

The research group at Politecnico di Bari gained fair experience in recent years in this concern. Some features proved to be critical during the carrying out of such initiatives, so requiring proper consideration when building operational environments (Khakee et al. 2002; Borri et al. 2006). A number of key issues can be singled out in particular, both contextually and in general terms.

(i) An initial issue, well known in organizational studies, is related to the management of IT-based interaction platforms, as particularly poor communities often suffer from difficulties in accessing and using such systems. This involves a frequent interaction and feedback delegation to groups of intermediate agents representing the community, so determining problems of legitimated representation of the community. Furthermore, there is the risk for even intermediate agents to be unaware of filter interactions, with consequent issues of misinterpretation, mystification and a general low level of knowledge transmission to/from the community (Forester 1988, 1999). Ad hoc methodologies and system approaches oriented to easing and broadening the access and the interactive capacity of all possible agents can significantly decrease the above problems and improve the efficacy of knowledge exchanging initiatives.
When setting up the IT-based learning environment focused on water questions, characters and technologies, a basic knowledge base built on relevant literature represents our initial ground to share and develop cognitive features among involved agents. Formalized as a normal hypermedia file, this basic repository is oriented to be searched, navigated, revised, integrated and

assimilated cognitively by agents using an ad hoc website, and finally aimed at setting up a complex, dynamic learning environment.

(ii) A second issue linked to the building up of the system concerns languages, idioms and expressions, that are particularly critical when indexing knowledge contents to carry out searching and navigation activities across documents (Khaled and Mohamed 2004). As a matter of fact, such agent types as researchers, professionals, scientific experts make use of their formal domain languages and expressions when searching over the web. In that case, the structuration of documents for search enhancing aims is relatively easy, as searching tags are actually formal tags, and therefore basic navigation and the related sharing of knowledge base is quite straightforward (King and Munson 2004). On the other side, non-expert agents have to rely on more commonsense, informal approaches, widely founded on conceptualizations, phrases, locutions and customized syntagms (for example, searching for techniques to "*...have less stinky water*", or devices to "*...carry out maintenance more easily*", etc.). Indeed, a concept-related approach is useful also to expert agents, particularly if they need to make more varied and combined search activities. Both agent categories eventually entail fuzzier and more articulated potentials of navigation, so needing more complex document indexing in the system structure, aimed at allowing a more diversified and effective learning environment.

(iii) Thirdly, the cognitive frames of agents involved in navigation activities do affect IT-based interaction systems. As a matter of fact, literature reports that each agent carrying out interactive reasoning tasks is characterized by distinct cognitive layout and organization in performing their knowledge-based activities (e.g., Shanahan 1997). The frame problem seems able to influence the ability of each cognitive agent to navigate without getting lost in massive 'problem spaces'. In memory studies, framing is exercised to prune dangerous and unproductive areas of problem spaces, basing on contextual situations. This is particularly true when there is a large number of knowledge agents involved—as occurring in internet-based multi-agent environments (Borri et al. 2004). In terms of search-based activities, searching itself is impacted by the frame problem, so suggesting the use of distinct indexing approaches when dealing with knowledge memories and deposits, to attain effective navigation. In this context, a more versatile, deep, semantic search indexing seems more effective in responding to diverse cognitive frames of agents than the usual synthetic tag indexing.

(iv) A final issue concerns the inferring of indexing criteria for searching activities. In fact, navigating agents may derive hints for new indexes either from interacting with other agents or/and from their substantial search activity. Searching itself may well suggest new conceptual items useful to refine navigation and gain access to new characters of technological knowledge. Should new navigation criteria unveil structural relations among concepts, then an ontological structure of interrelated criteria may emerge and finalize dynamic indexing frameworks useful for successive navigations. Dynamic ontological indexing means that the indexing feature evolves during navigation and/or interaction

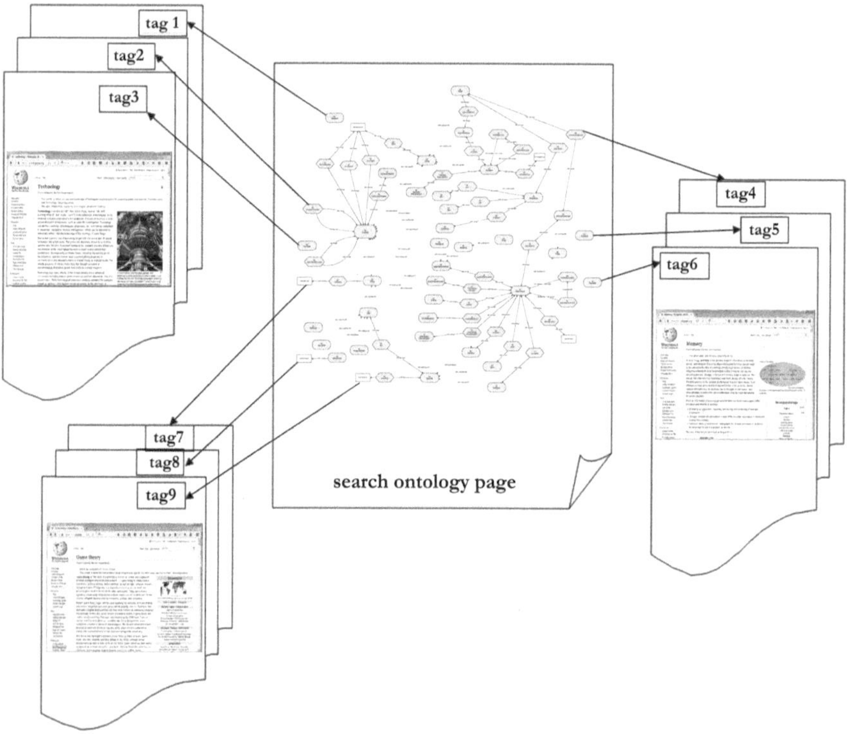

Fig. 3 An example of the ontology-based indexed search

activities. The system tries to keep memory of the indexing ongoing update, so determining a sort of self-learning intelligent systems, using an ontology-based approach (Abraham and Grosan 2008) (Fig. 3).

From the reasoning developed up to here, ontologies seem to suggest useful perspectives toward the setting up of indexing systems to support learning-oriented navigation environments. However, many suggested features need deeper reference to IT-based devices and in general to computer science, i.e., to subjects significantly out of our research aims. In the subsequent section we will give a synthetic account of an interactive, internet-based learning environment, loosely evoking semantic-based search indexing with ontological perspectives.

5 The Web-Based Interactive Learning Environment

In the implementation of the knowledge management system (KMS) software and structure, the previous general issues are combined, in order to meet demands more directly connected to the research project itself. The operational life of the system may be viewed as made up of a first phase, with mainly research suggestions, followed by a second disseminating and organizational phase, aimed at letting information be used by local communities in diverse developing Countries (Fig. 4).

The users involved in the first phase are mainly author-type agents and their number is relatively limited. Here the system tries to meet the requirements of partners from different sectors and Countries and with specialized, expert knowledge.

In the second phase, the number of users is quite greater and is mostly based on reader-type agents. The user is registered and her/his work submitted to a board controlling added and/or modified pieces of information. The requirements supported by the KMS are particularly relevant to the connection with global and local networks, as well as to the possibility of filling knowledge gaps and enabling the solution of possible differences in the vision of the subjects at hand.

While the operational time of the system can roughly be separated into two steps, other general characteristics are demanded to the KMS. Access to significant, real time and easy comprehensible information, easy using of such information for decision making, creation of an environment allowing learning but also knowledge generation are all features that the system should ensure throughout.

As a matter of fact, from a mere architectural viewpoint, the standard LAMP (Linux, Apache, MySQL, PHP) was the solution stack used. This environment was

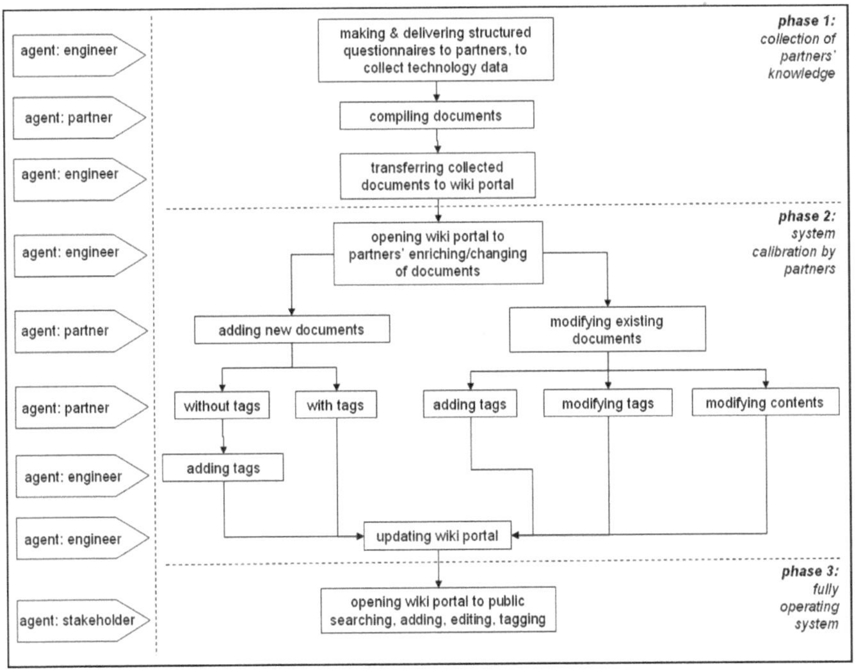

Fig. 4 Phases and agents participating to the building up of KMS

selected because of some well-known advantages, namely open source code, wide user-community support, easy-to-code, easy-to-deploy and develop-locally features, cheap and ubiquitous hosting (Gerner et al. 2006).

Furthermore, a collaborative web-based knowledge-sharing software was adopted, belonging to the known 'wiki' category. Particularly, the "MediaWiki" software was preferred, with the addition of 'semantic MediaWiki' extension. This choice was due to a number of interesting characteristics, such as, e.g., the easy creation and editing of a very large quantity of web pages through a browser, by using very simple markup language and WYSIWIG text editor. Most processing costs (usually very low when using wikis) are almost completely server-side supported. From the client's side, just a standard, not necessarily up-to-date web browser is required (Ebersbach et al. 2008) (Figs. 5 and 6).

An idea behind this KMS is the possibility of satisfying two different, sometimes conflicting, requirements, i.e., (i) being prominently collaborative and community-supporting and (ii) securing publishing rights, in order, on one hand, to safeguard the intellectual rights of different authors and, on the other, to get control of any off-topics.

The requirement of organizing knowledge according to different meaningful topics is met partly by creating linkages among different web pages via hyperlinks,

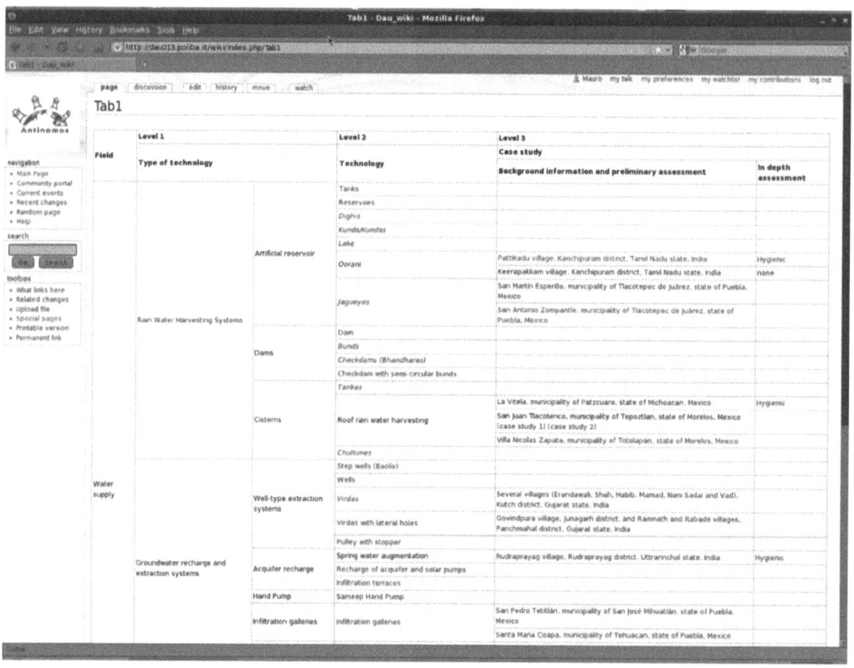

Fig. 5 The wiki portal of ANTINOMOS project

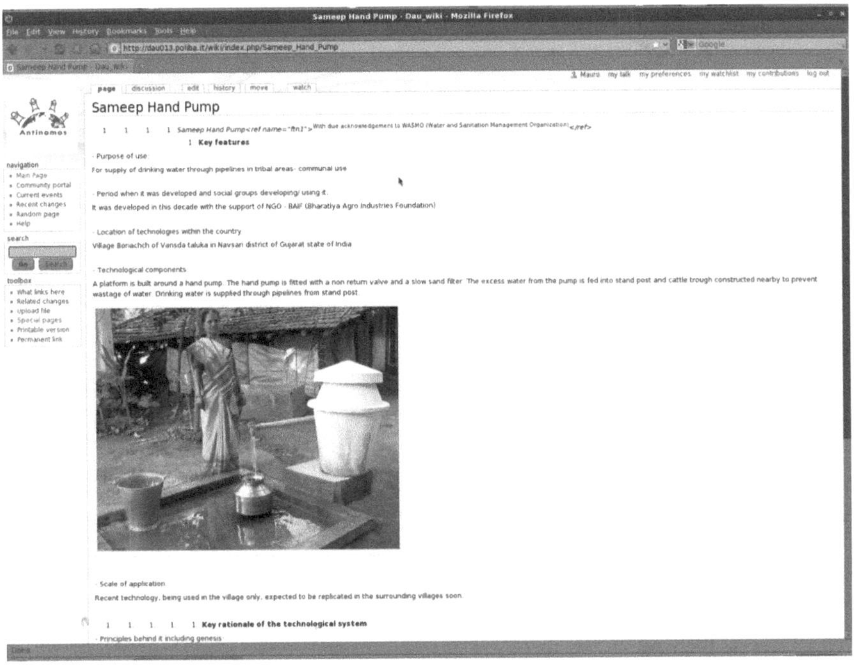

Fig. 6 An excerpt illustrating the informal technology of water supply

partly through the markup system of the semantic extensions installed. The ongoing creative and collaborative process oriented to change the 'cognitive landscape' from the viewpoints of both stored raw data and the classification meaning of information in its thorough complexity is supported by the KMS functions. The aim is to set up a cooperative system of knowledge management, able to foster creation, navigation, search features and, last but not least, attribution of meaning.

Some features of our KMS are implicit in the non-linear navigation support of wikis: users may create links, indexes and tables, with reference to any preferred knowledge organization. Yet, a more structured organization layout of contents is being implemented in the system, framed on semantic extensions and ontologies. Those extensions introduce some additional markup into the page text, which allows users to add "semantic annotations" to the wiki. In this way the structure of the wiki is simplified and the overall quality and consistency of the contents are improved. Other benefits of using "semantic extensions" are: automatically-generated lists, visual display of information, enhanced information searching, external reuse, integration and mash-up of data.

The completing of such structuring effort was intended as a peculiar activity of the research project. It represented an interesting research perspective that is still being considered, aimed at developing technology-learning environments in informal contexts based on multi-agent approach.

6 Brief Final Remarks

The present paper dealt with the importance and the perspectives of the multi-agent learning and sharing of water technologies in informal contexts, within the EU-funded ANTINOMOS project. In particular, the setting up of system architectures to support learning and knowledge interactions among diffused agents has been investigated, using web-based models and technologies. In this context, knowledge exchanges and technology learning are supposed to be highly dependent on the quality of the interaction environment. Therefore, the study has devoted a particular attention to the structuring of the searching and navigation features of the system, heavily highlighting the quest for efficient indexing approaches.

To this aim, the importance of semantic navigation has been emphasized, and the quest for searching features has been oriented to the potentials of ontological indexing, as a base to allow semantic tasks.

At present, the study is still at an early stage of complex, yet traditional, web searching tasks. This paper has shown some aspects of current features, also presenting the funding rationale, some potentials and some criticalities of the next semantic-based learning environment. The development of the system architecture of that environment represents an important legacy of research activities started off with the ANTINOMOS for the future.

Acknowledgements The research on which the paper is based is part of the project ANTINOMOS, "A knowledge network for solving real life water problems in developing countries", funded by the European Commission under the Sixth Framework Programme.

References

Abraham A, Grosan C (2008) Engineering evolutionary intelligent systems: methodologies, architectures and reviews. Stud Computat Intell 82:1–22

Agarwal A, Narain S (1997) Dying wisdom. rise, fall and potential of India's traditional water harvesting systems. State of India's Environment: A Citizens' Report, N. 4, Centre for science and Environment (CSE), New Delhi

Ahmad S (1966) On the theory of induced innovation. Econom J 76:344–357

Ahuja G, Lampert CM (2001) Entrepreneurship in the large corporation: a longitudinal study of how established firms create breakthrough inventions. Strateg Manag J 22:521–543

Archibugi D, Lundvall BA (eds) (2001) The Globalizing Learning Economy. Oxford University Press, Oxford

Barbanente A, Borri D, Grassini L (2012) Evolutionary technologies in knowledge-based management of water resources: perspectives from South Asian case studies. In: Higgitt D (ed) Perspectives on environmental management and technology in Asian River Basins. Springer, Dordrecht, pp 45–67

Bathelt H, Malmberg A, Maskell P (2004) Clusters and knowledge: local buzz, global pipelines and the process of knowledge creation. Prog Hum Geogr 28(1):31–36

Bergek A, Jacobsson S, Carlsson B, Lindmark S, Rickne A (2008) Analyzing the functional dynamics of technological innovation systems: a scheme of analysis. Res Policy 37:407–429

Bijker WE (1995) Of bicycles, bakelites and bulbs: towards a theory of sociotechnical change. The MIT Press, Cambridge/London
Bijker WE, Hughes TP, Pinch TJ (eds) (1987) The social construction of technological systems: new directions in the sociology and history of technology. MIT Press, Cambridge
Binswanger HP, Ruttan VW (1978) Induced innovation: technology, institutions, and development. John Hopkins University Press, Baltimore
Borri D (2011) Technological memories. Plurimondi. Int Forum Res Debate Hum Settl IV(9):81–91
Borri D, Camarda D, De Liddo A (2004) Envisioning environmental futures: Multi-agent knowledge generation, frame problem, cognitive mapping. Lecture Notes in Computer Science, vol 3190, pp 230–237
Borri D, Camarda D, Grassini L (2005) Complex knowledge in the environmental domain: Building intelligent architectures for water management. Lecture Notes in Computer Science, vol 3533, pp 762–772
Borri D, Camarda D, Grassini L (2006) Distributed knowledge in environmental planning: hybrid IT-based approaches in scenario-building contexts. Group Decis Negot 6:557–580
Borri D, Grassini L, Starkl M (2009) Technological innovations and decision making changes in the water sector: Experiences from India. In: Messerschmid C, El-Jazairi L, Khatib I, Al Haj Daoud A (eds) Water: values & rights. Palestine Academy for Science and Technology, Palestine, pp 577–590
Breschi S, Malerba F (1997) Sectoral innovation systems: technological regimes, Schumpeterian dynamics, and spatial boundaries. In: Edquist C (ed) Systems of innovation: technologies, institutions and organizations. Pinter, London/Washington, pp 130–156
Brodt SB (2001) A system perspective on the conservation and erosion of indigenous agricultural knowledge in central India. Human Ecol 29(1):99–120
Brown S, Kozinets RV, Sherry JF (2003) Sell me the old, old story: retromarketing marketing and the art of brand revival. J Customer Behavior 2:85–98
Buchanan B, Shortliffe E (1977) Production rules as a representation for a knowledge-based consultation prigram. Artif Intell 8(1):15–45
Carlsson B, Stankiewicz R (1991) On the nature, function and composition of technological systems. J Evol Econ 1(2):93–118
Cohen W, Levinthal D (1990) Absorptive capacity: a new perspective on learning and innovation. Adm Sci Q 35:128–152
Collins HM (1990) Artificial experts. social knowledge and intelligent machines. The MIT Press, Cambridge
Comelli E (2016) Tecnologie più veloci della società (Technologies faster than society). Il Sole 24 Ore 512:10
Dosi G (1988) Sources, procedures and microeconomic effects of innovation. J Econom Literature 26(2):1120–1171
Ebersbach A, Glaser M, Heigl R, Warta A (2008) Wiki: web collaboration. Springer, Berlin
Elzen B, Geels FW, Green K (eds) (2004) System innovation and the transition to sustainability: theory, evidence and policy. Edward Elgar, Cheltenham
Escobar A (1996) Constructing nature: elements for a post-structural political ecology. In: Peet R, Watts M (eds) Liberation ecologies. Environment, development, social movements. Routledge, London, pp 46–68
Farla J, Markard J, Raven R, Coenen L (2012) Sustainability transitions in the making: a closer look at actors, strategies and resources. Technol Forecast Soc Chang 79:991–998
Ferber J (1999) Multi-agent system: an introduction tio distributed artificial intelligence. Addison Wesley Longman, Harlow
Fleming L (2001) Recombinant uncertainty in technological search. Manage Sci 47(1):117–132
Fleming L (2007) Brekthroughs and the 'long tail' of innovation. MIT Sloan Manage Rev 49:68–74
Forester J (1988) Planning in the face of power. University of California Press, Berkeley

Forester J (1999) The deliberative practitioner: encouraging participatory planning processes. MIT Press, Cambridge
Gartner WB (1988) "Who is an entrepreneur" is the wrong question. American J Small Business 12(4):11–32
Garud R, Karnøe P (2003) Bricolage versus breakthrough: distributed and embedded agency in technology entrepreneurship. Res Policy 32:277–300
Geels FW (2002) Technological transitions as evolutionary reconfiguration processes: a multi level perspective and a case study. Res Policy 31:1257–1274
Geels FW (2004) From sectoral systems of innovation to socio-technical systems. insights about dynamics and change from sociology and institutional theory. Res Policy 33(6/7):897–920
Geels FW, Schot J (2010) The dynamics of socio-technical transitions—a socio-technical perspective. In: Grin J, Rotmans J, Schot J (eds) Transitions to sustainable development. Routledge, London, pp 9–101
Geels FW, Hekkert MP, Jacobsson S (2008) The dynamics of sustainable innovation journeys. Technol Anal Strateg Manag 20(5):521–536
Gerner J, Naramore E, Owens M (2006) Professional LAMP: Linux, Apache, MySQL, and PHP web development. Wiley & Sons, New Jersey
Giovannini E (2016) Innovazione per la sostenibilità (Innovation for sustainability). Il Sole 24 Ore 512:9
Grassini L (2011) Buone prassi per chi? Politiche di sviluppo e pratiche interpretative in India. Alinea, Firenze
Grassini L (2012) Water resources management and territorial development: technological changes in the Apulia region in the post-Unitary period. Plurimondi. Int Forum Res Debate Hum Settl 11:89–123
Grassini L (2013) Gestione delle risorse idriche e sviluppo territoriale. Percorsi di innovazione tra modernità e tradizione. Planum. J Urbanism 27(2)
Greenwald BC, Stiglitz JE (1986) Externalities in economies with imperfect information and incomplete markets. Q J Econ 101(2):229–264
Greenwald BC, Stiglitz JE (1990) Asymmetric information and the new theory of the firm: Financial constraints and risk behavior. Am Econ Rev 80(2):160–165
Griffith M (ed) (1983) Aeschylus' Prometheus Bound. Cambridge University Press, Cambridge
Griliches Z (1984) R&D, Patents and Productivity. University of Chicago Press, Chicago
Hekkert MP, Suurs RAA, Negro SO, Kuhlmann S, Smits REHM (2007) Functions of innovation systems: a new approach for analyzing technological change. Technol Forecast Soc Chang 74:413–432
Hendry C, Harborne P (2011) Changing the view of wind power development: more than "bricolage". Res Policy 40(5):778–789
Hicks J (1932) The theory of wages. Macmillan & Co, London
Howells JRL (2002) Tacit knowledge, innovation and economic geography. Urban Stud 39(5–5):871–884
Hughes TP (1987) The evolution of large technological systems. In: Bijker W, Hughes TP, Pinch T (eds) The social construction of technological systems. MIT Press, Cambridge, pp 51–82
Johnson-Laird P (1988) The computer and the mind. An introduction to cognitive science. Collins, London
Kamp LM, Smits REHM, Andriesse CD (2004) Notions on learning applied to wind turbine development in the Netherlands and Denmark. Energy Policy 32:1625–1637
Katila R (2002) New product search over time: past ideas in their prime? Acad Manag J 45 (5):995–1010
Katila R, Ahuja G (2002) Something old, something new: a longitudinal study of search behavior and new product introduction. Acad Manag J 45:1183–1194
Keeney J, Lewis D, O'Sullivan D (2007) Ontological semantics for fistributing contextual knowledge in highly distributed autonomic systems. J Netw Syst Manage 15(1):75–86
Khakee A, Barbanente A, Camarda D, Puglisi M (2002) With or without? Comparative study of preparing participatory scenarios using computer-aided and traditional brainstorming. J Future Res 6:45–64

Khaled MH, Mohamed SK (2004) Efficient phrase-based document indexing for web document clustering. IEEE Trans Knowl Data Eng 16:1279–1296
King P, Munson EV (eds) (2004) Digital documents: systems and principles. Springer Verlag, Berlin
Kubursi A, Borri D, Grassini L (2011) In praise of traditional technologies. Plurimondi. Int forum Res Debate Human Settl. Special issue on Antinomos: Dilemmas in water supply and sanitation technologies IV(8):21–50
Latour B (1987) Science in action: how to follow scientists and engineers through society. Open University Press, Milton Keynes
Mansfield E (1968) Industrial research and technological innovation. Norton, New York
March J (1991) Exploration and exploitation in organizational learning. Organ Sci 2:71–87
McCully P (1996) Silenced rivers: the ecology and politics of large dams. Zed Books, London
Messeni Petruzzelli A, Albino V (2012) When tradition turns into innovation. How firms can create and appropriate value through tradition. Woodhead Publishing Limited, Cambridge
Messeni Petruzzelli A, Rotolo D, Albino V (2012) The impact of old technologies on innovation: the case of US biotechnology industry. Technol Anal Strateg Manag 24:453–466
Mignolo W (2000) Local histories/global designs: coloniality, subaltern knowledges, and border thinking. Princeton University Press, Princeton
Minsky ML (1987) The Society of Mind. Simon & Schuster, New York
Minsky M (1988) The emotion machine: common sense thinking, artificial intelligence, and the future of human mind. Simon and Schuster, New York
Negro S, Hekkert MP (2008) Explaining the success of emerging technologies by innovation system functioning: the case of biomass digestion in Germany. Technol Anal Strateg Manag 20 (4):465–482
Nelson RR, Winter SG (1973) Towards an evolutionary theory of economic capabilities. Am Econ Rev 63:440–449
Nelson RR, Winter SG (1977) In search of useful theory of innovation. Res Policy 6(1):36–76
Nelson RR, Winter SG (1982) An evolutionary theory of economic change. Cambridge University Press, Cambridge
Nerkar A (2003) Old is gold? The value of temporal exploration in the creation of new knowledge. Manage Sci 49:211–229
Nonaka I (1999) The dynamics of knowledge creation. In: Ruggles R, Holtshouse D (eds) The knowledge advantage. Ernst and Young, Dover
Postel SL (2000) Entering an era of water scarcity: the challenges ahead. Ecol Appl 10(4):941–948
Raven RPJM, Geels FW (2010) Socio-cognitive evolution in niche development: Comparative analysis of biogas development in Denmark and the Netherlands (1973–2004). Technovation 30(2):87–99
Rinkus S, Walji M, Johnson-Troop KA, Malin JT, Turley JP, Smith JW, Zhang J (2005) Human-centered design of a distributed knowledge management system. J Biomed Inform 38 (1):4–17
Rip A, Kemp R (1998) Technological change. In: Rayner S, Malone EL (eds) Human choice and climate change—resources and technology. Battelle Press, Columbus, pp 327–399
Rosenberg N (1982) Inside the black box: technology and economics. Cambridge University Press
Rostow WW (1960) The stages of economic growth: a non-communist manifesto. Cambridge University Press, Cambridge
Scherer F (1986) Innovation and growth, schumpeterian perspectives. The MIT Press, Cambridge
Schmookler J (1966) Invention and economic growth. Harvard University Press, Cambridge
Schumpeter JA (1934) The theory of economic development: an inquiry into profits, capital, credit, interest and the business cycle (trans from the German by Opie R). Transaction Publishers, New Brunswick and London
Schumpeter JA (1942) Capitalism, socialism and democracy. Harper, New York
Severino E (1979) Legge e Caso (Law and chance). Milano, Adelphi
Severino E (1998) Il Destino della Tecnica (The destiny of the technique). Milano, Rizzoli
Shanahan M (1997) Solving the frame problem. MIT Press, Cambridge

Shiva V (2001) Water wars. privatisation, pollution and profit. Pluto Press, London
Silver D et al (2016) Mastering the game of Go with deep neural networks and tree search. Nature 529:484–489
Smith A, Jan-Peter Voβ J-P, Grin J (2010) Innovation studies and sustainability transitions: the allure of a multi-level perspective and its challenges. Res Policy 39(4):435–448
Solow RM (1957) Technical change and the aggregate production function. Rev Econ Stat 39:312–320
Sørensen J, Stuart T (2000) Aging, obsolescence, and organizational innovation. Adm Sci Q 45 (1):81–112
Stuart TE, Podolny JM (1996) Local search and the evolution of technological capabilities. Strateg Manag J 17:21–38
Torregrosa Armentia ML, Castro JE, Allen A, Gómez González Cosío R, Vera J, Kloster K (2006) Institutional development and political processes, baseline document for the cross-cutting perspective 2. National Water Commission, Mexico City
Unver OIH (1997) South-eastern Anatolian Project (GAP). Water Resour Dev 13(4):453–483
van de Kerkhof M, Wieczorek A (2005) Learning and stakeholder participation in transition processes towards sustainability: methodological considerations. Technol Forecast Soc Chang 72:733–747
van den Bergh JCJM, Gowdy JM (2000) Evolutionary theories in environmental and resource economics: approaches and applications. Environ Resour Econ 17(1):37–57
Voβ J-P, Smith A, Grin J (2009) Designing long-term policy: rethinking transition management. Policy Sci 42(4):275–302
Winograd T, Flores F (1987) Understanding computers and cognition. A new foundation for design. Ablex, Norwood, New Jersey
Zeleny M (2005) Human systems management: integrating knowledge, management and systems. World Scientific, New York
Zhang J (1998) A distributed representation approach to group problem solving. J Am Soc Inform Sci 801–809

Energy Supply, Thermodynamics and Territorial Processes as a New Paradigm of Sustainability in Planning Science and Practice

Antonio Leone, Federica Gobattoni and Raffaele Pelorosso

Abstract The world's population is growing constantly and, more importantly, the need for raw materials and food products is growing quickly, as a result of the western development model. The energy-consuming (energivorous) and consumerist nature of this model is being consolidated globally, ignoring both the issue of resource limitations, and the medium-long term environmental consequences (e.g. climate change, water pollution). This development model, in order to maintain its internal integrity and further develop (often at increasing rates of growth), needs to import energy and materials from the external environment and to produce waste and disorder (entropy) in an inexorable slide toward thermodynamic equilibrium. Sustainable development should focus on contrasting these processes as far as possible, and on developing suitable planning praxes. This paper aims to show how to achieve sustainable land-use through local resource evaluation, overturning the "linear" logic of acquisition-consumption-disposal of wastes, in search of circular processes, capable of reducing entropy growth in a social-ecological system. An analysis of the exergetic availabilities of the landscape mosaic demonstrates great potential for exploiting energy supplies from local and renewable sources, thus lessening the system's overall impact on the global environment.

1 Introduction

Thermodynamics is probably the most structured science of complex systems and many concepts developed in this discipline find applications in other fields, such as ecology (Naveh 1987), sociology (Mckinney 2012), economy (Georgescu-Roegen 1971;

A. Leone (✉) · F. Gobattoni · R. Pelorosso
DAFNE, University of Tuscia, Viterbo, Italy
e-mail: leone@unitus.it

F. Gobattoni
e-mail: gobattoni@unitus.it

R. Pelorosso
e-mail: pelorosso@unitus.it

R. Papa and R. Fistola (eds.), *Smart Energy in the Smart City*,
Green Energy and Technology, DOI 10.1007/978-3-319-31157-9_5

Annila and Salthe 2009; Von Schilling and Straussfogel 2008), industrial ecology (Stremke et al. 2011) and planning (Fistola 2011; Leone et al. 2014; Pelorosso et al. 2014; Vandevyvere and Stremke 2012). Energy and matter flows in the landscape are governed by the Laws of Thermodynamics (Chen et al. 2014; Gobattoni et al. 2011; Inostroza 2014; Marull et al. 2010; Pelorosso et al. 2011; Rees and Wackernagel 1996). In consequence, Thermodynamics (in particular the second law) can be considered the science of sustainability, or, at least, one of its milestones.

Despite these achievements, the Second Law of Thermodynamics (SLT or law of entropy) is not sufficiently considered in disciplines such as urban and landscape planning. Among the real applications to study cases, we find the South Limborg Province (Stremke and Koh 2011) and the Province of Groningen (Stremke et al. 2011) in the Netherlands. Energy Potential Mapping, i.e. the method of visualizing energy potentials and demands by making information regarding the quantity, quality and location of demand and supply accessible, has recently been elaborated and implemented throughout the Netherlands (Broersma et al. 2013).

While the First Law of Thermodynamics (FLT) states that energy is always conserved, the SLT states that during any process, exergy (work capacity) is destroyed and entropy (disorder) is produced. Work is defined as the useful final output derived from the expenditure of energy resources. Work can be used to move a train, to transform water into ice, or to heat a house. Work is made possible through many forms, such as the differences in pressure, electric charge, chemical bond energies, and temperature. The work available in a system is exhausted when the system reaches an equilibrium with the surrounding environment. In a process optimizing the exergy of a system, the entropy and the consequent environmental problems (e.g. CO_2 emissions) are reduced. Examples of exergetic optimizations are the reuse of residual thermal energy between an industrial process and a residential area, the employment of local biomass or industrial waste heat for heating houses and public offices etc.

In general a SLT planning approach, also defined as "second-law thinking", aims at increasing the energy component able to produce work (exergy) and, consequently, at reducing the production of pollutants (entropy) responsible for the alteration of ecosystem ecological functionality (e.g. climate change, freshwater degradation etc.).

To increase the sustainability, efficiency and resilience of the social-ecological systems, a new land planning paradigm is discussed here where, by mimicking nature and the structure and functioning of an ecosystem (Ferguson and Lovell 2014; Ho 2013; Mancebo Quintana et al. 2010; Stremke et al. 2011) and by looking at the principles of circular economy (Preston 2012) the following outcomes can be achieved: (a) the reduction of dissipative systems (b) the closure of cycles (c) the maximization of reciprocity, symbiosis, cooperative relationships and landscape diversity and complexity, (d) the enhancement of exchange of energy and matter.

This chapter focuses on these concepts, with particular reference to planning praxis, which needs a stricter reference to SLT thinking than currently adopted in order to pursue sustainability more effectively.

2 For a New Deal on Landscape: Linear and Circular Economy

The present modernity crisis could become an occasion for new development paradigms, oriented towards environmental and social sustainability. It is necessary to shift from the present thinking on development, centered on technology, where technology is not often the instrument but too frequently the goal, oriented towards business (finance), rather than economic development. The technology question is thus crucial; it is the main shifting factor from the pre-modern age (circular and complex) to the modern age, which is prevalently linear. The tumultuous technological development since the eighteenth century, and the related industrial revolution, has influenced humankind's feelings, not only the practical aspects of our lives. Undoubtedly, this progress has generated enormous advantages (healthiness, social development, democracy) even if prevalently in the world's northwest, but people are beginning to realize that we have thrown out the baby with the bathwater.

Thus, today, a stop to the use of technology for its own sake is perceived by many as a desirable goal. The reference to smart city praxis seems appropriate, because this approach is becoming a mantra, although too often it is reduced to a simple implementation of electronic devices in the city. Before being smart, a city should be clever, i.e. able to interpret the complexity of the built-up urban system, raising awareness of the need for a more robust and sustainable development (Leone, in press).

One consequence of the Modernity crisis consists in the simplification of the "linear" logic of acquisition-consumption-disposal of wastes. North-western world lifestyles are based on heavy consumptions of resources and energy, conceptually oriented on the presumption of an unlimited availability of resources. These are the milestones of the linear economy, where issues relating to resources simply translate as finding the cheapest supply, rather considering how long those supplies will last. The limitations of this concepts have long been clear (Meadows et al. 1972; WCED 1987), but they have remained at a niche level. However, today there is a pressing need to extend our awareness of the finite nature of resources to all aspects of our daily lives. There are several reasons, which make this change in attitude essential: the need to feed the growing (and impacting) world population; ongoing climate change; the energy question. Many answers to these problems have been proposed: sustainable development, convivial degrowth, and low carbon, green and blue economies.

Nevertheless, in day to day practice, such solutions are not satisfactory and the gap between environmental and social questions is growing, as the available data on the increase in atmospheric CO_2 and on land use change confirm. The Modern age simplification continues, the quintessential example being the new, vigorous lease of life given to fossil fuel exploitation thanks to shale gas/oil development in the USA. Moreover, landscape evolution is also following the simplification line, as can be inferred from the two examples illustrated in Fig. 1 showing an urban case (Rome's surroundings, Fig. 1a, after Recanatesi 2014) and a rural case (a typical Italian mountain area, Fig. 1b). A worldwide overview of this situation is given in Table 1 (Hillel 1998).

(a)

URBAN (ROME) SITUATION

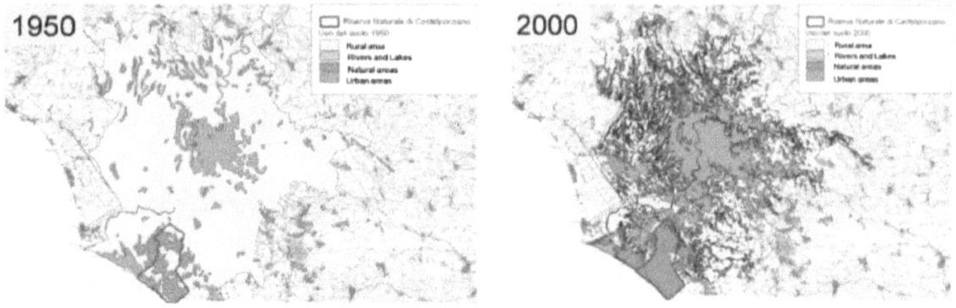

Main land use/land cover: Yellow=arable land/meadowlands; Green=woods; Grey=urbanization; Red line=protected area

	2000				
1950	rural areas	natural areas	rivers and lakes	urban area	tot. 1950
rural areas	**56852**	17716	1197	28809	104575
natural areas	3673	**6728**	28	1845	12274
rivers and lakes	318	63	**163**	62	606
urban area	264	1466	129	**9375**	11234
tot. 2000	61107	25973	1518	40090	**128688**

(b)

RURAL (MOUNTAIN) SITUATION

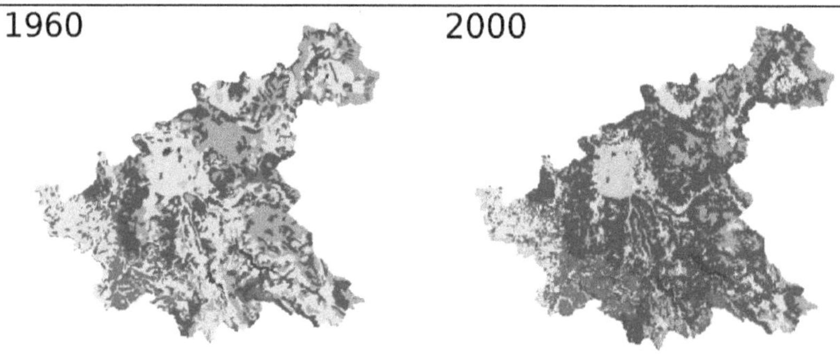

Main land use/land cover: Yellow=Non-irrigated crops; Green=woods; Violet=Orchards; Orange=meadowlands; Cyan= Irrigated crops

	TCI/CNR 1960													
	Woodlands		Meadowlands		Non-irrigated crops		Orchards		Irrigated crops		Urban		Water	
LUM 2000														
Woodlands	**62,993**	*1126*	**43,554**	*3403*	**27,435**	*5391*	**12,720**	*4738*	**276**	*19*	**149**	*5*	**313**	*11*
Meadowlands	**6536**	*1114*	**21,748**	*3373*	**12,939**	*3191*	**3178**	*1786*	**59**	*23*	**116**	*19*	**41**	*11*
Non-irrigated crops	**2026**	*133*	**2691**	*123*	**26,668**	*5601*	**8132**	*5510*	**403**	*12*	**106**	*18*	**78**	*4*
Orchards	**1765**	*115*	**818**	*110*	**6613**	*3111*	**16,006**	*2974*	**42**	*9*	**72**	*15*	**15**	*1*
Irrigated crops	**629**	*6*	**325**	*1*	**8627**	*2310*	**2374**	*2310*	**189**	*0*	**39**	*3*	**310**	*1*
Urban	**50**	*2*	**32**	*2*	**1219**	*640*	**810**	*640*	**58**	*0*	**274**	*4*	**3**	*2*
Water	**14**	*0*	**52**	*1*	**231**	*28*	**35**	*28*	**0**	*0*	**14**	*0*	**1048**	*1*

Fig. 1 Land use/land cover change in two typical Italian situations and related transition matrices (values in hectares): **a** Rome Municipality (urban case); **b** Rieti Municipality (rural case) (Color figure online)

There is an evident polarization between artificial areas (urbanized grey areas) and green areas (above all new woodlands). In the urban case, protected areas grow as a sort of fig leaf to hide large urbanization processes; in the rural case, green areas

Table 1 World land use/land cover changes between 1966 and 1980 (in million hectares)

	1966	1980	Variation %
Agriculture	1381	2362	+71
Rangeland	3198	3117	–6
Forest	4236	1756	–141
Other (urban, marginal lands etc.)	5995	7675	+78

are represented by "new" woods, which occupy the space of less remunerative agricultural and livestock land uses. In other words, the landscape is becoming simplified: it is either artificial (urban), or semi-natural (new woods, not necessarily characterized by increasing biodiversity). The long-lasting metastable equilibrium between agricultural and forest landscape components has disappeared and the landscape's connectivity, i.e. its capacity to exchange material, energy, and organisms, and the long term persistence of biodiversity, has been reduced (Pelorosso et al. 2011, 2015). Rural areas have also lost their identities and cultural heritages (Gobattoni et al. 2015a, b). This contradicts the concept of a landscape as the product of the complex synthesis between nature and culture, as perceived by the population (European Landscape Convention; Florence, 2000).

These are the roots of the present landscape crisis, which no plan and no landscape architecture design can solve without solving the fundamental nodes, because the present landscape crisis is the crisis of the present historical age.

A profound change of the socio-economical system is necessary, a shift from a linear to circular economy, whose concepts derive from waste management (see, for example, the recent European Directive COM/2014/0397) whose cycles are necessarily closed. For a circular economy, nurturing local resources is the focus of all technological applications and this is basis for re-discovering complexity in landscape evolution. This approach should be largely reproduced in territory governance, in order to achieve an actual sustainable development. The Territorial Engineering approach, in which the problem solving approach typical of engineering is based on local resources and closed cycles, could prove useful in the effort to attain these sustainable objectives.

Linear technological development has not been followed by a similar cultural development, resulting in the breaking of ancient closed cycles (Fig. 2), simplifying and linearizing resource exploitation, as a result of the cheapness, the enormous energy and the availability of fossil fuels, the milestone of the industrial revolution (Leone, in press). In the scheme of Fig. 2 all pre-modern systems are functional:

- Agriculture is not simply the "banal" production of food, it is also a place of resource recovery and restoration. Here the waste concept is unknown, because agriculture is not only the primary economic sector, but it is also the final destination of various "wastes", where "wastes" in quotation marks has a special meaning, with the same implications as it has in ecology, i.e., what is discarded by one system becomes a resource for another. The challenge for landscape management consists in re-discovering this circularity, with agriculture as the

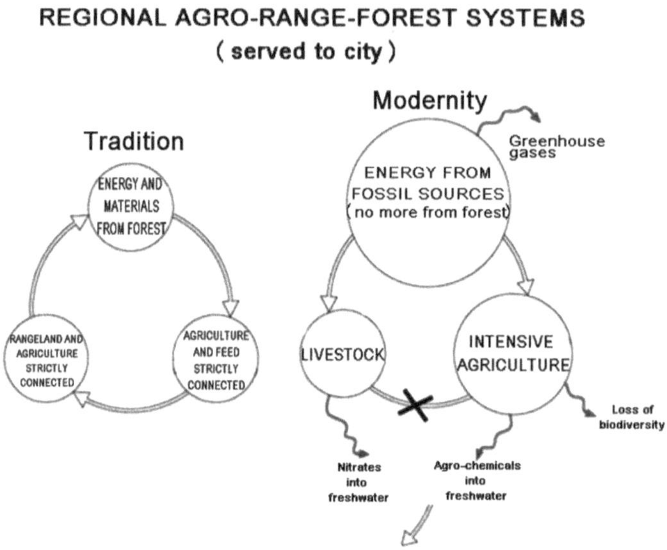

Fig. 2 Schemes of regional systems

primary and final sector, which implies struggling against the global systems' increasing entropy.

- Woodlands are not the impenetrable forest of modern citizens' imagination, seen as undeveloped and marginal zones. In the pre-modern circular economy, they were not only a source of building materials and an energy supply, but they also provided food for both livestock and humans[1] in periods when other food sources failed. Modernity has marginalized the functionality of woodlands and, paradoxically, their growth is the signal of a landscape simplification process. New forms of circular economy should re-discover their role in the landscape evolution, for example through ecosystem service evaluation (Hermann et al. 2011). Livestock produce proteins, the noblest part of human diet, but this is not the only, nor the most important function of livestock: two other important roles of livestock are as work power for tilling and the production of high quality fertilizers for agriculture.

In synthesis, in traditional rural systems, everything is nested and multifunctional and the consequence is a complex, robust and resilient system. Doubtlessly, modernity has allowed humankind a development, which would have been unthinkable in the past, but, in the present historic phase, its propulsive push is turning off. Technological development has deceived humankind, in a sort of

[1]Woodlands have been an insurance against famine, providing forage for livestock, but, in extreme cases, forage for humans too.

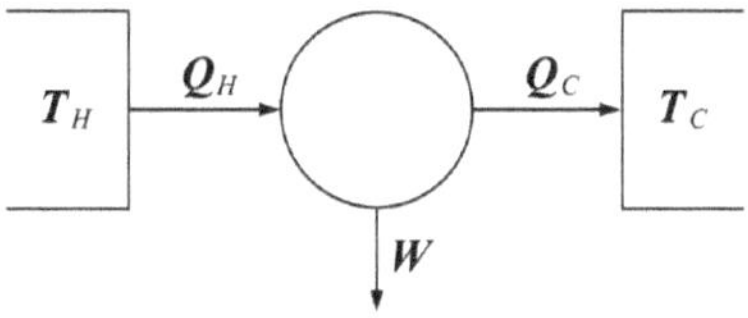

Fig. 3 One of the many expressions of first thermodynamic law: Carnot's Principle

Promethean myth, but it should be recognized that, globally, society and civilization are not yet adapted to technological development.

A re-thinking is necessary, saving all the development brought by modernity. In this sense, the contemporary debate on landscape management is significant: consider, for example, the European Landscape Convention focusing on the dynamic character of landscape, deriving from the interaction among natural and cultural factors.

To pursue this aim, Thermodynamics is fundamental, because it is the science of complex systems. The usefulness of Thermodynamics when approaching sustainability scientifically is well documented (see Georgescu-Roegen 1971 for economic; Scandurra 1995 for planning). It is now time to take into account these concepts in order to identify best management practices for landscapes.

Considering the Thermodynamic principles, the first law (FLT) can be considered the fundamental law at the basis of the Industrial Revolution, i.e. the starting point of Modernity. One formulation of this law is Carnot's Principle,[2] represented in Fig. 3:

$$\Delta U = Q - W$$

where ΔU is the system's internal energy variation, Q the exchanged heat, W the work done by the system.

This equation synthesizes the Modern Age, with its advantages (the heat engine and related blind faith in progress) and its weaknesses (the long-term unsustainable linearity of development).

The engine derived from Carnot's theorem consumes resources (Q_H) to obtain work, with the consequent standardization of products, rise of the merchant class, and tumultuous development of cities. This is the "thermodynamics" bill, corresponding to Q_C in FLT and this is the key of unsustainability: sooner or later the bill must be paid; it is only a dangerous game of Russian roulette amongst generations in the developed world,[3] whereas in "undeveloped" countries the bill has already been paid. The Second Law of Thermodynamics (SLT) allows this bill to be quantified, showing how blind faith in linear growth contrasts with the laws of

[2]Sadi Carnot was the son of Lazarus Carnot, known as the "Organizer of victory" of the French Revolution and an archetypal modern man: engineer, mathematician, politician and freemason. It can be said that Sadi and Lazarus Carnot personify the Modern Age.

[3]Indeed, it seems that this shrinking process is started and even rich northwest young people heredity a minor living level.

nature and it is measurable in terms of resource degradation (increasing Entropy). This scientific evidence has, to date, scarcely influenced economic and technological practice.[4]

Above all, this is due to:

- the fact that, doubtlessly, the industrial revolution "worked" well for people of the world dominating northwest, whose life has evolved in a very impressive way.
- the fact that the SLT is valid for closed systems, while the industrial revolution involved, until few years ago, only the northwest of the world, an open system that compensated its thermodynamics bill by appropriating resources from undeveloped countries and colonies. Nowadays, technological development itself is tending to close the system, the economy is becoming global and Earth is "smaller", as the Gaia metaphor evocates. The present crisis is not only economic; it is a system crisis too: the common attitude of modernity must undergo a fundamental change. Climate change can be considered as the tangible sign of this situation. Gaia's entropy is increasing unsustainably and Prometheus starts to pay for his arrogance, in terms of irreversible resource degradation. It is time that SLT thinking became a commonly accepted approach, no longer limited to intellectual elites. A circular economy and consequently local resource engineering and planning is hence necessary.

Figure 3 shows how simple FLT thinking works, which involves not worrying about resources and energy degradation, with the illusion of unlimited fuel (Q_H) availability and unlimited space for waste (Q_C) or with the thaumaturgical power of technical development, capable of solving any problems related to resource supplies and waste management. This "Futurist" attitude, typical of beginning of the twentieth century, has almost run its day and a new humanist way of thinking is necessary, whose technical basis should be the integrated FLT-SLT thinking. In the Modernity-linear approach, the focus is on the work produced W, that should be as high as possible without any other considerations. It is perfectly consequential, because the aim of Carnot's law is to build and exploit the work of "simple" machines. This was a great achievement because machines offered humans the opportunity for a cheap, vast empowerment. This capacity was truly a revolution, with relevant and positive consequences for western civilisation in all fields.

But Q_C will never be zero, even in theory, and, furthermore, Q_C accumulation progressively degrades resources and accelerates the system's collapse. Here is the sustainability key: the shift in thinking and acting, taking into account the Second Law of Thermodynamics (SLT). The Laws of Thermodynamics are absolute values, which cannot be bypassed and which have to be the ideological basis of a new, clever, technical age. FLT implies technological and smart development; SLT implies clever awareness of the world's complexity and resource limits, which is

[4]The Georgescu-Roegen (1971) economic theory has to date scarcely influenced praxis and economic mainstream thinking.

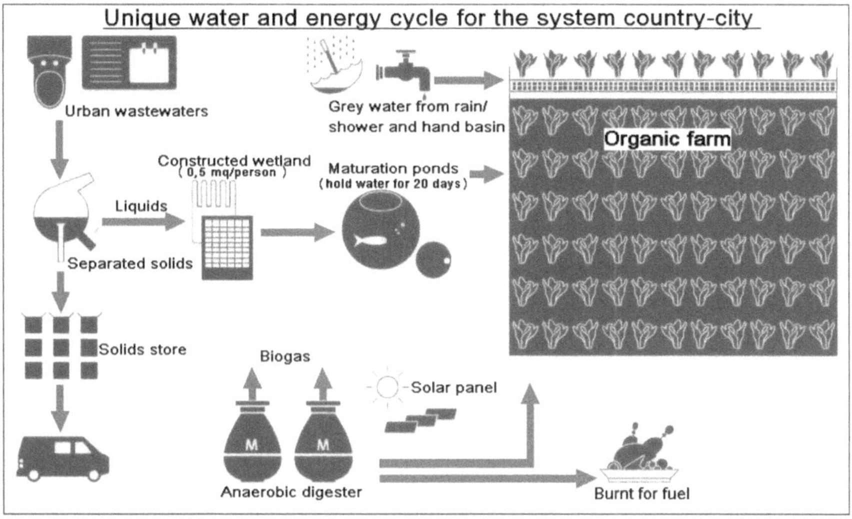

Fig. 4 The creation of a virtuous cycle of connections with urban agriculture (Rydin et al. 2012, modified)

translated into wiser behaviour. The greater challenge of contemporary applied sciences is to find how to transfer it into practice, and environmental sustainability is the keystone of this new thinking.

Figure 4 (modified after Rydin et al. 2012) shows a scheme of rural and urban systems integration. It shows how the use of waste and local resources can build a complex landscape.

As a consequence, for "circular-future" systems the focus is on Q_C, in particular on the system's ability to utilize Q_C, transforming it into a resource, while, for Carnot's machine, it is only a waste, a factor of efficiency reduction.

Generalizing: "smart-simple" is represented by physical infrastructures, also encompassing information and communication technologies (ICTs); "clever-complex" is represented by the strategies that increase human social and environmental capital, also thanks to physical infrastructures and ICTs.

3 A Proposal

The present situation appears to be an impasse, but the current crisis should be considered an opportunity to overcome this impasse and the environmental question is the focus. Obviously, the solution is not a slavish return to the past: technology's conquests are not at issue. The solution is in FLT thinking (saving the modernity tradition) innovating it by integration with SLT thinking, i.e. in building more complex, socio-ecological systems.

More awareness in development choices needs to be re-discovered, linking choices to the landscape and local resource availability. This is easy to say, but not so easy to do. Indeed, mainstream thinking affirms that there is no alternative to a linear economy, that waste is inevitable and that the only solution consists in managing wastes with a blind faith in technology: dumping grounds or incinerators, in other words FLT, but no SLT thinking. On the other hand, the alternative, current ecological thinking is weak, bound to the anchor of conservation, but not completely able to meet the challenge of landscape dynamism. The consequence is the polarization previously debated and defenders of traditions are skeptical about the economic sustainability of any alternative to a linear economy.

Greening is then only cosmetology, fig leaves that separate the "real" economy (i.e. really functioning things of the linear economy) from elite niches, where some ecologists accept this compromise, maybe due to the lack of alternative pragmatic proposals. The ecologist world seems to be fixed on the first phase of the ecological movement, i.e. the nature conservation phase of the 1950–1960's, but it seems not completely able to face the present day challenge of sustainable development in complex and dynamics social-ecological landscapes. Although sciences such as ecology have studied in depth, and clarified, several environmental issues, more effort is necessary to realize sustainable and efficient actions in practice. As a consequence, what is really needed today is to transform theory in practice, to extend what is now confined to small intellectual niches to common technical procedures and, in particular, to urban and landscape planning.

The modernity crisis is the landscape crisis and simple conservation is not sufficient because it fails from the dynamic nature point of view. Green areas, natural reserves, ecological nets etc. are not yet structural components of the landscape, they are ghettoized into the polarized green areas of Fig. 1.

On the contrary, real, high quality landscape is dynamic, at the temporal scale of human history[5]; it is the fruit of interactions and contaminations, deriving from the *artialisation* processes of Roger (2009). The pre-modern development was based on *artialisation*, on building and acting considering local resources and, hence, was more able to build appreciated landscape. This is what the modern age lacks, this is the future challenge, which is a titanic challenge, because it needs a profound renewal of our way of thinking. Planning by processes, i.e. technical procedures aimed to revalue local resources, stressing genius loci, have to meet this challenge, demonstrating the possibility to shift from the linear economy to circular economy in the praxis (by local resources engineering). Not many cases of genius loci development strategies, with opportune adaptations to the current socio-economic situation and cultural/technological advancements, have been proposed in north-western landscape management plans. Recently, Gobattoni et al. (2015a, b) have suggested several strategies to support and maintain traditional farming

[5]Also the environment has a dynamic nature, if on a larger temporal scale, as Darwin demonstrated.

systems and other activities such as craftsmanship (e.g. manufacturing activities, local food production), as an example of sustainable human integration with nature in a rural context in Central Italy.

4 An Example: Energy from the Landscape

4.1 The Exergy Concept

In Thermodynamics, a system's exergy is the maximum first species (mechanical, electrical) energy that brings the system to equilibrium with reference to the surrounding environment. The FLT says that energy is never destroyed during a process, while the SLT says that exergy is always destroyed. The exergy concept is synonymous with available energy, maximum useful work and, hence, it is applied in industrial ecology, with the aim of using energy in the most efficient way.

The concepts of energy and exergy are useful in engineering to compute what comes in and out of several possible designs before a factory is built. Energy input and output will always balance according to the FLT or the energy conservation principle. Exergy output will not balance the exergy input for real processes since a part of the exergy input is always destroyed according to the SLT for real processes. After the input and output have been completed, the engineer will often want to select the most efficient process. An energy efficiency or *first law efficiency* will determine the most efficient process based on wasting as little energy as possible with respect to the energy inputs. An exergy efficiency or *second-law efficiency* will determine the most efficient process based on wasting and destroying as little available work as possible from a given input of available work.

It is time that these concepts were extended to landscape planning and design (Stremke and Koh 2011), with the aim of planning and designing with local resource evaluation.

A milestone of this topic is the issue of efficiency in resource exploitation. It is here that the main shift in thinking and acting must take place. The Prometheus effort, till now, has consisted in increasing W, because today's machines have a maximum efficiency of only 40–45 %, despite more than 200 years of technological development.

Circular thinking allows this question to be overcome, because the goal of increased efficiency is a problem only when the energy comes from external systems. On the contrary, when local resources of complex systems are considered, the maximum efficiency question disappears. The comparison with the ecosystem working is clear: in nature, the only energy supply derives from photosynthesis, whose efficiency is "only" 1 % (Blankenship et al. 2011; Oakley and Rao 1999). For Modernity thinking it is surprisingly low, because Prometheus, until now, has only considered FLT. On the contrary, for nature, efficiency in energy supply is not a problem at all, thanks to the complex organisation reached by evolution, whose driving force is the reduction of increasing entropy, i.e. SLT considerations. A complex organization is aimed at optimizing exchanges and symbiosis among subsystems, which involves cancelling the concept of waste. In complex-symbiotic systems wastes are absent, because what is waste for one subsystem is a resource for another. The problem, in consequence, consists in creating opportunities for exchange, in the building of the maximum possible number of synapses into the landscape.

Modernity contributed to human evolution by linear technology, present and future challenge consists in getting circular this evolution, in increasing complexity. The waste question is at the basis of this evolution.

This would be the aim of the new landscape management.

4.2 Planning of Local Renewable Energies Following SLT Thinking

The depletion of fossil fuels in combination with climate change necessitates a gradual reduction of today's fossil energy dependence. The need for a low carbon energy system, where local renewable energies (RE) substitute external fossil fuel, is seen as an essential pillar of sustainable development and energy self-sufficiency. In the context of sustainable energy transition, designers and planners cannot limit the discussion to the location and design of individual projects such as wind turbines or photovoltaic parks. Rather, an energy-conscious (re)organization of the landscapes appears to be necessary. "Energy-conscious planning and design" implies increasing the proportion of energy generated from renewable sources and enhancing the efficiency of energy use through the spatial organization and design of the larger physical environment (i.e. the landscape) (see also Magnaghi and Sala 2013). Indeed, generally, renewables have a lower energy density than fossil fuels and consequently they have much larger spatial footprints. Bulky biomass, for example, may have an energy density as little as 7 megajoules per liter (MJ/L) whereas diesel has a density of 39.0 MJ/L. The low energy density of biomass means that it should only be transported over short distances to be economically feasible. The energy density of warm water/steam is even lower (0.02–0.2 MJ/L) and limits the spatial extent of heat networks. The relationship between energy density and possible transport in space has direct implications for the design of sustainable energy landscapes. Moreover, temporal variability (seasonal and daily) of the RE (e.g. wind, solar radiation and biomass availability) may induce planners and designers to re-think processes and features at landscape scale. This means that, in a society based on renewables, space equals energy, and source and sinks must be spatially connected in smart ways. For example, although some uses (transport and heavy industry) will continue to require high-density energy carriers, residential, commercial and public buildings (accounting for 35 % of the total global final energy consumption) will need a significant amount of low and medium-quality energy (e.g. heating and cooling, hot water and cooking). The latter demand is currently satisfied by the high-quality energy forms of petroleum products and natural gas, thus the employment of local RE in this case would be an efficacy strategy for a more rational use of resources.

To face the problem of spatio-temporal variability and the lower energy density of RE, several strategies have been proposed to date, such as the interconnection of energy systems (sources and sinks of energy), intelligent management systems for blocks of buildings, heat-cascades (i.e. the multi-stage reuse of residual thermal energy by temperature level), the storing of hot and cool energy by means of aquifers, retention basins or green areas. Through this spatial and temporal reorganization of the landscape mosaic, it is also possible to satisfy other people's needs (ecosystem services), increasing the multifunctionality of the territory: an example could be the re-use of areas destined to biomass production for recreational purposes after the harvest.

Obviously, all these interventions can (and must) be supported by other strategies for energy consumption reduction at the level of industries, services, residential textures, single buildings and the community. Among these interventions, systems for building insulation, LED lamps or environmental education aimed at energy consumption reduction through virtuous behaviors are worthy of mention.

4.3 A Case Study

The study area (Fig. 5) is located southeast of Foggia (South Italy) and includes the industrial zone ASI-Incoronata where several companies have their corporate headquarters (e.g. Barilla for pasta production, Princes-A.R. for peeled tomatoes

Fig. 5 Study area

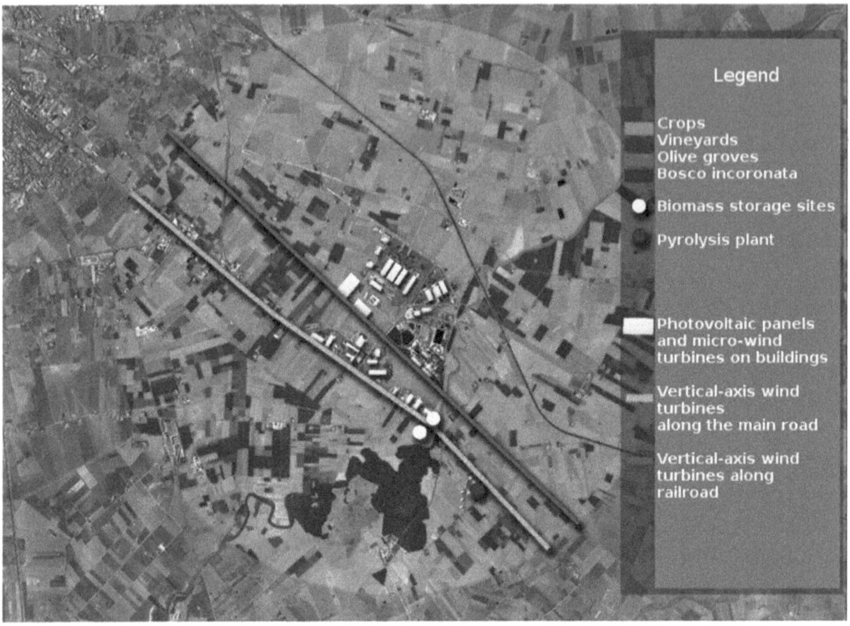

Fig. 6 Planned actions within the study area (Modified from Pelorosso et al. 2014) (Color figure online)

manufacturing, Alenia Aereonatutica for carbon-resin processing and assembly, FPT-Sofim who produce diesel engines for Iveco Motors).

The SCI (Site of Community Importance) "Valle del Cervaro-Bosco Incoronata" (IT9110032) falls within the study area with about 280 ha of forest and the residential area named Borgo Incoronata established in 1928 following land reclamation. In order to evaluate the available energy resources from agricultural and forest biomass, a portion of territory included in a range of 7.5 km from the industrial zone has been taken into account. With the aim of a sustainable land use development through RE implementation, an accurate territorial analysis was carried out in several phases. An analysis of the technologies and productive processes of each company located in the study area followed an initial inventory of the available data. Through interviews and consultation of scientific literature and technical reports, the potential energy uses and amounts obtainable from each landscape element and productive area, were evaluated. The analysis took the following elements into consideration (Fig. 6):

(a) Installation of vertical-axis wind turbines on buildings and along the main streets
(b) Installation of photovoltaic panels on existing buildings
(c) Biomass from agriculture and forestry productions
(d) Biomass from industrial waste

(e) Installation of a pyrolysis plant PiroDistill for the production of Syngas, Synoil and active carbon and a Syngas cogeneration plant 200 KW (planned and awaiting funding)
(f) Realization of a district heating plant for the residential area of Borgo Incoronata
(g) Plantation of 10 ha with Harundo donax for energy purposes
(h) Energy requalification of buildings
(i) Biomass storage

For each intervention, the productive potential and the energy potential requirement was calculated for each unit of occupied surface. The cost per KWh produced was also evaluated to quantify both the energy production expenses and the income which can be obtained through the extraction of exergy from the system using the different technologies considered. Then a multicriteria analysis based on Analytical Hierarchy Process (AHP) was used to assess three sustainable scenarios with reference to the six objectives to pursue which are reported in order of importance below (the weight assigned in the multicriteria analysis is reported in brackets):

1. increasing exergy recovery (0.31)
2. increasing the number of sources involved in energy provision (0.24)
3. minimizing land-use changes (0.14)
4. reducing energy requirements (0.12)
5. minimizing social transformations (i.e. the impacts that variations had on family and company balances and on their view of energy saving) (0.10)
6. minimizing intervention costs (0.09)

The scenarios analysed differ in their intensity of intervention on the territory and in the way they use the energy produced within the system. The optimal scenario was then designed in detail and the correspondent energy balance was evaluated.

The optimal scenario derived from the multicriteria analysis, envisages the usage of the Syngas produced from industrial, agricultural and forest wastes for the production of electric energy through a cogeneration plant and for the district heating of the residential area Borgo Incoronata (scenario B in Fig. 7).

This alternative involves the greatest number of renewable energy sources. The energy balance of this sustainable landscape design showed an energy recovery of 53.4 % with respect to the base scenario.

The final balance of all the interventions proposed up to now is reported in Fig. 8, taking into account the actual annual consumption and the actual annual recovery, measured in MWh/year, a unit which gives a realistic and comprehensive account of consumptions and recoveries in the study area.

From Fig. 8, it is possible to infer that the industrial zone and the agricultural system are the subsystems, which mainly consume energy, and the agricultural system is the most energivorous. In the agricultural system, indeed, not all the processes contributing to the final product can be energetically controlled as

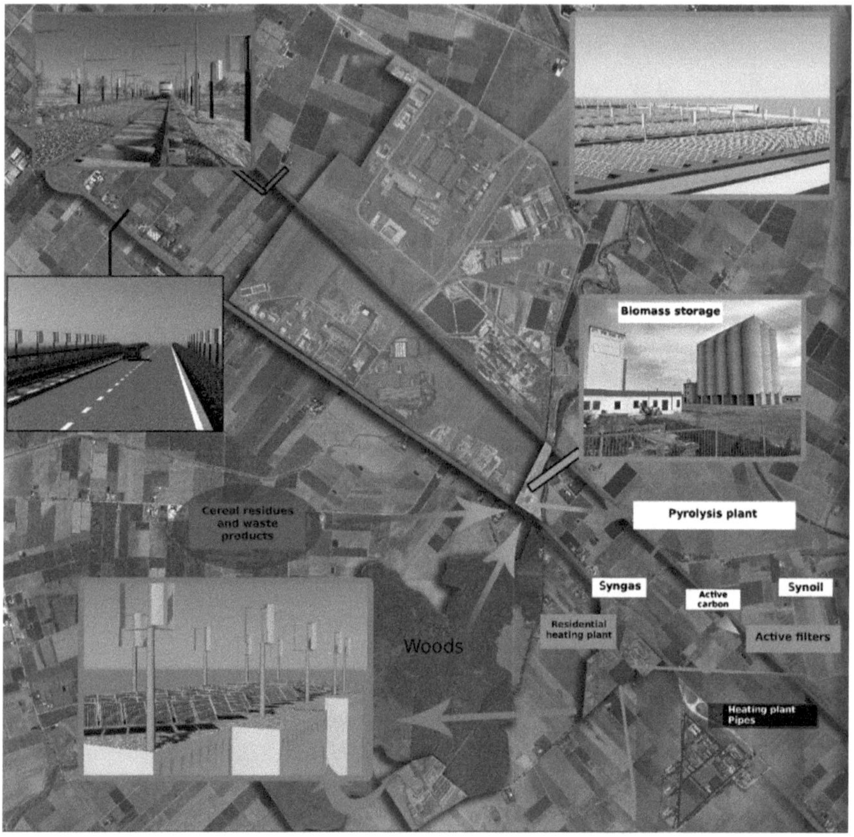

Fig. 7 Scenario B: detail of planned actions (Modified from Pelorosso et al. 2014)

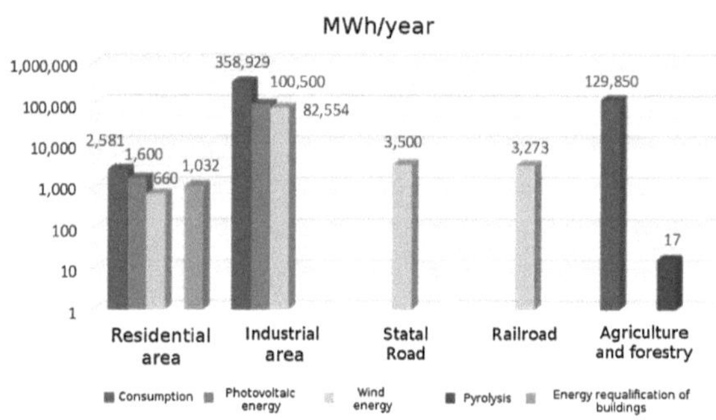

Fig. 8 Energetic Balance of scenario B (Modified from Pelorosso et al. 2014)

happens in an industry. As an example, to produce olive oil, several processes use electric engines or fossil fuel engines, which are not energetically efficient thus inducing a great difference between consumption and recovery in terms of energy due to the low energetic density of biomass.

The subsystems from which it is possible to recover the greatest amount of energy to redistribute throughout the territory considered, are the industrial zone and the residential area. In particular, the residential area could cover all its energy consumption and produce an energy surplus with some small interventions of energy requalification.

The results underline that the agricultural system needs actions aimed at the energy requalification in all its processes, from cultivation to final product transformation.

Differentiated and highly efficient systems, in terms of exergy, have a greater perspective of life and income and, in this view, the low energetic efficiency of agricultural production and transformation systems could be seen as one of the causes of the crisis in that sector. The inclusion of agricultural wastes and by-product in the energetic system could contribute to keep the primary sector alive, since it represents the real motive force of rural society and of Italian landscape identity.

5 Conclusion

The present paper is aimed at demonstrating concretely how it is possible to find alternatives in the landscape that, normally, are remitted to "allochthonous", impacting solutions. This is the basis for shifting from linear to circular economy. More examples demonstrate that this change is a *forma mentis* problem, rather than a technical one.

In the specific case, the aim is improving energy supply, which is evaluated first of all from a local resources point of view.

The focus is on the Laws of Thermodynamics: our common way of thinking about this issue only takes into consideration the first law, the needful change in our way of thinking requires considering the second law too. In technology development terms, this means that attention should shift from the Modern Age tradition of work production to a new age, based on waste management. With the Industrial Revolution, humankind met the challenge of easily multiplying work production, but today's task is local resource engineering, whose milestone, rather than work, is waste optimization and re-use. To do this, the focus must be on territorial organization, aimed at promoting synapses among landscape components and all possible symbiotic and exchange processes.

In consequence, the new landscape challenge consists in finding a new complexity, based on ecosystem services. At present, the theory appears clear, what is needed is praxis. The change consists in the will to rise to the challenge, in social and political structures.

The case study illustrated in this contribution demonstrates how it is possible to evaluate the energy incidental to each land-use, obtaining significant productions, respecting landscape identity and increasing system resilience. Indeed, a conscious spatio-temporal organisation of the landscape based on the SLT, local RE and smart energy systems increases the ratio of energy self-sufficiency and the resilience of the socio-ecological system. It allows local populations to have a greater capacity to persist and develop in the territory to which they belong through the mitigation of the negative effects derived from the fluctuations of the energy availabilities external to the system both on the productive processes and on the essential vital activities.

References

Annila A, Salthe S (2009) Economies Evolve by Energy Dispersal. Entropy 11(4):606–633. doi:10.3390/e11040606

Blankenship RE, Tiede DM, Barber J, Brudvig GW, Fleming G, Ghirardi M, Sayre RT et al (2011) Comparing photosynthetic and photovoltaic efficiencies and recognizing the potential for improvement. Science (New York, NY) 332:805–809. doi:10.1126/science.1200165

Broersma S, Fremouw M, Dobbelsteen A (2013) Energy potential mapping: visualising energy characteristics for the exergetic optimisation of the built environment. Entropy 15(2):490–506. doi:10.3390/e15020490

Chen S, Chen B, Fath BD (2014) Urban ecosystem modeling and global change: Potential for rational urban management and emissions mitigation. Environ Pollut (Barking, Essex: 1987) 190C:139–149. doi:10.1016/j.envpol.2014.03.032

Ferguson RS, Lovell ST (2014) Permaculture for agroecology: design, movement, practice, and worldview. Rev. Agronomy for Sustain Dev 34:251–274. doi:10.1007/s13593-013-0181-6

Fistola R (2011) The unsustainable city. Urban entropy and social capital: the needing of a new urban planning. Procedia Eng 21:976–984. doi:10.1016/j.proeng.2011.11.2102

Georgescu-Roegen N (1971) The entropy law and the economic process. Harvard University Press, Cambridge

Gobattoni F, Pelorosso R, Lauro G, Leone A, Monaco R (2011) A procedure for mathematical analysis of landscape evolution and equilibrium scenarios assessment. Landscape Urban Plan 103(3–4):289–302. doi:10.1016/j.landurbplan.2011.08.011

Gobattoni F, Pelorosso R, Leone A, Ripa MN (2015a) Sustainable rural development: the role of traditional activities in Central Italy. Land Use Policy 48:412–427

Gobattoni F, Pelorosso R, Ripa MN, Leone A (2015b) Percezione e identità paesaggistica: un'opportunità di sviluppo per i territori rurali. Italian J Reg Sci 14(3):49–70

Hermann A, Schleifer S, Wrbka T (2011) The concept of ecosystem services regarding landscape research: a review. Living Rev. Landscape Res 5(1):1–37

Hillel D (1998) Environmental soil physics. Academic Press

Ho M-W (2013) Circular thermodynamics of organisms and sustainable systems. Systems 1 (3):30–49. doi:10.3390/systems1030030

Inostroza L (2014) Measuring urban ecosystem functions through "Technomass"—A novel indicator to assess urban metabolism. Ecol Ind. doi:10.1016/j.ecolind.2014.02.035

Leone A, Gobattoni F, Pelorosso R (2014) Sustainability and planning: thinking and acting according to thermodinamics laws. J Land Use, Mobil Environ (Special Issue Eighth International Conference INPUT)

Magnaghi A, Sala F (2013) Il territorio fabbrica di energia. Wolters Kluwer italia

Mancebo Quintana S, Martín Ramos B, Casermeiro Martínez MA, Otero Pastor I (2010) A model for assessing habitat fragmentation caused by new infrastructures in extensive territories—evaluation of the impact of the Spanish strategic infrastructure and transport plan. J Environ Manage 91(5):1087–1096. doi:10.1016/j.jenvman.2009.12.013

Marull J, Pino J, Tello E, Cordobilla MJ (2010) Social metabolism, landscape change and land-use planning in the Barcelona Metropolitan Region. Land Use Policy 27(2):497–510. doi:10.1016/j.landusepol.2009.07.004
Mckinney LA (2012) Entropic disorder: new frontiers in environmental sociology. Sociological Perspectives 55(2):295–317. doi:10.1525/sop.2012.55.2.295.296
Meadows DH, Meadows DL, Randers J, Behrens WW (1972) I limiti dello sviluppo. Rapporto del System Dynamics Group, Massachusetts Institute of Technology per il Club di Roma. Mondadori, Milano
Naveh Z (1987) Biocybernetic and thermodynamic perspectives of landscape functions and land use patterns. Landscape Ecol 1(2):75–83. doi:10.1007/BF00156229
Oakley Hall D, Rao KK, Institute of Biology (1999) Photosynthesis. Cambridge University Press, 6 p. Accessed in November 2014. ISBN 978-0-521-64497-6
Pelorosso R, Della Chiesa S, Tappeiner U, Leone A, Rocchini D (2011) Stability analysis for defining management strategies in abandoned mountain landscapes of the Mediterranean basin. Landscape Urban Plan 103(3–4):335–346. doi:10.1016/j.landurbplan.2011.08.007
Pelorosso R, Gobattoni F, Geri F, Monaco R, Leone A (2015) Evaluation of Ecosystem Services related to Bio-Energy Landscape Connectivity (BELC) for land use decision making across different planning scales. Ecol Ind. doi:http://dx.doi.org/10.1016/j.ecolind.2015.01.016
Pelorosso R, Gobattoni F, Lauriola D, Bari P, Civile I, Leone A (2014) Pianificazione territoriale e termodinamica: nuova declinazione della sostenibilità. In Atti della XVII Conferenza nazionale SIU, L'urbanistica italiana nel mondo (pp 1–8). Planum
Preston F (2012, Mar) A global redesign? Shaping the circular economy. Energy Environ Resour Governance 2
Recanatesi F (2014) Variations in land-use/land-cover changes (LULCCs) in a peri-urban Mediterranean nature reserve: the estate of Castelporziano (Central Italy). Rendiconti Lincei. doi:10.1007/s12210-014-0358-1
Rees W, Wackernagel M (1996) Urban ecological footprints: why the cities cannot be sustainable —and why they are a key to sustainability. Environ Impact Assess Rev 16:223–248
Roger A (2009) Breve trattato sul paesaggio, Palermo, p 141
Rydin Y et al (2012) Shaping cities for health: complexity and the planning of urban environments in the 21st century. Published Online, http://www.smartcities.eu. Accessed in July 2013
Scandurra E (1995) The man's environment. Towards the design of a sustainable city. Etas Books editions, 304 p (in Italian)
Stremke S, Koh J (2011) Integration of ecological and thermodynamic concepts in the design of sustainable energy landscapes. Landscape J 30(2):194–213. doi:10.1353/lnd.2011.0036
Stremke S, Van den Dobbelsteen A, Koh J (2011) Exergy landscapes: exploration of second-law thinking towards sustainable landscape design. Int J Exergy 8(2):148–174
Vandevyvere H, Stremke S (2012) Urban planning for a renewable energy future: methodological challenges and opportunities from a design perspective. Sustainability 4(12):1309–1328. doi:10.3390/su4061309
Von Schilling C, Straussfogel D (2008) Introducing the research. Entropy debt: a link to sustainability? In: 52nd Annual Meeting of the International Society for the Systems Sciences
WCED (1987) Rapporto Brundtland, WCED–Commissione mondiale sull'ambiente e lo sviluppo. Scaricato dal web il 3 novembre 2014

Managing Mobility to Save Energy Through Parking Planning

Maurizio Tira, Silvia Rossetti and Michela Tiboni

Abstract Energy saving is the most sustainable solution in the long run to achieve the 2020 goals and mobility is one of the highest energy-consuming activities in our towns. The way people move in the urban environment is manageable through several policies, strategies and actions. Parking management is an important strategy in most planning activities, those addressing land use management as well as traffic plans. Private parking lots (usually at the origin of trips) are planned with a minimum standard quantity in most countries, but there are some good examples where a maximum amount of parking is ruled by plans, in order to discourage residents from possessing too many cars! The availability of public parking places (at the destination of the trip) and their fees have a direct influence on modal choice and so on inter-modality, as the economic sustainability of private motorised mobility is also influenced by economic estimations. Several options in managing public parking regulations can influence mobility patterns, such as regarding location, parking fees and time-related policies. The paper proposes a methodology for the analysis of the space-time relations between public parking and individual travel choices. The methodology has been assessed in the case study of Brescia, in northern Italy. First of all, the location and the density of parking areas within the city have been mapped to show the spatial coverage of car parking supply. Then, the time variable has been considered, to illustrate the degree of use of each parking area during the day—thus showing modal choices and their variation in time and space. The results of the analysis can be extended to similar situations as the methodology has a broad application. The final goal of the research, rather than simply monitoring parking use, is to encourage sustainable mobility through the management of parking supply in urban areas and so to foster energy saving policies.

M. Tira · S. Rossetti (✉) · M. Tiboni
DICATAM, University of Brescia, Brescia, Italy
e-mail: silvia.rossetti@unibs.it

M. Tira
e-mail: maurizio.tira@unibs.it

M. Tiboni
e-mail: michela.tiboni@unibs.it

R. Papa and R. Fistola (eds.), *Smart Energy in the Smart City*,
Green Energy and Technology, DOI 10.1007/978-3-319-31157-9_6

1 Managing Mobility to Save Energy: Passenger Mobility, Negative Externalities and European Policies

Energy and mobility are deeply related issues: mobility is the second highest energy-consuming activity and energy saving is probably the most suitable solution in this field: alternative fuels can only partly contribute to reach the 2020 goals. The paper will focus on passenger mobility and related policies, namely in urban environment.

Passenger mobility is nowadays a very critical issue as it is vital for the quality of life of citizens (European Commission 2011: 3). Mobility plays an essential role in economic and social development in every society but it has several negative consequences, both on the environment and on the liveability of towns and cities.

Current passengers' modal split is highly overbalanced towards road transport and private cars. In 2012, total passenger transport activities in the European Union (EU-28) by any motorised means of transport were estimated to have amounted to 6391 billion passenger kilometres (or on average around 12,652 km per person), and passenger cars accounted for 72.2 % of this total.[1] Non-motorised means of transport, such as cycling and walking, account only for a very marginal share of road transport: the average person in the European Union cycles about 0.5 km, walks about 1.0 km, and travels 28 km by car per day (WHO Europe 2004: 12). In 2012, the European Union (EU-28) accounted 487 passenger cars per 1000 inhabitants (European Commission 2014). Italy has one of the highest rates in EU countries[2]: 621 passenger cars per 1000 inhabitants (European Commission 2014).

The massive diffusion of individual motorised vehicles started in Europe after the Second World War, and the growth of mobility demand has reduced the quality of urban public spaces and highly affected the environment. Increasing land and energy consumption, congestion, greenhouse gas (GHGs) emissions, air pollution, noise and road accidents are the negative externalities of such a system. Indeed, car traffic contributes to climate change, consumes land and natural resources, is a burden on public and private budgets, and causes environmental and health problems.

The European Commission estimated that in the EU-27 31.7 % of all energy consumption derives from the transport sector (European Commission 2014: 116). Furthermore, transport is responsible for a significant share of total emissions of greenhouse gasses (GHGs): from 1990 to 2010, total EU-27 GHG emissions fell by 15 % with decreases in all sectors but transport. The increase of transport emissions in the same period was of 20 %. In 2010, 80 % of the total GHG emissions were energy related, and transport accounted for 20 % of this share (Eurostat 2012: 143).

---

[1]These figures include intra-EU air and sea transport but not transport activities between the EU and the rest of the world. As for the other modes: powered two-wheelers accounted for 2 %, buses and coaches for 8.2 %, railways for 6.5 % and tram and metro for 1.5 %. Intra-EU air and intra-EU maritime transport contributed 9 and 0.6 % respectively (European Commission 2014).

[2]According to the "*EU transport in figures*", only Luxembourg shows a higher motorisation rate, such as 663 passenger cars per 1000 inhabitants in 2012.

Global awareness on climate change grew since the '90s: in 1992 the United Nations Framework Convention on Climate Change (UNFCC) was defined, and five years later, in 1997, the Kyoto Protocol was adopted.

At European level, the White Paper "*Adapting to climate change: Towards a European framework for action*" was published in 2009 (European Commission 2009), to set objectives and actions aimed at integrating adaptation to climate change and resilience into EU policies. Among the objectives, the document states that "*a more strategic and long-term approach to spatial planning will be necessary, including transport*" (European Commission 2009: 4).

Therefore, spatial planning should deal with transport issues in a strategic and long-term vision: mobility must be primarily Environment and Climate Friendly.

According to the White Paper of the European Commission,[3] there is a "*…need to drastically reduce world Greenhouse Gas emissions, with the goal of limiting climate change below 2 °C…*" and "*…a reduction of at least 60 % of GHGs by 2050 with respect to 1990 is required from the transport sector. …by 2030 the goal for transport will be to reduce GHG emissions to around 20 % below their 2008 level…*".

For this reason, the current continuous growth in private car ownership and usage is not sustainable in the long term.

2 The Role of Parking Policies and Planning Towards an *Energy Saving* Mobility

2.1 A Short Introduction

The way people move in an urban environment is manageable through several policies, actions and measures, but parking policies are often underestimated. How can parking planning and management contribute in curbing energy consumption and promoting environment friendly transportation modes?

Car ownership depends on several pre-conditions, mainly economic and social, but one of them is the availability of a parking space (or garage) near or at the residence. As a matter of fact, in densely urbanized areas of big towns a smaller motorization rate is observed, as a consequence of both high level of public transport services and high real estate costs, where land is too valuable to "waste" on garages.

In Italy, the motorization rate has diminished by 5.4 % in Rome between 2008 and 2012 and by 5.2 % in Milan.

The other side of the coin is the progressive abandonment of inner town centers in small and medium sized settlements, where public transport is poor and inefficient, due to the lack of parking places, both private and public.

---

[3]"Roadmap to a Single European Transport Area—Towards a competitive and resource efficient transport system" (2011).

At the destination, private parking supply is one of the key factors for attracting customers. Shopping malls are the prototype of a large parking supply in order to accomodate many various shops in one place, but elsewhere, most large companies and employers provide large spaces for the cars of their customers and employees.

Public policies then influence public parking facilities which could potentially change the modal choice, as the cost of parking is one of variables in the function of transport costs. As an example, the use of public transport instead of private cars can be promoted by increasing the parking fees in the inner town parking, which in turn could help reduce car dependence. In sum, by providing a disincentive to drive, parking management can support the use of alternative modes, it can improve walkability, and influence land uses.

According to some authors, parking planning can also help in curbing Heat Island Effects, by reducing the total amount of pavement in an area (see, i.e., Gorsevski et al. 1998; Selicato and Cardinale 2014).

2.2 *Setting* Maximum *Parking Requirements*

The minimum parking requirement is a measure to avoid the inapt use of public parking from residents and customers of commercial activities.

Urban planners guided by City ordinances typically set the minimum parking ratios (number of spaces per unit of development) surface to devote to parking lots for every land use (Shoup 1999), but those parking requirements represent a tacit policy for automobile use and sprawl (Willson 1995). Private parking lots are foreseen with a minimum standard in most countries, including Italy, where the D.M. 1444/68 suggests a surface area of parking of at least 2.5 m^2 per inhabitant in residential areas, and a minimum of 40 % of the gross floor surface in commercial areas (now most often raised to 100 %). Providing parking lots within private developments should avoid passing off to local administrations the cost of those facilities in public spaces (Fig. 1).

In contrast, there are some good examples where a maximum quantity is specified instead, in order to discourage residents from possessing too many cars!

Already in the 70's Boston, Portland and New York City removed minimum parking requirements, and established the so called "parking caps" or maximums requirements in downtown areas (Weinberger et al. 2010).

Later on, since the 1990s, Berlin has had no car parking requirements, except for the disables.

In London, minimum parking standards for residential developments were replaced with maximum standards in the early 2000s (Greater London Authority 2002). Then, the Greater London Master Plan of 2011 reaffirmed this approach, retaining maximums for both residential and non-residential developments. For retail, those requirements also depend on the Public Transport Accessibility Level (PTAL) of the area (as mapped in the plan): locations with a low PTAL are penalised also in terms of maximum parking requirements (Table 1), so to

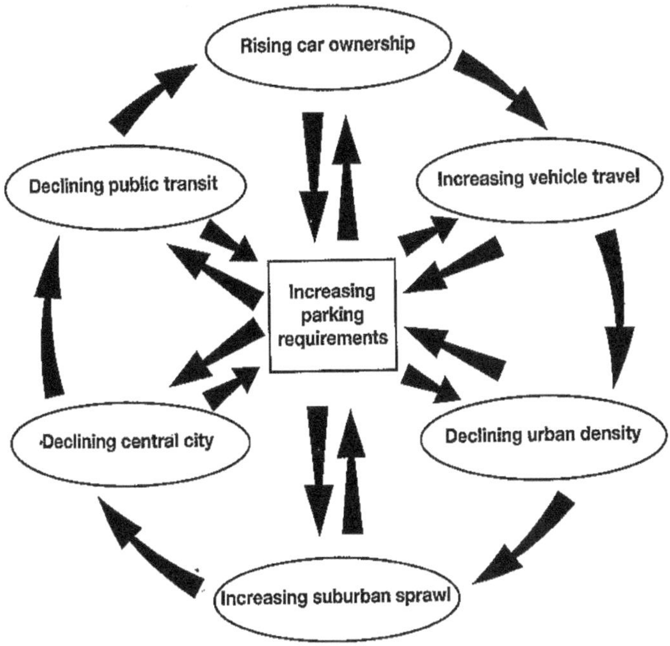

Fig. 1 The cycle of parking dependency. *Source* Shoup (2013)

Table 1 Parking for retail in the Greater London master plan

USE	PTAL 6 and 5	PTAL 4 to 2	PTAL 1
Food			
Up to 500 m^2	75	50-35	30
Up to 2500 m^2	45-30	30-20	18
Over 2500 m^2	38-25	25-18	15
Non food	60-40	50-30	30
Garden centre	65-45	45-30	25
Town centre/shopping mall/dept. store	75-50	50-35	30

Maximum standards for retail users: space per m^2 of gross floor-space (Greater London Authority 2011)

discourage new commercial developments where Public transport connections are scarcely available (Greater London Authority 2011).

The City of Seattle currently has a maximum amount of parking allowed per square foot for office and commercial developments in the city centre ("parking caps") (City of Seattle 2015).

A well-developed public transport service is the common condition in all the above mentioned examples, together with inter-modality between private cars and other transportation modes (public transport, cycling, walking…) (see, i.e., Simićević et al. 2012; Maternini and Ferrari 2014).

Car ownership cannot decrease without an efficient alternative for mobility.

2.3 *Other Policies*

There are more options in managing public parking standards in order to influence mobility patterns regarding both location, fares and time-related policies. Several studies have shown that parking demand greatly changes during the day, and that it also depends on land use (Fig. 2).

A fundamental dilemma arises when trying to make urban development less dependent on cars: the inability of most alternatives to match the quality of accessibility provided by private motorised transport (Bertolini and Le Clercq 2003). The assessment of the accessibility level provided by car parking areas and their use is therefore only a first step towards the implementation of sustainable mobility policies, aimed at providing competitive alternatives to car use. Therefore, there is a need to investigate the relationship between local accessibility to urban activities by collective transports and the location of current parking areas.

The results could be used to better locate parking garages and lots, but also to eliminate them when a public transport means could be an efficient alternative.

Also parking fee systems can be assessed through the analysis, in order to coordinate the fees and the daily distribution of costs. It has been estimated that charging motorists directly for parking tends to reduce automobile trips by 10–30 % (VTPI 2015).

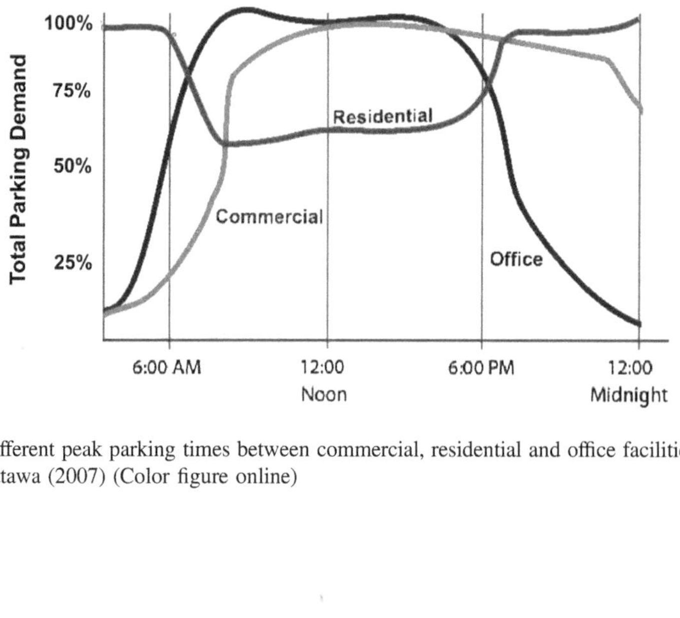

Fig. 2 Different peak parking times between commercial, residential and office facilities. *Source* City of Ottawa (2007) (Color figure online)

3 Some Analysis on the Parking System in Brescia

An analysis of the parking system in the City of Brescia is presented below (see also Bonotti et al. 2014). That analysis involves the temporal dimension (how car parking use varies during the day), and it is complementary to some space-time accessibility assessments that were developed for the public transport system in Brescia, with a special focus on the new light metro line. Those further assessments are presented in Bonotti et al. (2015).

Car spatial accessibility was analysed with regard to the availability of car parking and was compared to public transportation accessibility.

The parking Plan for the city centre was designed so that any destination within the centre from the nearest car park is equal to or less than 300 m. Considering the entire city, i.e. outside the city centre, all areas within 350 m of a parking structure or lot were considered accessible, corresponding to the average walkable distance in 5 min. The accessibility level was then rated using a linear function varying from 1, corresponding to the centre of the parking area, to 0 at the edge of the area of influence. The results are shown in Figs. 3 and 4 compared to the public transport accessibility.

Car parking is an indirect indicator of the degree of use of services. Despite the fact that the accessibility by private modes can generally be guaranteed 24 h a day, the filling rate of parking lots varies according to the time slot, as well as, in some cases, the opening hours of the car park.

Three thematic maps have been drawn. They focus on three significant time slots during a spare day:

- 10.00 a.m., critical time for toll parking;
- 01.00 p.m., lunch time;
- 4.00 p.m., afternoon rush hour.

In the following map (Fig. 5), the degree of use of toll parking places has been represented by using different colours for each filling rate. The green colour identifies parking lots with more than 75 % vacancy, yellow between 75 and 50 % vacancy, orange between 51 and 25 % vacancies, red with less than 25 % vacancy, and finally black identifies completely filled parking areas. Figure 4 shows the results for a limited area of downtown Brescia, including the historical centre.

Finally, according to accessibility theories, the availability of car parking should be linked with the location of the activities and opportunities located in the space (see, i.e., Hansen 1959).

Processing available data, it is possible to evaluate the average number of parking lots available for each service and to what extent they are filled during the day. To give an example, this analysis is provided with reference to two kinds of facilities in Brescia: public offices and sport facilities.

Concerning the data available on the location and on the hours of operation for all the services in the municipality, some critical aspects can be highlighted,

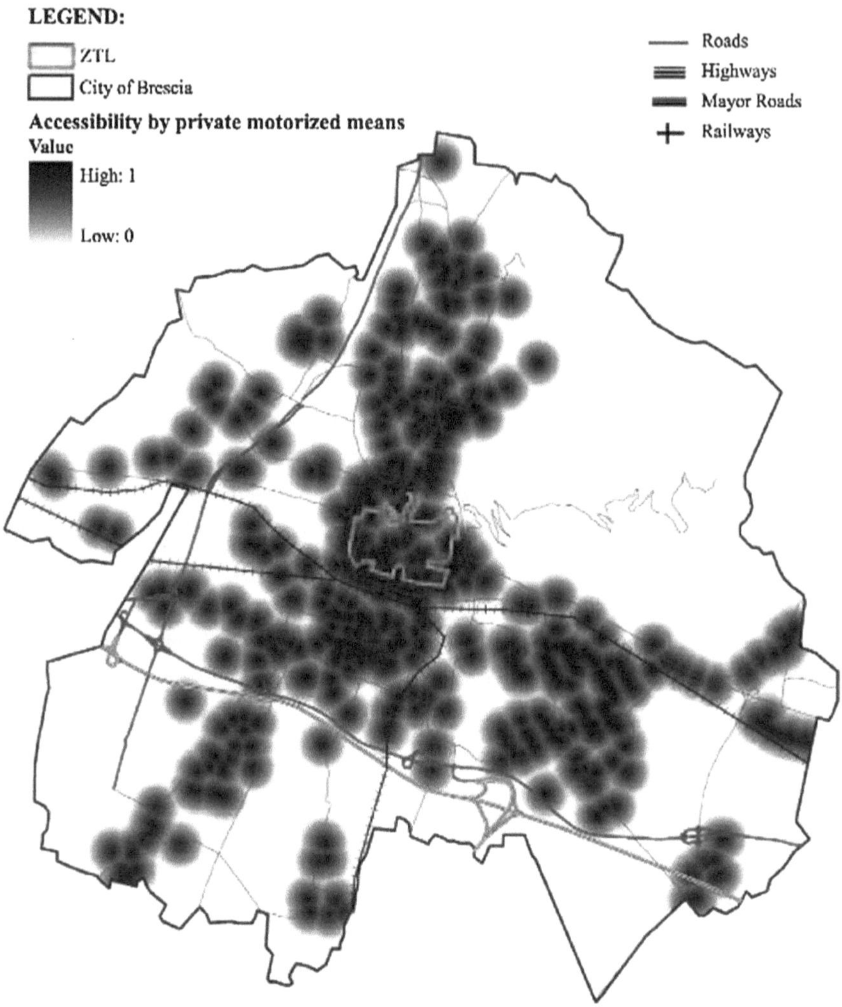

Fig. 3 Brescia and the accessibility rate of cars, described by the *grey* intensity (Color figure online)

as some services are open when the public transport system does not operate, or when the parking supply is not able to meet the demand (Fig. 6).

Comparing these data and the land use plans it is possible to optimize the location and management of parking.

The development of the Brescia metro line (opened in 2013) led to a renewed use of big central parking structures. The liveability of the downtown is now possible leaving the car in peripheral lots, being the metro ticket is far more affordable when compared to the city centre parking fees.

The diachronic analysis can rather influence the operation hours of the metro.

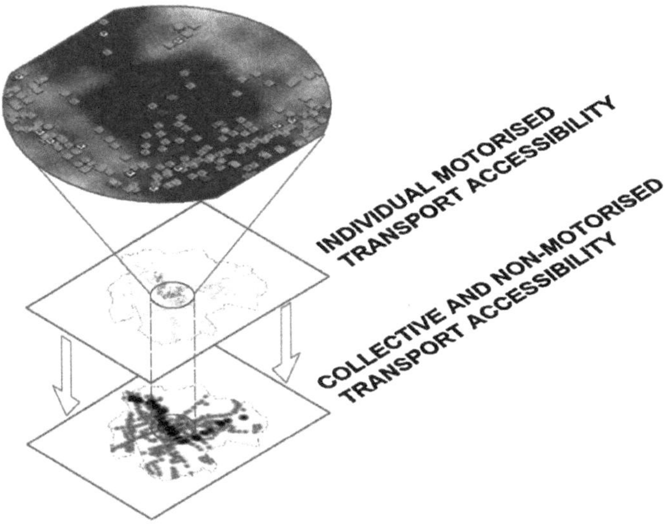

Fig. 4 The link between the accessibility by collective transit and the availability of space for parking vehicles

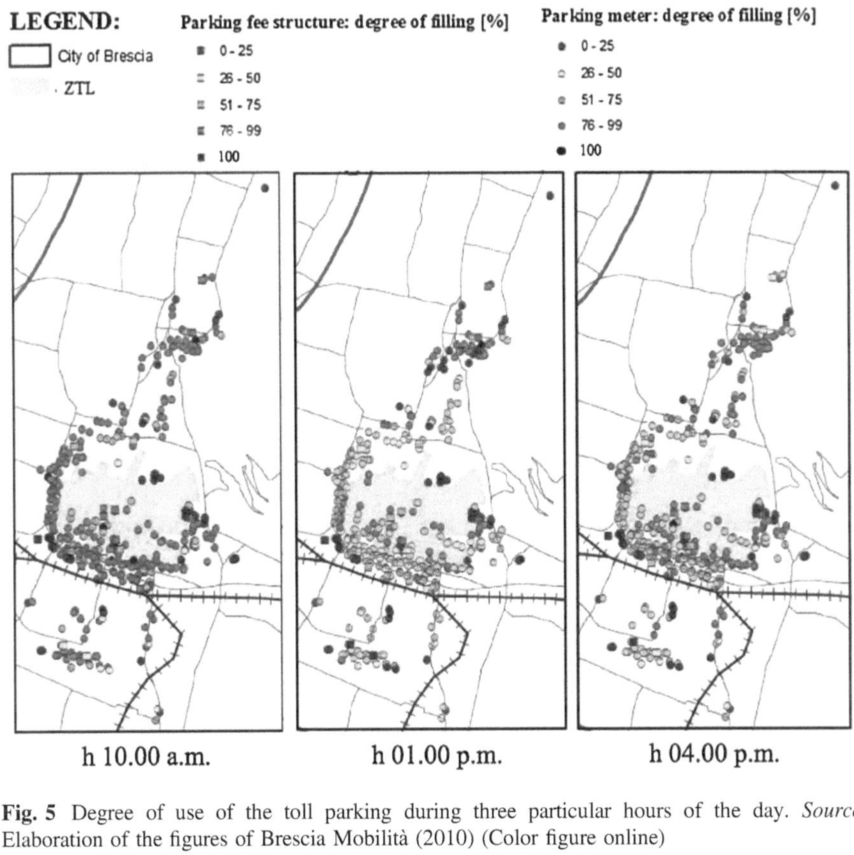

Fig. 5 Degree of use of the toll parking during three particular hours of the day. *Source* Elaboration of the figures of Brescia Mobilità (2010) (Color figure online)

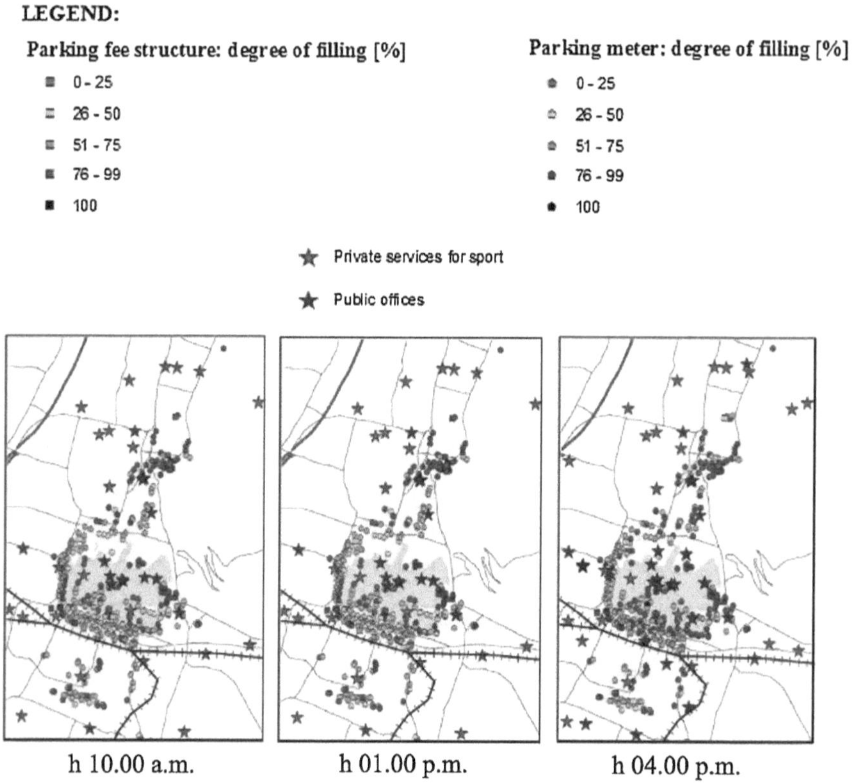

Fig. 6 The relationship between the degree of parking filling in time and public services opened at the same time (Color figure online)

4 Conclusive Remarks

There is no doubt that car parking planning policies and strategies can support alternative and more sustainable modes of transport, and therefore promote energy saving. Car parking management policies and strategies can help in promoting alternative and more sustainable modes of transport. Traditionally increased parking demand was addressed by enlarging the parking supply. However, in the last few years several cities all over the world have set the goal of reducing private traffic and increasing public transport, especially in metropolitan areas. So, the recent trend is to optimise the use of the existing parking supply by management strategies (Maternini and Ferrari 2014).

Furthermore, parking planning and management represent only one aspect that should be considered when dealing with mobility and energy saving issues.

Low carbon mobility is "a mobility that disincentivizes car use, along with the promotion of non-motorised modes and good public transport, and the introduction of alternative fuels infrastructures" (UNECE 2011). Some authors argue that transport

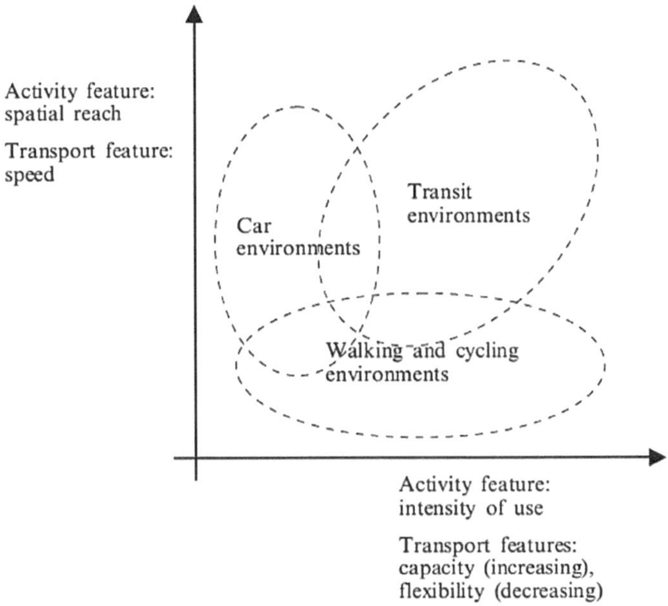

Fig. 7 Principles of multimodal urban and regional development. *Source* Bertolini and Le Clercq (2003: 578)

planning is currently at a critical point, and that transport planners underestimate the key challenges facing urban planners (see, i.e., Banister 2005; Wickham 2006).

The need for integrating urban and mobility planning is even greater as environmental concerns increase (Tira 2015).

For example, Bertolini and Le Clercq (2003) propose an accessibility-oriented multimodal urban development that considers two dimensions: the spatial reach of an urban activity or function, and its intensity of uses. The available transport modes, with their features determine the preferred location of an activity (Fig. 7).

References

Journals and Articles

Bertolini L, Le Clercq F (2003) Urban development without more mobility by car? Lessons from Amsterdam, a multimodal urban region. Environ Plann A 35(4):575–589. doi:10.1068/a3592

Bonotti R, Rossetti S, Tiboni M, Tira M (2014, June) Diachronic analysis of parking usage. The case study of Brescia, TeMA J Land Use Mobil Environ (Special Issue):77–85. doi:http://dx.doi.org/10.6092/1970-9870/2505

Bonotti R, Rossetti S, Tiboni M, Tira M (2015) Analysing space-time accessibility towards the implementation of the light rail system: the case study of Brescia. Plann Pract Res. doi:10.1080/02697459.2015.1028254
Gorsevski et al. (1998) Air pollution prevention through urban heat island mitigation: An update on the urban heat island pilot project. 1998 ACEEE Summer study on energy efficiency in buildings Proceedings, 9. Report Number: LBNL-42736. Washington, D.C.: American Council for an Energy-Efficient Economy
Hansen WG (1959) How accessibility shapes land use. J Am Inst Plann 25(2):73–76
Maternini G, Ferrari F (2014) Innovative parking strategies through the application of variable pricing techniques. The case of San Francisco. CSE J 2:49–60. doi:10.12896/cse20140020019
Selicato F, Cardinale T (2014) Energy aspects of urban planning. The urban heat island effect. CSE J 1:79–91. doi:10.12896/cse2014001008
Shoup DC (1999) The trouble with minimum parking requirements. Transp Res Part A 33:549–574
Simićević J, Milosavljevi N, Maletić G, Kaplanović S (2012) Defining parking price based on users' attitudes. Transp Policy 23:70–78
Tira M (2015) Verso nuove mobilità sostenibili. Sentieri urbani: 14–22. ISSN:2036-3109
Wickham J (2006) Public transport systems: the sinews of European urban citizenship? European Societies, 8(1):3–26
Willson RW (1995) Suburban parking requirements: a tacit policy for automobile use and sprawl. J Am Plann Assoc 61(1):29–42

Books and Chapters

Banister D (2005) Unsustainable transport. City transport in the new century, Routlege (Taylor & Francis), London and New York
European Commission (2009) White paper. Adapting to climate change: towards a European Framework for action. EU Publications office, Brussels
European Commission (2011) White paper. Roadmap to a single European transport area. Towards a competitive and resource efficient transport system. EU Publications office, Brussels
European Commission (2014) EU transport in figures. Statistical pocketbook 2014. Publications office of the European Union, Luxembourg
Eurostat (2012) EU transport in figures. Statistical pocketbook 2012, Publications office of the European Union, Luxembourg.
Greater London Authority (2002) SDS maximum parking standards: derivation of PTAL-based parking restraint. SDS Technical Report Twenty. Greater London Authority, London. ISBN 1 85261 412 9
Greater London Authority (2011) The London plan. Spatial development strategy for Greater London. Greater London Authority, London. ISBN 978 1 84781 451 7
Shoup D (2013) The high cost of free parking. American Planning Association, Chicago-Washington DC
United Nations Economic Commission for Europe (UNECE) (2011) Climate neutral cities. How to make cities less energy and carbon intensive and more resilient to climatic challenges, United Nations, Geneva
Weinberger R, Kaehny J, Rufo M (2010) U.S. parking policies: an overview of management strategies. Institute for Transportation and development Policy, New York
World Health Organisation (WHO), Peden M, Scurfield R, Sleet D, Mohan D, Hyder A, Jarawan E, Mathers C (eds) (2004) World report on road traffic injury prevention. WHO, Geneva

Online Publications

City of Ottawa (2007) Transit oriented development guidelines. Approved by City Council on 26 Sept 2007. http://ottawa.ca/en/city-hall/planning-and-development/community-plans-and-design-guidelines/design-and-planning-0-1-8pdf. Accessed 18 Apr 2015

City of Seattle (2015) Best practices in transportation demand management. Seattle urban mobility plan. http://www.seattle.gov/transportation/briefingbook.htm. Accessed 18 Apr 2015

Victoria Transport Policy Institute (2015) Parking management strategies for more efficient use of parking resources. TDM Encyclopedia. http://www.vtpi.org/tdm/tdm28.htm. Accessed 18 Apr 2015

ISUT Model. A Composite Index to Measure the Sustainability of the Urban Transformation

Roberto Gerundo, Isidoro Fasolino and Michele Grimaldi

Abstract The urban transformation is the result of several decisions made in a variety of temporal and spatial scales, which concern the field of urban planning regarding the location of buildings and activities, position of transport networks, that in fact are more persistent (Morris 1994). The urban transformation generates environmental impacts in terms of soil sealing, energy consumption and urban heat island (Ewing and Rong 2008), which result in consequences on stormwater runoff, on raising the temperature and increasing of CO_2 emissions. From a representation of the state of the art, the present paper defines a composite index called ISUT (index of sustainability of the urban intervention). The ISUT covers all the essential components of the urban design that condition the energy balance equation (Parham and Fariborz 2010). The composite index value is used to propose better strategies to guide the development of local area plans in conjunction with the City's Plan using an easy data-set which derives from the knowledge of the territory, which is usually already acquired for urban planning tools on a municipal basis.

1 Introduction

Cities represent a dramatic manifestation of human activities on environment (Ridd 1995). This artificial organism degrades natural habitats, simplifies species compositions, interrupts hydraulic systems and modifies energetic flows (Alberti 2005), with health and socio-economic consequences in the long term too. In urban planning, the renewed interest in the environmental-energy sustainability criteria is

R. Gerundo (✉) · I. Fasolino
DICIV, University of Salerno, Fisciano, Italy
e-mail: r.gerundo@unisa.it

I. Fasolino
e-mail: i.fasolino@unisa.it

M. Grimaldi
DCB, University of Salerno, Fisciano, Italy
e-mail: migrimaldi@unisa.it

R. Papa and R. Fistola (eds.), *Smart Energy in the Smart City*,
Green Energy and Technology, DOI 10.1007/978-3-319-31157-9_7

relatively new, likewise the recent scientific approaches of coding and logging of procedures, parameters and sustainability indicators used in this area of study (Jaeger et al. 2010; De Wilde and Van Den Dobbelsteen 2004; Schwarz 2010). It is quite obvious how essential is the alignment between the architectural design and sustainable urban design, because strategic choices, in matter of regulation and urban planning, if not well calibrated on the sustainability, could make partially ineffective the results in the area, also if those are modeled on principles of environmental sustainability and respectful of the parameters of any protocols (IISBE 2012).

Urban plans generally have little consideration of environmental aspects apart from, in fact, that they provide some reference, like, for example, to the obligation to respect permeability thresholds.

Sustainable urban development is a broad and multi-dimensional concept, a theoretical framework is needed to deal with what is meant by sustainability of the urban project and which kind of tools should be developed for the evaluation (Carraro et al. 2009).

The impact of the urban structure and therefore the actions of urban transformation on the environmental components has a formal expression in the surface energy balance equation (Parham and Fariborz 2010; Stull 1988). For a surface energy in equilibrium the following is a valid relation:

$$\mathrm{Rn} = \mathrm{Rg}(1 - \mathrm{A}) - \mathrm{RL} = \mathrm{QG} + \mathrm{Q} \tag{1}$$

which expresses how energy inputs, that is the net radiation (Rn), are used to heat the air in contact with the ground (triggering a flow of sensible heat QH); to evaporate, or transpire water, if there are any plants, (triggering a flow of latent heat, QE); to heat the interior of the soil (triggering a flow of heat in the soil, QG). Rn is the result of the balance between incoming global solar radiation (Rg) net of albedo (A), and the emission of long wave radiation at the earth's surface (RL); greenhouse gases, along with clouds and atmospheric dust, act on RL term, partly intercepting and re-radiating it to the ground. Thereby they limit the cooling of the earth's surface by radiation into space.

The presence of urban area changes the terms of the energy balance and introduces two new terms (Oke 1987; Oke et al. 1991), the anthropogenic heat flow QF, and the flow of heat stored in the urban structure ΔQS (WMO 2008).

QF is an additional source of heat that comes from human activities and the energy they involve, such as traffic, heating, combustion industrial, production and consumption of electricity.

ΔQS depends on the materials and geometry of the urban context, but above all on the density of buildings.

As a first approximation ΔQS can be expressed as a function of green spaces and built-up areas (Oke 1981).

But even QF, indirectly, is a function of the built-up areas, as urban areas increase with increasing human activities. Overall, therefore, the urban areas compared to country areas are rich in energy and this imbalance is even more fragile because of extra heat sources (Gerundo et al. 2012b).

Through the analysis and the considerations about the experiences as a system, a new model has been developed that faces the urban planning and projects in an integrated and organic way that includes the different impacts.

The aim of this contribution is the construction of an index, defined by sustainability of the urban transformation (ISUT).

This index allows us to control, in a quantitative way, the sustainability of both the individual building project and a sub-urban plan project to implement, likewise also to check on, the existing urban structure, the efficiency of possible adaptations, energy and environmental, of existing settlements, however always desirable. This method allows investigating the impact of human development on the natural environment through a model that evaluates direct and indirect effects on an urban ecosystem through some indicators that are measured on the specification of the context. This method monitors the distribution and organization of surfaces, seeking the protection of soil, of water, of air and guarantees a better life's quality for the users of urban spaces (Gerundo et al. 2012a).

Using, in fact, a set of selected indicators, it was assumed that we would combine these linearly in a sustainability index that supports the processes of territorial government at an urban scale, both in the analysis of the existing design and verification. The screening at the base of the construction of ISUT index has allowed us to expand the macro-fields of study, to extract an appropriate set of indicators of sustainability in urban areas. Moreover, it allows us to define a framework that takes the form of a model of eco-sustainable urban regulations, possibly to be included in the instrument of municipal planning building rules.

2 Methodology

The methodological framework we have proposed consists of the two following macro-phases:

- macrophase 1: selection of a set of indicators to measure the influence of the urban structure on ΔQS and QF;
- macrophase 2: construction of the ISUT relative to each minimum unit, in which the urban structure has been split.

The proposed methodology requires a level of knowledge of the area, ordinarily taken for the preparation of an instrument of structural urban planning, on a municipal basis. Specifically, for its proper application, the methodology requires a minimum level of information consisting of a cartographic support of vector type at a 1:2000 scale, made by an aerial digital survey. This last one has associated an alpha-numerical database containing information about the volume and floor areas of individual buildings, as well as information about the materials and the prevailing urban land use.

2.1 Macrophase 1

In the first phase, according to categories of green spaces (G) and built-up area (B) of the urban structure, we identify the following components:

- vegetation;
- permeability of soil;
- materials of urban construction;
- urban morphology.

Afterwards, a set of indicators have been selected according to available knowledge, which has allowed us to measure the morphometric characteristics of the urban settlements that affect QF and ΔQS respectively (Table 1).

In order to consider the presence of vegetation on a land area (Sf) or territorial area (St) (Saito et al. 1991; Honjo and Takakura 1991), as in the literature, we refer to the coefficient of vegetation occupation (CVO) (Arnofi and Filpa 2000).

The CVO is an indicator that aims to assist the other parameters, related to built-area; this indicator is a measure of the presence of vegetation, and it is used to approximate the visual perception of different structures of urban green, also depending on the permeability of the soil and its water retention.

The potential direct supply of the aquifer is taken into account through the ratio between the sum of the individual surfaces exposed to rainwater, considered

Table 1 Indicators selected

Cat.	Energy factors	Indicator		Formula	Description
G	ΔQS	I1	Coefficient of occupation of the vegetable land (CVO)	|(Salb * Halb) + (Sav * Hav)|/St	Salb = area covered by trees Halb = tree height Sav = occupied area by lower beds and green roof Hav = height of flower beds St = land area
	ΔQS		Permeability ratio (RP)	S_p/St	S_p = the sum of the permeable areas exposed to rain water, considered according to their degree of permeability St = land area
B	ΔQS; QF	I2	Average reflection coefficient (ICRM)	∑Ci * Ai/∑Aì	Ci = coefficient or reflection of a given material Ai = surface area of a specific material
	ΔQS; QF	I3	Index of local volumetric density (IV)	V/St	V= sum of the volumes of buildings St = land area
	QF	I4	Thermal dispersion index (IDT)	S/V_h	S = dispersive surface area V_h = heated volume

according to their degree of permeability measured as the inverse of the runoff coefficient of the material of which is built the ith permeable surface. Both the indicators are linearly combined in an overall indicator (I1).

The average reflection coefficient, of all the surfaces exposed to radiation related to the building or urban project, is used to calculate the albedo factor control's indicator for urban construction materials (I2), considering the totality of the building structures on single lot or isolated.

The application of this indicator requires identification of the type of material of its extension and the definition of its reflection coefficient.

As regards the characterization of the existing urban structure, for the detection of different types of material we use a direct survey, through the interpretation of a photographic survey, or by using the potential of the observation's technology by thermal remote sensing.

In the case of an urban project during the design phase, we impose the adoption of materials with a known reflection coefficient in order to check in advance the Cm of the project.

Regarding the urban morphology component, we take into account morphological indicators and the indicator related to passive energy efficiency of the urban project. About the morphology aspects, in order to automate the procedure in a GIS environment, the choice of an indicator of local volumetric density (I3) has been considered significant as a ratio between the sum of the volumes of the buildings and the reference area. This indicator, on the same reference area, takes into account both the effects of the distance between the buildings and the ratio of unobstructed views. This assumption loses the information related to the orientation of buildings, however, the calculation of this indicator, extended to the block or to the lot, would require that we strictly consider also the phenomenon of the correlated shadows, which is difficult to model in the GIS environment. About the energy aspects, attention must be paid to the compactness of the building (cfr. Dlgs 192/05, Dlgs 311/06, Dir 2002/91/CE, energy performance of buildings). Specifically we consider the ratio S/V (I4) between the volume and the heated dispersing surface, which provides positive information, as having a lower is resulting value. We must also strictly consider the orientation with respect to the solar path (cfr. Regione Campania, Dgr 659/07 "Obiettivo D1"). In this local law the East-West axis is taken as the optimum arrangement for buildings, with decreasing values down to the last one corresponding to the angular variation of +/−45° with respect to the ideal).

2.2 Macrophase 2

In the second phase of the methodological approach, we proceed to synthesize the various indicators that we previously described in a complex index, in order to identify a measure that enables us to monitor, in an expeditious way, the energetic and environmental outcomes of urban transformation (Fig. 1).

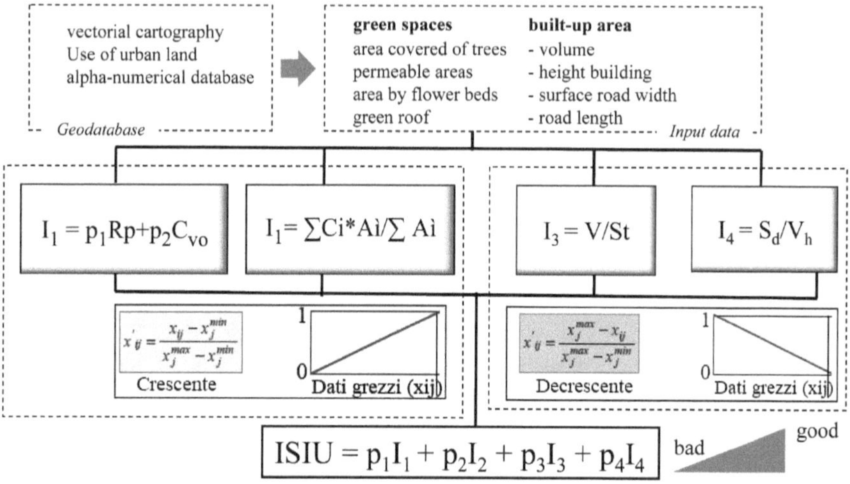

Fig. 1 Methodological framework for the definition of the ISUT

First, we have proceeded with normalization of the various selected indicators. The procedure of normalization is an operation that allows us to make a comparison between the statistical distributions, which are expressed in the respective scales of measurement, in relation to a single scale, that typically includes zero and one. The transformation should not alter the informative content of the initial data. There are several features to provide the above-mentioned transformation. In particular, we have proceeded with the normalization of the selected indicators in order to make comparable the different distributions expressed in the respective scales of measurement. The overall index, ISUT, has obtained by using a linear combination of these indicators.

In this case, we refer to the types of changes which maintain the direction of preferences (i.e., the maximum value is the maximum and the minimum is minimal even after processing) for the values of the indicators pertinent to the thematic areas of the vegetal cover of soils and the permeability (I1) materials of urban construction I2 (higher value is better), while we refer to the type which reverses the direction of preferences for the indicators relating to the local volumetric density (I3) and thermal dispersion index (I4) (lower value is better).

This choice is due to the fact that the searched index of sustainability, as a linear combination of those parameters with equal weights, expresses positive outcomes for higher values of the result. In order to read the phenomenon on urban scale, we have made the spatialization of the various indicators for unit of urban structure, in relation to the type of intervention that corresponds to the lot or the block. This series of factor maps let to express the spatial distribution of ISUT, through operations of map algebra.

The map algebra means the use of basic operators, arranged in sequence, in order to solve complex spatial problems. It is substantially similar to an algebraic expression in which the result is a combination of several characteristics. It involves the use of logical and mathematical relationships, in a geographic information system, applied to spatial data, which allows with two or more layers (maps), of similar size, to produce a new layer.

The result of the map algebra operation is a new map that is representative of the spatial variability of the ISUT. We need to estimate the intensity of the phenomenon, through the definition of a suitable scale of values that must be divided into classes in order to obtain appropriate evaluations. Due to the lack of values of reference, in an absolute sense, for the individual indicators that make up the ISUT, and in order to limit arbitrariness in the definition of the thresholds, we have chosen the method of natural breaks (Jenks 1967) as our method of classification. This method of classification is based on the procedure of optimization that provides the identification of break points in the distribution of values, given the number of classes.

Through this method the variation of the elements is minimized within each class. In this way the amplitude of the classes varies as a function of the distribution of data within the considered groupings (Boffi 2004).

Having defined the number of classes, the concerned thresholds are depending on the only values related to the fabric of settlement that has been analyzed. In this way the methodology is generally valid and applicable to different blocks, regardless of their extension We have divided the evaluation range into five classes (Carraro et al. 2009), corresponding to different degrees of intensity, which have been labelled as: very high (VH), high (H), medium (M), low (L) and very low (VL).

3 Case Study

This methodology has been tested on the urban fabric of the City of Baiano (Campania, Italy), in order to assess the changes brought about by the choices of urban transformation planned by the Municipal Urban Plan (MUP) (Fig. 2).

3.1 Definition of Control Indicators in Relation with the Degree of Knowledge

On the first phase of the methodology, that is the measurement of the indicators related to the selected components, we have made several operations of spatial analysis in GIS environment, taking into account the degree of the available knowledge.

Specifically, with reference to the first two components, namely permeability and the vegetation cover of the soil, we have considered appropriate to synthesize,

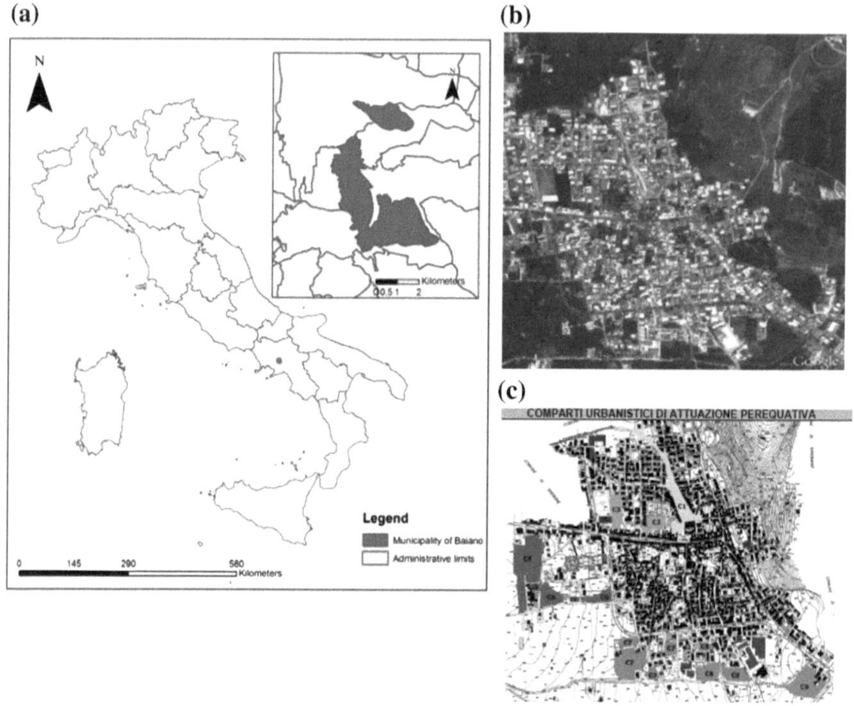

Fig. 2 Geographic location Baiano, Italy (**a**); base scenario (**b**); design scenario (**c**)

in a single indicator, both information, thus constructing the controlling index of vegetation cover and permeability. First of all, the permeability ratio has been measured. It is intended as the ratio between the unsealed surface and the reference surface that, depending on the considered unit of processing, corresponds to the surface of the lot or of the block. Subsequently, this ratio was increased, through an appropriate multiplier ratio, according to the qualitative component of the specific vegetable cover, classified as a first approximation, in tree or meadow (I1). With reference to the phenomenon of the island of urban heat, in order to construct the index for the control of the albedo factor of the urban construction materials, we have proceeded to a first subdivision of the components of the urban fabric in horizontal and vertical surfaces (I2). The horizontal surfaces include the totality of urban open spaces, which are divided, in turn, in converted areas, in particular road surfaces (asphalt or similar) and squares, and in unsealed surfaces. Moreover the surfaces of the flat roofs of the buildings belong to this category. In addition, we take into account an analysis of the use of urban land for the definition of the average reflection coefficient of such areas, derived from the prevailing law. The vertical surfaces include the totality of the constituents of the buildings. Since the reflecting walls of buildings, in particular of the most recent ones, are composed by a large number of faces and made by materials of a different nature, it was considered allowable to adopt, for each individual building, an average reflection coefficient with respect to the extension of the vertical surfaces of the same. In this way, the operation of association of the material to a single vertical surface is simpler. First of all, from an operational point of view, we proceeded to the definition of a topological relationship for the construction of the vertical walls. This relationship is necessary to transform the polygonal feature in a linear feature that is representative of the plan view of the ith wall to which the relative height can be associated. Instead, the attribution of the average coefficient of reflection of the vertical walls to each building was carried out through a their visual analysis, that has been made with the aid of information provided by google-maps, where possible, or through the outcomes of the direct investigations This simplification has been particularly useful for the urban fabric, which could be complex from the morphologic point of view. This one has adopted in order to streamline the input provided to the algorithm and speed up work. Finally, as regards the construction of the index for the control of the factors of urban morphology, we have considered the local volumetric density (I3) and the Thermal dispersion index (I4), in order to automate the procedure in a GIS environment.

3.2 Results and Discussion

After the database was implemented with the measured values, we proceeded with construction of factor maps representing the block's spatial distribution of the index of control of permeability and vegetation and the index of control of its albedo factor (Fig. 3a, b).

Moreover we built the factor map of the local volumetric density and the range of the index of compactness on buildings (Fig. 3c, d).

Looking at the category of green spaces, the corresponding factor map records the spatial variability of the components of permeability and vegetation that are related to the variations of the rates of evapotranspiration and surface runoff about the different types of land cover. In particular, the map shows how the settlement fabric is characterized by values of the index totally placed in the middle class of intensity range, presenting however, especially within the mainly dense fabric, blocks characterized by medium to high levels.

Concerning the class built-up areas, that allow us to investigate the effect of the urban island of heat determined by the urban structure, the factor map is representative of the component of urban materials and shows values of the index, i.e. medium-high, on over all urban fabric, with the exception of the suburban area.

The factor map I3, that is representative of the component of urban morphology, shows a concentration of density of buildings with intensity mostly medium-high, as shown also by the factor map I4, bringing out a substantial correlation between the two values.

We subsequently built the factor map ISUT, by a linear combination performed with operations *mapalgebra* of a factor's map for each indicator.

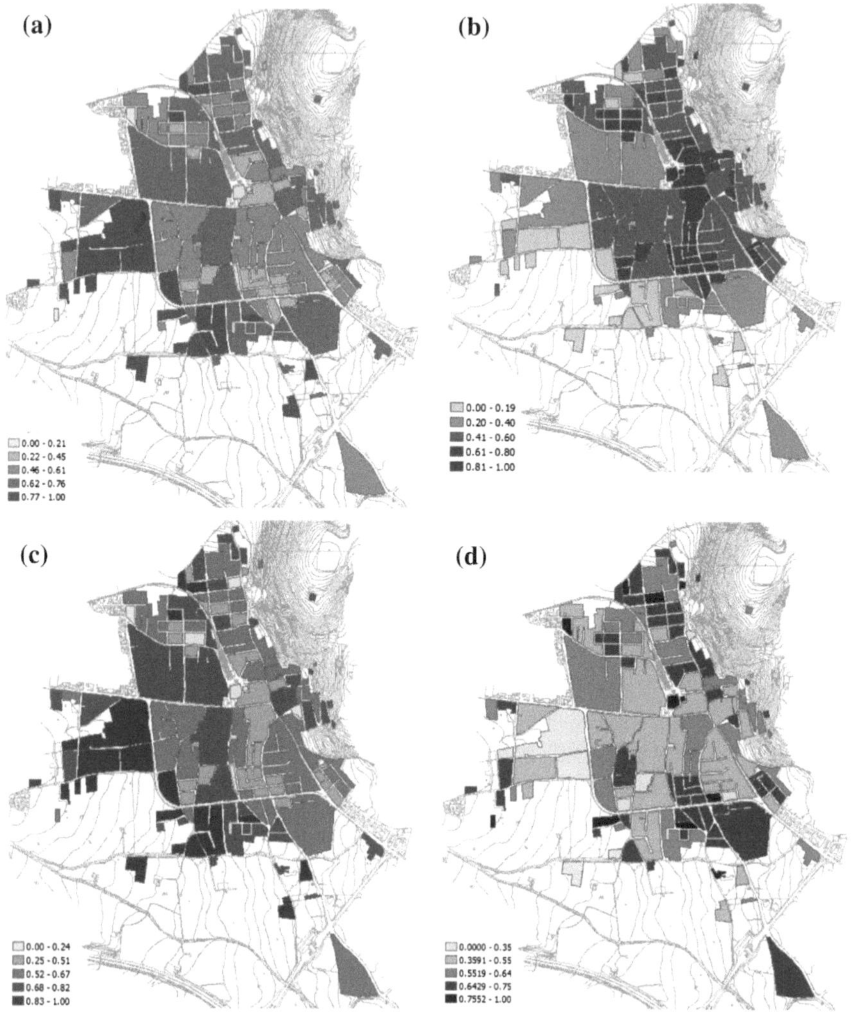

Fig. 3 Factor map representative of the spatial distribution of each indicator: I1 (**a**); I2 (**b**); I3 (**c**); I4 (**d**)

As described in the methodology, the spatialization and the classification of the obtained values, using the "natural breaks" classification's method identifies a scale of intensity of the phenomenon, for the urban fabric, considered divided into five classes according to the literature.

The operation was carried out for both the *base scenario* and the *design scenario*.

Looking at the base scenario, the map highlights the critical points of the settlement fabric investigated, represented by blocks characterized by an ISUT value belonging to medium-low and low classes (Fig. 4c).

Fig. 4 Sector C7 (**a**); Factor map design scenario: I1, I2, I3, I4, (**b**); factor map representative of the spatial distribution ISUT for basic scenario and design scenario (**c**)

The results obtained for the ISUT value in the design scenario, namely for the sectors planned as new areas of settlement development of the MUP, let us obtain the factor map of all areas of the project. This also allows us to make judgments about the sustainability of the changes brought by the choices of spatial planning that need to individually reach a programmed level of acceptability, namely a greater ISUT value than the maximum range present in the intensity scale identified.

Placing, in fact, the focus on the sector C7 (Fig. 4c) shows that the planned transformation, according to the assumptions used, is placed in the upper class of the scale of intensity defined.

Having established what can be considered acceptable and, therefore, sustainable, the medium-high and high thresholds, it is clear that the transformation in question exceeds the average level of sustainability of the fabric where it is integrated.

Finally, we have evaluated the impact of all the changes on the level of sustainability of existing tissue, measured by comparing the average value of the ISUT of each block, calculated for the base scenario and the design scenario.

In particular, for the case in question, there is an increase of percentage of the average ISUT's value and, therefore, it is evident that, basically, the new interventions, although of reduced dimensions, contributes, overall, to improve the energy and environmental quality of the existing settlement fabric to which are added.

The results show how the index ISUT has the potential to be used for a comparative analysis of the sustainability performance for the production of information of land use, reliable and detailed, at the urban scale.

The spatial scale is an important aspect of assessment in detecting the impacts of urbanization on natural resources and ecosystems.

In fact, the scale of the assessment influences both the identification of the actors of environmental issue and the range of possible actions and policy responses (Weins 1989; Levin 1992; Millennium Ecosystem Assessment 2003).

The set of indicators has been selected considering the characteristics of land use, the sustainability issues, the strategies of environmental planning and the availability of data.

Moreover, the indicators, at the urban scale, are effective tools in monitoring the complex phenomena, increasing transparency and accountability with the provision of widespread access to information, engaging stakeholders, supporting policy-making and allowing comparisons across time/space with other municipalities/regions.

The ISUT is basically a static index of sustainability, since it provides information for a specific moment and allows us to understand the wider impacts of urbanization.

However in the application ISUT has been used like a dynamic index, between the base scenario and the design scenario, because, to support appropriate actions of planning, it is necessary to understand and explain the relationships between the different factors that change over time.

As regards the aspects evaluation, the ISUT provides a comparative evaluation between these scenarios.

However, sustainability should be judged by some threshold standards on a scale of absolute values, because sustainability is not a relativistic concept (Fischer et al. 2007).

We should note that it is difficult to fix such absolute limits in decision-making, based on ecological and environmental thresholds, although they are essential for assessing the sustainability of the city.

The result is that it is essential to provide an absolute evaluation with a certain standard of thresholds until you can.

4 Conclusion and Future Developments

The explained methodology shows that appropriate designing criteria helps to enhance global energy and environmental quality of the existing urban frame, namely the object of urban planning.

The ISUT helps to support the processes of land's government, both in the design and testing of new settlements, and for planning projects of ecological and energetic redevelopment of the existing urban fabric.

Specifically, it permits (1) a definition of the benchmark of the current situation, the strengths and weaknesses; (2) an evaluation of the efficiency of the required implementation plans, and (3) measurement of advancements towards sustainable development.

The future developments of our methodology consist in the sensitivity analysis of the model with reference to the procedures of weighting. The application has been implemented with an Equal weighting which uses the measurement of each indicator with the same degree of importance.

With regard to the overcoming of an assessment only based on the comparison between scenarios, we should prepare, gradually, large records of data to be able to locate and consolidate the analysis's values, especially of the partial indexes, such as to make significant results of the individual application.

We need a wider trial of the ISUT, which necessarily has to be applied in multiple and different settlements in order to compare the different intensity classes and validate threshold values of reference. This validation requires the definition of standard urban structure, which is an ideal reference and that should be extracted using a parametric comparative analysis of the different components that contribute to the formation of the ISUT. In this way we will be able to identify possible correlations between parameters in play and see which of them influence, in a more sensible way, the value of the ISUT.

By working in this way, we can further refine the judgments of value in terms of energetic and environmental sustainability, both for the existing urban fabric and for the new ones.

References

Alberti M (2005) The effects of urban patterns on ecosystem function. Int Reg Sci Rev 28(2):169–192

Arnofi S, Filpa A (2000) L'ambiente nel piano comunale. Guida all'eco-management nel Prg. Milano: Il Sole 24 Ore

Boffi M (2004) Scienza dell'informazione geografica. Introduzione ai GIS. Zanichelli, Bologna

Carraro C, Ciampalini C, Cruciani S, Giove L (2009) The FEEM sustainability index methodological report. Retrieved from: www.feemsi.org. Accessed 04 June 15

De Wilde S, Van Den Dobbelsteen A (2004) Space use optimisation and sustainability—environmental comparison of international cases. J Environ Manag 73:91–101

Ewing R, Rong F (2008) The impact of urban form on residential energy use. Hous Policy Debate 19(1):1–29
Fischer J, Manning AD, Steffen W (2007) Mind the sustainability gap. Trends Ecol Evol 22 (12):621–624
Gerundo R, Fasolino I, Grimaldi M (2012a) Dimensioni della trasformazione. Edizioni scientifiche Italiane, Napoli
Gerundo R, Grimaldi M, Falcone I (2012b) Methodological proposal for the optimization of energy retrofit strategies in urban planning. In: Zhang et al. (eds) Proceedings of the international conference on green buildings and optimization design, Asme Press, New York
Honjo T, Takakura T (1991) Simulation of thermal effects of urban green areas on their surrounding areas. Energy Build 15:443–446
IISBE SB (2012). Method and SBTool. Retrieved from: http://www.iisbe.org/sbmethod. Accessed 04 June 15
Jaeger JAG et al (2010) Suitability criteria for measures of urban sprawl. Ecol Ind 10(2):397–406
Jenks GF (1967) The data model concept in statistical mapping. Int Yearb Cartogr 7:186–190
Levin S (1992) The problem of pattern and scale in ecology. Ecology 73:1943–1967
Millennium Ecosystem Assessment (2003) Ecosystems and human well-being: a framework for assessment. Island Press, Washington DC
Morris A (1994) A history of urban form: before the industrial revolutions, 3rd edn. Longman, London
Oke TR (1981) Canyon geometry and the nocturnal urban heat island: compareson of scale model and field observations. J Climatol 1:237–254
Oke TR (1987) Boundary layer climate. Routledge, London
Oke TR, Johnson GT, Steyn DG, Watson ID (1991) Simulation of surface urban heat islands under ideal conditions at night: diagnosis of causation. Bound-Layer Meteorol 56:258–339
Parham AM, Fariborz H (2010) Approaches to study urban heat island—abilities and limitations. Build Environ 45:2192–2201
Ridd M (1995) Exploring a V-I-S (vegetation-impervious surface-soil) model for urban ecosystem analysis through remote sensing: comparative anatomy for cities. Remote Sensing 16(12): 2165–2186
Saito I et al (1991) Study of the effect of green areas on the thermal environment in an urban area. Energy Build 15:493–498
Schwarz N (2010) Urban form revisited. Selecting indicators for characterising European cities. Landsc Urban Plann 96(Issue 1):15, 29– 47
Stull RB (1988) An introduction to boundary layer meteorology. Kluwer Academic Publishers, Norwell
World Meteorological Organisation (2008) Guide to meteorological instruments and methods of observation. WMO-N.8
Weins J (1989) Spatial scaling in ecology. Funct Ecol 3:385–397

The Landscape Assessment for New Energy Facilities. An Integrated Approach for Innovative Territorial Policies

Donatella Cialdea and Luigi Mastronardi

Abstract The landscape impact of renewable energy facilities requires an approach designed to create a new infrastructure network, compatible with the landscape features and with the agricultural production. It is necessary, therefore, to consider the impact of the energy infrastructures, caused by the large size of the involved area, and consequently their influence on the fruition of the surrounding area and, as a result, on its land value. Our thesis aims to highlight the importance of the analyses of the visibility undertaken on a portion of land characterised by its strongly rural features, and to show, through some study cases, how the control and the management of these performances can be transformed into control and management of the visual-perspective relationship between landscape and observer. It is clear that the visual control alone cannot be considered as a total control of the landscape but it can become an important tool for monitoring transformations in specific areas and a useful tool in the territorial planning process.

1 Introduction

Our study concerns the energy infrastructures that have the most impact on rural landscape. To test our methodology we analyze some sample areas in the Molise Region in the South of Italy, which is characterized as a "Region without cities" because of the small dimensions of its towns. The analysis focuses on the systems for wind power and ground–mounted photovoltaic plants.

The main aim of this research is to investigate the renewable energy plants phenomenon in this Region and to relate it with other cases. Looking to the most recent legislation about renewable energy and landscape, the research wants to give a project's approach to introduce new facilities in the Regional Landscape. The evolution of land use in order to establish renewable energy installations is

D. Cialdea (✉) · L. Mastronardi
L.a.co.s.t.a. Laboratory D.I.B.T. Department, University of Molise, Campobasso, Italy
e-mail: cialdea@unimol.it

R. Papa and R. Fistola (eds.), *Smart Energy in the Smart City*,
Green Energy and Technology, DOI 10.1007/978-3-319-31157-9_8

evaluated by comparing the evolution of the land use more than fifty years apart. This experimentation is the result of research, at an international level, carried out by the L.a.co.s.t.a. Laboratory of the University of Molise.

This paper aims to illustrate the construction process for the visibility analyses undertaken using a GIS platform in a sample area, located on the coast of Molise. An innovative methodology was used: it is based on consolidated scientific examples creating a new approach which can be applied to other case-studies.

For this purpose, a multi-criterial approach of territorial analysis was developed: it is meant to summarize interaction modes between the investigated phenomena, for the control of new infrastructures that involve rural areas creating new forms of landscape impacts.

1.1 Territorial Features and Energy Consumption

In an overview of medium and small size cities, very often the typical features of urban settlements are not recognizable. In such contexts, the city's relationship with its surrounding land plays an important role. In these cases, the topic of energy infrastructures and their impact assume relevant connotations, the urban settlements being closely correlated with the rural areas.

Landscape constantly requires attention in order to insure its sustainable conservation, because it is the result of a process of natural environment creation and modification. The uniqueness of the landscape consists of its close relationship with social systems and local contexts (Cialdea and Mastronardi 2014a).

Landscape could represent a model of integrated and sustainable territorial development (Mastronardi and De Gregorio 2012) because it can be considered a factor of reorganization of territorial competitiveness (Daugstad 2008), which creates ecological, environmental, cultural, social and economic values (Cialdea et al. 2006), through which it is possible to recognize and recover the identity of places (Antrop 2005; Mastronardi et al. 2012).

Landscape is also a place in which productive activities thrive, such as agriculture and tourism. Rural tourism, for example, is an inherently territorial activity, really connected to the specific territory, in terms of environmental diversity, architectural texture, cultural and social wealth.

In this context, we analyze the utilization form of energy, that is the prime mover of economic growth that depends on the long-term availability of energy from renewable sources (Popescu and Mastorakis 2010), that could be considered affordable, accessible and environmentally friendly (Ram and Selvaraj 2012).

Thanks to oil prices, the renewable resources can be seen as possible and economically acceptable alternatives. Renewable energy technologies are indicated as an appropriate alternative for providing a considerable portion of future energy demand. In fact, renewable energy supplies can play an important role in providing energy to the vast population in developing countries who as yet have no access to clean energy (Gabbasa and Sopian 2013). During the last few decades all countries started to increase their focus on clean and renewable energy sources (Cialdea and Mastronardi 2014b).

In Italy, the contribution of renewable energy sources grew by 32 % in the decade 1990–2000 (GSE 2012). It is particularly significant because the Italian energy system is characterized by a strong dependence on other countries. The supply data show that—in the period 1980–2010—the national primary energy imports more than 80 % of total consumption: Italy is in the last position in Europe for energy self-sufficiency.

Especially photovoltaic plants increased: at the regional level, they are concentrated to 43.5 % in the North, 37.5 % in the South and 18.9 % in the Central Italy.

This phenomenon is particularly interesting because these infrastructures have an impact on the aesthetic quality of the landscape (EEA-FOEN 2011). Renewable energy supply requires large parts of the land, involving arable and pasture land with suitable soils (Cialdea and Mastronardi 2014c). As a consequence of these growing demands, "the remaining unfragmented areas are under an enormous pressure. Therefore, much higher efforts are now required to conserve unfragmented landscapes" (Haber 2007).

1.2 The Regional Context in Molise

The Molise Region is located in the Central-South part of Italy, overlooking the Adriatic Sea. Even in this relatively small Region, numerous arrangements for the creation of infrastructure networks from renewable energy have been created, especially in recent years. Similarly, laws and guidelines for their implementation have been developed.

The *Molise Regional Environmental Energy Plan* was approved in 2006 (Regione Molise 2006): it was oriented to optimize energy conservation, and to enhance clean energy sources with particular attention to hydro-power and wind-power plants.

The *Energy Plan* enhances the potential of the production of energy from wind and assumes, in the 2015 scenario, the installation of wind turbines for over 1700 MW. In addition, it identifies the Biferno river basin as containing sites of great interest for wind energy areas within the only river completely inside the Region. This 2015 scenario completely ignores the contribution of photovoltaic plants connected to existing buildings, while it is not exactly considered achievements of the ground-mounted photovoltaic plants (Cialdea and Quercio 2014). This Plan also sets out Regional Guidelines for the evaluation of projects.

However, impacts of wind farms will be implemented and regulated by successive regional laws.

Two years later a new regional law was enacted (Regione Molise 2008a) in relation to the localization of wind and photovoltaic settlements in its territory. The subsequent Resolution no. 167 (Regione Molise 2008b) approves the related guidelines. Anyway, the Regional Law no. 15/2008, and therefore its subsequent guidelines, identifies a number of areas not suitable for the installation of wind and photovoltaic plants. In particular these areas are the regional parks and the reserves; the core zones in the national parks; the areas of "full protection and preservation" defined by the Territorial Landscape Plans, the Sites of Community Importance and the Special Protection Areas. Moreover it includes a particular area that is the "Valley of the Tammaro River" with mountains that surround it.

It is necessary to underline that this Law (Art. 2, paragraph 1) forbids construction of wind farms even in a buffer zone of not less than 3 km from the perimeter of parks and archaeological areas, in a buffer zone of not less than 1 km from the urbanized perimeter, in a buffer zone of 500 m from residential or rural houses, in a buffer zone of 200 m from the perimeter boundary of the neighbouring Municipalities, in a buffer zone of 5,000 linear meters from the coastline, and in a buffer zone of 1 km from the river shores, wetlands, lakes and dams.

In fact these points of paragraph 1 of article 2 were declared illegal by the Constitutional Court Judgment no. 282/2009 (Corte Costituzionale 2009) because the not suitable areas must be identified by the Region only with the support of technical reasons, that were not declared.

This Judgment also declared that wind offshore plants are forbidden (in Molise there was a project for a big settlement in front of the coast), because the administrative functions relating to the use of the maritime domain belong to the National Government. Article 3 has also been declared illegal because it makes the issue of permits to specific limits for wind turbines (set in 545 wind turbines, each of the minimum power of 2 MW) and for the ground-mounted photovoltaic plants (set in 500 MW of total power).

Really at the same time, the Molise Region already was changing the regional norms: in fact the Regional Law 15/2008 was replaced with the Regional Law 22/2009 (Regione Molise 2009a) with its new Guidelines (Regione Molise 2009b). This law takes into account emendation by the Constitutional Court Judgment and includes into the list of not suitable areas the Important Bird Areas and into the list of suitable areas the Sites of Community Importance (even if after successful completion of the Assessment Incidence Evaluation) and the "Valley of the Tammaro River" area, with mountains that surround it.

Moreover article 3 paragraph 1 of the Regional Law 22/2009 provides that plants —not exceeding 1 MW—were authorized directly by the Municipalities through the DIA ("Dichiarazione di Inizio Attività—Starting Construction Report").

Also in this case another Constitutional Court Judgment (Corte Costituzionale 2010) stated that this exception would be introduced only by Ministerial Decree and not directly by Municipalities; so the Molise Region amended its Guidelines stating that, it is necessary the procedure for a single authorization issued by Regional Norms even for systems not exceeding 1 MW (Regione Molise 2010a). In 2010 further modifications and integrations appear in a new Regional Law (Regione Molise 2010b): in particular the "Valley of the Tammaro River" area is replaced into the list of not suitable areas, as one of the most important archaeological values context in the Molise Region.

Subsequently a Resolution of the Ministers Council (Consiglio dei Ministri 2011) declares the unconstitutionality of this Regional because the already mentioned Legislative Decree no. 387/2003 provides that Regions can proceed to the identification of unsuitable areas, but in accordance with National Guidelines. In fact, according to Ministry Decree (D.M. 10/09/2010) not suitable zones could be identified with regard to specific sites after the completion of a depth investigation that would identify specific areas particularly sensitive or vulnerable.

Finally the Molise Region approved last guidelines for authorization to build and to manage plants producing electricity from renewable sources (Regione Molise 2011), that reproduces verbatim the DM 10/09/2010.

At present, the Evaluation Energy Study for the Molise Region is in progress: it is aimed at preparing the New Regional Plan.

2 The Major Critical Area: Energy Supplies Requirements and New Landscape Assessments

The invasion of the Adriatic coastal area by ground-mounted photovoltaic plants and by the large number of wind farms is very clear, especially in the eastern part of the province of Campobasso, located between the Biferno River Valley and the Fortore River Valley (Cialdea and Maccarone 2014).

In detail we will analyze the territorial assessment of the Coastal Zone whose main matrix appears to be agricultural-productive activities voted, because of its orographic characteristics, that are favourable to agricultural practices (Cialdea 2012). In this prevalent matrix there are also environmental excellences such as the coastal landscape, the waterways designed by the main watercourse of the Region, as the above mentioned Biferno and Fortore Rivers, and the high natural values areas such as the Sites of Community Importance.

At the same time this territory is one that has been affected by the major territorial changes that occurred in the last fifty years as a result of the settlement of the major regional infrastructures—all made along the coast—of the great transformations due to the processes of agrarian land reform, of the implementation settlements tourist resorts and, finally, of the insertion of new energy infrastructures.

2.1 The Land Overview

The primary aim of our experimental methodology is to provide useful information for the creation of a network of energy infrastructure that will be compatible with the distinctive features of the landscape and of the already existing agricultural activities and that will not constitute a handicap to the development of tourism.

This methodology, therefore, takes on previous and current data in the first instance about land use (which then is analyzed in its diachronic asset observing the changes of the last 50 years) and in a timely manner focusing on use due to agricultural activities.

The purposes of the landscape plans are then analyzed in detail, in order to understand how the current landscape plans identify areas by major perceptive and productive-agricultural features.

On the basis of these data that identify the actual state, the plants for energy from an alternative source were included, focusing in particular on wind and ground-mounted photovoltaic plants, which have been identified as the principal producers of impact on the landscape. Therefore they have been located and geo-referenced with the creation of a database of including their energy characteristics and physical properties.

However we have implemented a dedicated G.I.S. project, in the first instance voted for the collection of information; in this regard, we discussed issues related to the need to combine data already spatialized with data whose information was gathered only from the level of the entire Municipality for which new solutions for their spatialization were implemented (Cialdea and Maccarone 2014).

Finally, particular attention was paid for refinement of the methodology of visibility to allow us to have an overlooking of the territorial ranges of influence of energy facilities, existing or in arrangement.

2.2 The Visual Aspects

In the analysis of the evolution of land use we wanted particularly to highlight the great territorial evolution related to changes in agricultural zones and in urbanized areas. However in Fig. 1 part a the Land use evolution is showed by highlighting the major increases and decreases in the Molise Coastal Region by a diachronic analysis.

We used data from the vegetation map of the Molise Region (drafted in 1954) and data from the Corine Land Cover 2012. To do this analysis the keys were standardized and then a comparative vector informative layer was used (Cialdea and Maccarone 2012).

It can be seen that the greatest changes in the coastal zone have been determined by the urbanization and the expansion of the town of Termoli situated in the middle part of the coast. The land use predominance, especially in the pre-coastal zone, remains mainly its agricultural nature.

Moreover, this part of the figure shows the wind farms and the ground-mounted photovoltaic plants, already existing and in project. The data shows that in the coastal areas there is a substantial and exclusive concentration of photovoltaic plants, while in the pre-coastal territories there is a higher concentration of wind farms.

The Map of Coastal Zone, moreover, defines in part b, the main characteristics of the area analyzed by dividing it into different fields of interest. For each of these areas, depending on the prevailing interest, modes of transformation were shown.

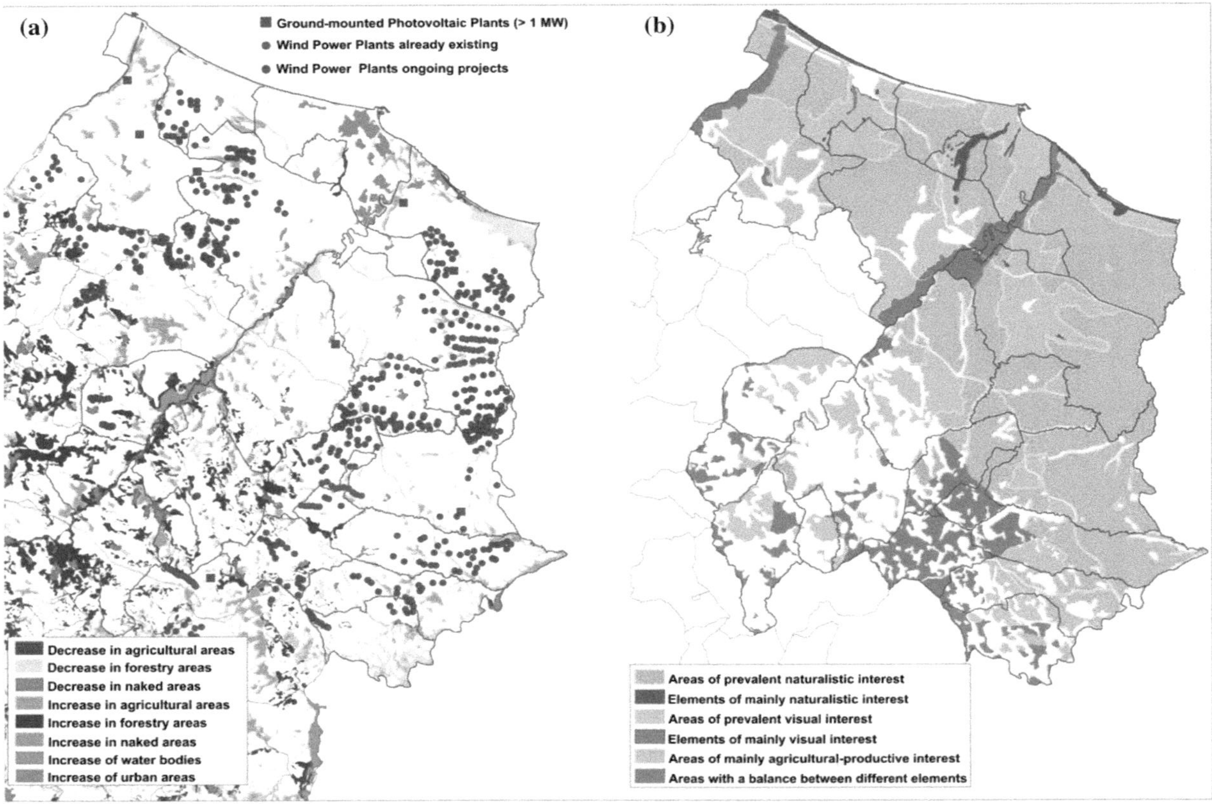

Fig. 1 **a** The Molise coastal and pre-coastal zones with the land use evolution and the wind power and ground-mounted photovoltaic plants; **b** Mainly interest for the actual landscape plans. (*Source Lab.lacosta elaboration*) (Color figure online)

In order to gain a complete picture of the transformability of the Region, all the Transformability Maps of the eight Landscape Plans in force have been analyzed: our sample area is included in two Plans, Plan no. 1 for the coastal zone and Plan no. 2 for the pre-coastal area.

Finally, our methodology wants to correlate the peculiarity landscapes related to the phenomenon of perception with the real knowledge of the land use also examined in an evolutional meaning. It is therefore necessary to examine them not only by their perceptual aspects but also by elements related to their utilization from the production point of view.

2.3 The Productive Aspects

The area shows a good level of development and of socio-economic and territorial integration, which involved all the regional landscape, except the inner areas. The area also shows a significant level of tourism development, even if it is tourism of domestic and hiking type (Mastronardi 2008; Mastronardi and De Gregorio 2012; Cialdea and Mastronardi 2014c, d).

Industrial activities and services play an important role in the already named town of Termoli. This condition is significant also in the industrial Map of Italian Localism. In the industrial area of Termoli, that reaches the top of specialization with the participation of the mechanical industry, diversification of production grew up in the direction of agro-food and chemical industries.

Moreover agriculture suffers a drastic restructuring and tends to specialization and intensification (Fig. 2a). Data reveal, however, a clear decrease of arable land and permanent grassland. In contrast, permanent crops showed a significant increase. Moreover the forest increase appears smaller. The weight of the arable land unchanged, the incidence of meadows and pastures reduced, while there is a significant increase in permanent crops. With particular reference to wine production, it is appropriate to emphasize the special vocation of agriculture in the area and the importance of the production of quality.

Irrigation played a very important role for the development of this area, where rainfall is scarce and poorly distributed over the course of the agricultural year.

Irrigation developed in the 80s, thanks to the use of water resources of the artificial Guardialfiera Lake creation, which has a usable capacity of about 137 million cubic meters of water. In 2010, irrigated areas total more than 5,000 ha and affect significantly all the Municipalities in the area: it is a phenomenon that must be carefully assessed in terms of farming but also of the environmental impact.

Almost all the Municipalities in the district are within the promotional circuits of "Oil City" and/or within itinerary of the "Wine Road".

In conclusion, our study area reveals the presence both of local foods and the organic ones (Fig. 2b): it would thus become an area of excellence in the framework of regional foods.

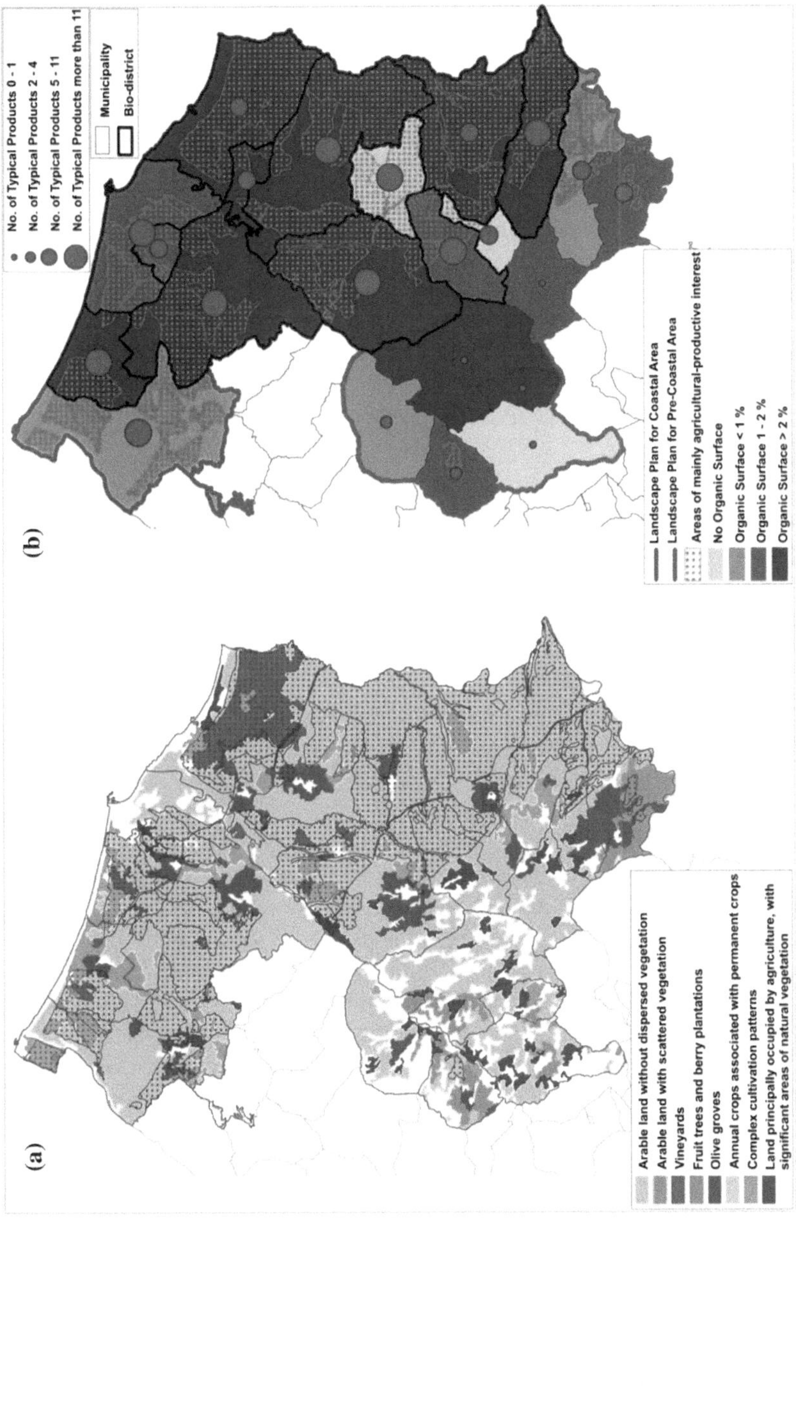

Fig. 2 **a** The Molise coastal and pre-coastal zone with the actual land use characterized by the agricultural production; **b** Mainly specification productions compared with the organic surfaces (*Source Lab.lacosta elaboration*) (Color figure online)

The presence of organic surfaces and the largest concentration of local products in this area allows us to characterize it as a typical district where the bio-agriculture and the environment tend to become more integrated than in the rest of the region. In fact, this area is defined by the Region as able to "promote the conservation and enhancement of biodiversity of agricultural areas and to create an innovative model for the integrated management interventions of public and private bodies." The proposed framework is therefore the ideal basin for experimenting with sustainable agricultural practices (Carabba 2011) through the implementation of systems of certification process and product, and it can aspire to the role of regional example in agriculture activity because combining the criteria of preservation of local traditional food with the application of advanced methods for the protection of the natural environment and rural.

3 A Possible Agreement Between Energy Requirements and Local Productive Needs

The methodology's assessment was carried out on a specific Municipality, that is Campomarino, in which our previous analyses show the must important proximity to an extensive park of new energy infrastructures. It is a witness to a passage through the importance of the "new" vocation of the territory linked to tourism too.

Moreover, it presents a specialized agriculture, with a high percentage of irrigated Agricultural Surface which must be carefully assessed in terms of environmental impact. In the last decade, the utilized Agricultural Surface and the total Agricultural Surface have showed an increase to underline its strong agricultural vocation.

At the same time, however, this Municipality has experienced a strong population growth (+40 % over the last thirty years), which had an impact on the housing stock and consequently on the regional asset. This consideration must be added the analysis of its strong tourist.

Campomarino has a clear attitude for the seaside tourism, that is a source of many environmental impacts, related in particular to the land consumption, to the biodiversity alteration, to the pollution of sea water, to the water and energy uses, and to the waste (Cialdea and Mastronardi 2014a, b). There are a lot of second homes and residences for tourism. The tourist demand is quite seasonal, with substantial territorial density. Tourism growth has not been accompanied, therefore, by a suitable hotel accommodation and it was developed thanks to an excessive number of buildings, very often illegal too (Cialdea 2007).

In particular the city of Campomarino, similarly to many other coastline cities, is organized in two urban centres, the Historical one—that is more set back from the coast—and the Lido area—that is not an integral part of the urban system. In fact, in winter it is considered as a "ghost town" while in summer it is overcrowded by the not very profitable tourism.

This phenomenon is also taken into account in the dimensioning of the General Town Plan, which is divided into two parts: the size of the residential part and the other one referring to the tourism activities.

The main problem in urban development, found more in the Lido area but also in other areas of the Municipality surface, is the lack of services, both related to permanent residency and to the exploitation of tourism. The plan foresees a reorganization of services and tourist activities available, as well as the stimulation of new plants that can revitalize globally the area. Among the objectives of the Plan is also an attempt to reverse the trend of growth of unoccupied dwellings compared to those employed, a phenomenon that strongly characterizes the Lido of Campomarino. However, our study can be a valuable contribution to the achievement of a possible happy marriage between the energy needs and the characteristics of local production, with them meaning both the preservation of traditional agricultural activities, and the incentive of those related to tourism development. For this reason, we have developed and implemented an appropriate methodology for data tourism spatialization (Cialdea et al. 2015). This methodology provides an evaluation of the distribution of tourist facilities according to localization factors related to the tourism supply, by including landscape features.

Landscape features, therefore, are calculated with reference to the elevation gradient and land coverage, while the proximity analysis uses G.I.S. functions.

Through a multivariate regression model, it was possible to extrapolate the localization factors by administrative units to subunits of equal dimension and estimate the accommodation density associated with each subunit (Fig. 3).

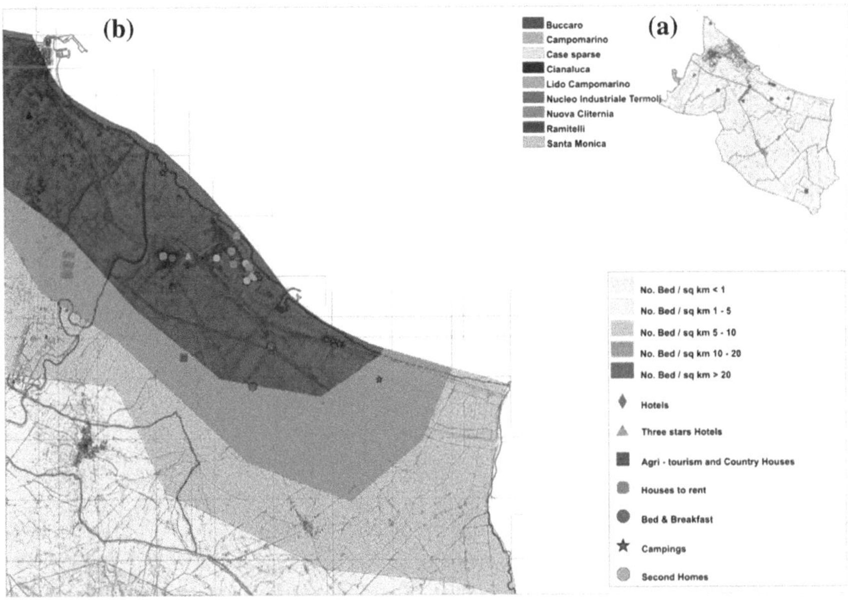

Fig. 3 **a** Different census areas in the Municipality of Campomarino; **b** The main tourism facilities in the coastal area of Campomarino (*Source Lab.lacosta elaboration*) (Color figure online)

3.1 The Intervisibility Methodology Input

Visual control can be considered an important tool for monitoring transformations on land: it also can be considered as a useful tool in the territorial planning process, even if it cannot describe the total control of the landscape features.

For this reason in our methodology we consider the reciprocal visibility (intervisibility) existing between diverse points in the territory in order to underline the aspects characterising the landscapes and the elements of which it is formed (Cialdea and Sollazzo 2012; Cialdea and Quercio 2014).

Now we expose a synthesis of different factors. There are physical-psychological laws of visual perception and evaluation of historical, anthropological and cultural characteristics which can be analyzed in the landscape.

Therefore, perceptive analysis does not only involve visual perception, but also involves the mental processing which constitutes cultural perception, that is a cultural interpretation of what people see.

Today the visual impact of the wind turbines is calculated with the simplified formula created by the Ministry of Heritage and Culture Activities (Di Bene and Scazzosi 2006; Casarotto et al. 2009).

Our research considers the Ministerial guidelines, but inserts more elements. This Ministerial calculation, the aim of which is to establish at what distance a wind turbine can be perceived, has two basic limits. That is, it considers the territory to be uniform and does not take into consideration the elements, and their inter-relationships, which are an integral part of the landscape.

First of all the perception of the landscape is linked to the observation point and its distance. Therefore the visual analysis is defined summarizing visibility and distance analyses. The first one is directly correlated to the height of each single element and its geographical position (including its height above the sea level), while the second is correlated to the distance within which it is possible to perceive its presence.

Moreover, our study area (that is in particular the portion of the Molise Region along the Adriatic Sea Coast) is made up of a series of elements which together form the image of the landscape.

These elements have been synthesised into two large groups, Values and Detractors. The values are elements which confer worth to the total landscape system (such as architectural, naturalistic, historical elements).

They include elements of great naturalistic worth, that is the presence of particular species of flora and fauna. They are: the Important Bird Areas and the Special Protection Zones (at the sense of the Directive 79/409/EEC), the Sites of Community Importance (at the sense of the Directive 92/43/EEC "Habitat" adopted in Italy by the D.P.R. 357/1997), the Beaches free (Data are required by the Plan for Use of State Coastal Property for Tourism and Leisure of Molise Region), the Areas restricted under Law 1497/39 and under Law 431/85 (Cialdea 2012). What remains of areas of great natural worth and are the historical memory of these places are also included. They are Residual of dunes areas; Residual of green pathways, Residual of woods' areas.

However, if values highlight landscape qualities, detractors penalize them.

The detractors list, even if it is not exhaustive, was prepared after detailed land use features analyses. They are: Airport, Industrial Areas, Quarries, Purifiers, Landfill sites, Electrical Infrastructures, Wind farms, Rail networks, Photovoltaic Plants, Dewatering Pumps, Port, Roads, Gas turbines.

We undertook our analysis in ArcGis field: the detractors have been represented through graphics primitives with its geographic coordinates.

Later, shapes for each detractor typology were realized: each wind turbine was represented with a point datum and areal datum were associated for photovoltaic plants. Then, we realized that using only the Digital Elevation Model datum results were uncertain as the detractors resulted as visible from numerous parts of the territory, when this is not the real situation.

We also realized that territory elements interfere with visibility.

These elements are not only of orographic origin. In fact, we have also woods, industrial and urban areas.

Therefore we correct the original datum of Digital Elevation Model using the Land Use datum, attributing them with an average height.

For the assessment of two different grids we realized the *Grid visibility* (directly related to the heights of the individual wind turbines and their geographical position and then to the portion above sea level) and the *Grid distance* (directly related to the distance within which it is possible to perceive the turbine presence.

Then they were combined to form a final grid, where values are between 0 and 5 (0 corresponding to invisibility while 5 corresponds to maximum perception) (Cialdea and Sollazzo 2012).

The calculation of the density of the detractors was carried out using a single data point of detractors (with appropriate conversions from linear and polygonal data point) to which was added an increasing weight.

For each detractor (in this case there is an example for wind farms in the coastal zone) type class we counted the number of points (Cnt_detrat); then we were given the weight (Max_peso) with the function 1/(number of detractors per class). Finally, we added a Corrective Factor (Correttivo) based on the importance of the detractor (Table 1).

The analysis of visibility represents a very important tool for the analysis of the landscape context. This research looked at all of the identified detractor elements; however the case of the wind farms is emblematic (Fig. 4).

3.2 Concluding Remarks

Our research has allowed us to highlight the difficulties that arose from recognition of the need for protection of numerous areas of the Region.

Table 1 The analysis of all detractors

Detractor	Cnt_detrat	Max_Peso	Correttivo
Airport	144	0.00694444444	0
Industrial areas	799	0.00125156446	4
Quarries	382	0.00261780105	1
Purifiers	3	0.33333333333	0
Landfills	1	1.00000000000	5
Electrical infrastructure	3226	0.00030998140	3
Wind farms	51	0.01960784314	4
Railway	1840	0.00054347826	1
Photovoltaic plants	756	0.00132275132	2
Dewatering pumps	3	0.33333333333	0
Port	84	0.01190476190	0
Roads	8168	0.00012242899	2
Gas turbine	27	0.03703703704	5

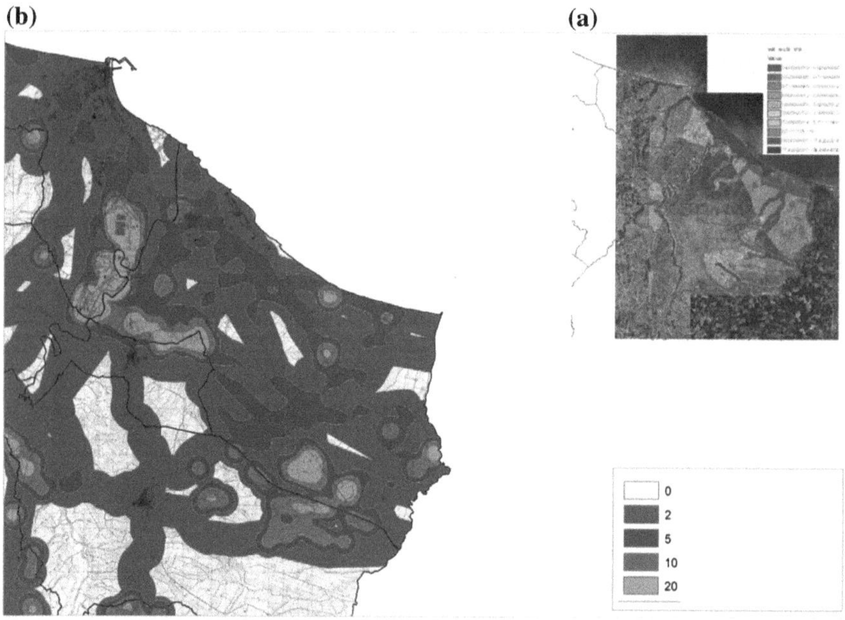

Fig. 4 **a** Visibility and Distance analyses in the sample area; **b** The final Intervisibility Grid (*Source Lab.lacosta elaboration*) (Color figure online)

Our research was then organized according to the following methodology for surveying the territory that could substantially support legislative strategies for the plants installation control. This activity was realized in collaboration with the Molise Region in order to create the "*New Regional Landscape Plan of Molise*" that has enabled us to put a strong emphasis on interdisciplinary approach that characterizes the study of the energy plants location.

The National Legislation (and consequently also the Regional Legislation) for the protection of the landscape provides a great importance for the maintenance of landscape features (landscape characteristics and its various natural and anthropogenic components) in connection with the historical systems (historical centres, historical widespread buildings), agrarian landscapes (typical cultural landscape), weavings historical territorial, scenic routes and presence of areas to strong symbolic significance.

Our methodology is based on the intervisibility and it is realized through a comparison of different factors by a matrix. Therefore, our final result is precisely the realization of this matrix, shared by the Molise Region and by the Regional Authority responsible for the Protection of Cultural Heritage and Landscape, in order to support the ongoing regional regulation.

Our analytical approach, which is included in a larger landscape analysis, underlines the importance of visual perception, in order to highlight areas in which already exists a detractor's strong impact. Moreover, it could be very useful in planning stage and in the policies determination for implantations of the renewable sources energy production.

Taking into account the landscape visibility, it will be possible to select more appropriate locations so that the alternative sources energy production (important for environmental protection) will not cause negative repercussions on portions of territory of high landscape quality.

Moreover this territorial analysis has universality and repeatability characters that make it applicable at all kinds of territory, unless changes related to the nature of places. It is really relevant especially in those areas in which, as in our case-study, there are excellence in food production, due to the presence of local products and organic surfaces.

Moreover we took into account several aspects, as the analyses of territorial transformations (in order to underline the local vocation), the production aspects (especially related to agricultural and tourism activities), the valences underlined by the actual Landscape Plans and finally the surveying of the energy plants from alternative sources. It was particularly oriented to underline the relationship between these new energy infrastructure systems and the tourist tendency that could be invalidated by them. It is necessary, anyway, to evaluate diverse components.

Moreover, the final goal is to understand for each area the best future, in a situation in which several components plays their role.

In order to prepare the New Regional Landscape Plan, we organized a table of the new landscape quality aims as listed in art. 135 of the Code (L. D. 42/2004) which contains for each homogeneous area specific requirements and provisions, oriented to the conservation of the constituent elements, to the rehabilitation of compromised or degraded areas; to the protection of landscape features; to the identification of the lines for the development, with particular attention to the preservation of rural landscapes and sites included in the UNESCO World Heritage List. Moreover, about the costal area in which our sample test is, we organized our information in five

Table 2 The landscape quality aims declined for the five Resources Systems in which spatial data analyses were organized

Resources systems	General aims		Specific aims	
Physical-environmental system	1	Promote the preservation of the integrity of areas of high naturalness and high ecosystem value	1.1	Safeguard geological-geomorphological systems with high integrity (geological formations, ravines, cliffs, crags)
			1.2	Safeguard protected areas and areas of high environmental value such as those covered by the Nature 2000 Network
			1.3	Safeguard and improve environmental functionality of river and lake systems of Molise
			1.4	Safeguard and rebuild coastal marine habitats of Molise (coastal wooded areas, dune systems, river mouths)
			1.5	Safeguard woods and forests of mountainous and hilly areas of Molise
			1.6	Redevelop and redesign the coastal landscapes of Molise
Landscape-visual system	2	Promote improved integration of landscape and the quality of infrastructures	2.1	Define territorial and landscape quality standards in the settlement of new network infrastructure
			2.2	Define territorial and landscape quality standards in the settlement of new energy infrastructure
			2.3	Define territorial and landscape quality standards in the settlement of new productive activities
Historical-cultural system	3	Promote the preservation of cultural values	3.1	Preserve cultural value and witnesses of settlements and historical manufacts
			3.2	Preserve cultural value of traditional rural buildings
			3.3	Preserve the visible cattle-tacks rest
			3.4	Redevelop the historic rural landscapes

(continued)

Table 2 (continued)

Resources systems	General aims		Specific aims	
Agricultural-productive system	4	Promote the conservation of agricultural landscapes	4.1	Develop the agricultural landscape of Molise, recognize and promote its social functions
			4.2	Preserve open landscapes of the reclamation as a characteristic aspect of identity of coastal landscape of Molise
			4.3	Redevelop the agricultural landscape of Molise
Demographic-touristic system	5	Promote the improvement of the quality of the settlements	5.1	Improve quality of urban settlements and their environmental performance, for greater well-being of the population
			5.2	Redevelop degraded contemporary urbanization landscapes
			5.3	Improve urban quality of and touristic settlements
			5.4	Improve urban quality of agricultural and productive settlements
			5.5	Improve soft mobility quality (walking, cycling, trekking on horse) and its interconnection with the traditional mobility

Resource Systems, as they are in Table 2: the Physical-Environmental, the Landscape-Visual, the Historical-Cultural, the Agricultural-Productive and the Demographic-Tourist Systems, which has enabled us to formulate targets for the Molise landscapes (Cialdea and Maccarone 2012, 2014).

The targets identified for the entire regional area are related to the conservation, protection, management and planning of exceptional, ordinary, and degraded landscapes with particular reference to typical natural landscapes such as rivers, lakes, hills, mountains, coastal and rural landscapes, forestry and agro-pastoral, not to mention historic, rural, urban, industrial and infrastructure sectors.

The objectives identified are also related to the government of the processes of urbanization and abandonment of the territory and to the preservation of material cultural values and intangible values such as the traditions and history of the Region.

The general objective was subdivided into specific objectives, as shown in Table 2, in which these objectives were finally associated with landscape quality directions that indicate policies to adopt and those who have an interest in achieving these objectives, as well as the measures required to adapt the urban planning instruments to the indications of the new Regional Landscape Plan.

Therefore, the landscape quality targets in this area aim to safeguard the surviving heritage in the area, to recover and improve the landscapes altered and degraded by human activity and to define quality standards for the correct insertion of new energy infrastructure in the coastal territory.

References

Antrop M (2005) Why landscapes of the past are important for the future. Landscape Urban Plan 70:21–34

Carabba P (ed) (2011) Il Progetto DINAMO per la Biodiversità del Molise. ENEA, Roma

Casarotto A, De Guio A, Ferrarese F (2009) Action Gis: un modello predittivo del movimento antropico in un paesaggio antico. Il caso di studio della Val d'Alpone (VR). Archeologia e Calcolatori 20:291–307

Cialdea D (2007) Definition of landscape quality aims and ongoing research. In: Cialdea D (ed) Interreg reports. Materials for adriatic cross border project report N. 3. environmental assessment GES.S.TER./Interreg IIIA. Arti Grafiche La Regione, Campobasso, pp 154–173

Cialdea D (2012) A new landscape planning approach in areas with a strongly rural matrix. In: Campagna M, De Montis A, Isola F, Lai S, Pira C, Zoppi C (eds) Planning support tools policy analysis: implementation and evaluation. Franco Angeli, Milano, pp 453–464

Cialdea D, Maccarone A (2012) Territory diachronic maps for the Regional Landscape Plan. In: Campagna M, De Montis A, Isola F, Lai S, Pira C, Zoppi C (eds) Planning support tools policy analysis: implementation and evaluation. Franco Angeli, Milano, pp 386–398

Cialdea D, Maccarone A (2014) The energy networks landscape. Impacts on rural land in the Molise Region. TEMA J Land Use, Mobil Environ 223–234 (Special Issue June 2014)

Cialdea D, Mastronardi L (2014a) Integrated approach in the planning stage for landscape conservation in the Coastal Italian Areas. Int J Des Nat Ecodynamics 9(4):296–306

Cialdea D, Mastronardi L (2014b) New land use in rural marginal areas. Renewable energy vs landscape preservation. In: Mastorakis N, Sokolov V, Neck R, Pedro Juliao R, Wang J, Karymbalis E et al (eds) Advances in environmental sciences, development and chemistry. WSEAS Press, pp 468–474

Cialdea D, Mastronardi L (2014c) Renewable energy resources and their impact on the rural land. WSEAS Trans Environ Dev 10:423–433

Cialdea D, Mastronardi L (2014d) A multi-criteria methodology oriented to the preservation of the coastal areas landscape. In: Batzias F, Mastorakis N, Guarnaccia C (eds) Recent advances in energy, environment and financial planning. Series: mathematics and computers in science and engineering, No. 35. pp 237–243

Cialdea D, Quercio N (2014) An overview on the Molise (Italy) renewable energy law: the conflict between the landscape protection and the territory management. Renewable Energy Power Quality J 12

Cialdea D, Sollazzo A (2012) Visibility analysis for identifying detractors in rural areas. In: Campagna M, De Montis A, Isola F, Lai S, Pira C, Zoppi C (eds) Planning support tools policy analysis: implementation and evaluation. Franco Angeli, Milano, pp 498–500

Cialdea D, Vitiello M, Saio G (2006) The GIS architecture elements for the coastal areas along the Adriatic Sea. In: Proceedings of the Ersa 2006 Volos Greece, 46th congress of the European Regional Science Association, Enlargement, Southern Europe and the Mediterranean. Paper n. 348, www.sre.wu-wien.ac

Cialdea D, Mastronardi L, Giannelli A (2015) Data-tourism Spatialization: a new methodology useful for landscape planning assessment. In: Nijkamp P, Kourtit K, Bucek M, Hundec O (eds) Proceedings of the 5th Central European conference in regional science. Technical University of Košice, Košice, pp 114–124

Consiglio dei Ministri (2011) Delibera Consiglio dei Ministri del 23 febbraio 2011- Impugnativa della legge n.23 del 23 dicembre 2010. Roma

Corte Costituzionale (2009) Sentenza della Corte Costituzionale n. 282/2009 del 2 novembre 2009. Roma

Corte Costituzionale (2010) Sentenza della Corte Costituzionale n. 194/2010 del 26 maggio 2010. Roma

Daugstad K (2008) Negotiating landscape in rural tourism. Annals Tourism Res 35(2):402–426

Decreto Legislativo 22 gennaio (2004) n. 42 recante il "Codice dei beni culturali e del paesaggio ai sensi dell'articolo 10 della legge 6 luglio 2002, n. 137 (in Gazzetta Ufficiale 24 febbraio 2004) e successive modifiche ed integrazioni. G.U. n. 45 del 24/02/2004"

Di Bene A, Scazzosi L (2006) Gli impianti eolici: suggerimenti per la progettazione e la valutazione paesaggistica, issued by MIBAC (**MI**nistero per i **B**eni e le **A**ttività **C**ulturali—Ministry of Heritage and Culture Activities. Gangemi Editore, Roma

EEA-FOEN (2011) Landscape fragmentation in Europe. Report 2011. Copenhagen

Gabbasa M, Sopian K (2013) Potential and status of the renewable energy technologies for sustainable energy development in OIC. WSEAS Trans Environ Dev 9(4):253–262

GSE (2012) Rapporto statistico 2012. Impianti a fonti rinnovabili. www.gse.it

Haber W (2007) Energy, food and land. The ecological traps of humankind. Environ Sci Pollut Res 14(6):359–365

Legge 29 giugno (1939) n. 1497, Sulla protezione delle bellezze naturali e panoramiche. G.U. n. 241 del 14/10/1939

Legge 8 agosto (1985), n. 431, Conversione in legge, con modificazioni, del decreto-legge 27 giugno 1985 n. 312, recante disposizioni urgenti per la tutela delle zone di particolare interesse ambientale. G.U. n. 197 del 22/08/1985

Mastronardi L (2008) Le analisi delle trasformazioni del paesaggio rurale: discussione dei risultati. Il sistema socioeconomico e demografico: Il Basso Molise. In: Tassinari P (A cura di) Le trasformazioni dei paesaggi nel territorio rurale: le ragioni del cambiamento e possibili scenari futuri. Gangemi, Roma, pp 65–80

Mastronardi L, De Gregorio D (2012) Il fenomeno della stagionalità e le conseguenze sull'ambiente. In: Meini M (A cura di) Il turismo al plurale. Una lettura integrata del territorio per un'offerta turistica sostenibile. Franco Angeli, Milano, pp 162–180

Mastronardi L, Battaglini E, Carranza LM, Giannelli A, Stanisci A, Truglia FG (2012) Analisi delle relazioni tra il mosaico paesaggistico-naturalistico ed il turismo nelle aree rurali: prospettive metodologiche ed applicative. Annali del turismo 1:99–118

Ministero dello Sviluppo Economico di concerto con Ministero dell'Ambiente e della Tutela del Territorio e del Mare e con Ministero per i Beni e le Attività Culturali (2010) Decreto 10 settembre 2010 – Linee guida per l'autorizzazione degli impianti alimentati da fonti rinnovabili. Roma

Popescu MC, Mastorakis N (2010) Aspects regarding the use of renewable energy in EU countries. WSEAS Trans Environ Dev 6(4):265–275

Ram BS, Selvaraj M (2012) Entrepreneurship in the environmentally friendly and economically sound renewable energy conversion system. WSEAS Trans Environ Dev 8(1):13–22

Regione Molise (2006) Delibera Consiglio Regionale Molise n. 117 del 10 luglio 2006 - Approvazione del Piano energetico-ambientale regionale. Campobasso

Regione Molise (2008a) Legge Regionale Molise 15 del 21 maggio 2008 - Disciplina degli insediamenti degli impianti eolici e fotovoltaici sul territorio della Regione Molise. Campobasso

Regione Molise (2008b) Delibera Consiglio Regionale Molise n. 167 del 10 giugno 2008- Prime linee guida per lo svolgimento del procedimento unico di cui al comma 3 dell'articolo 12 del Decreto Legislativo n. 387/2003, relativo all'installazione di impianti per la produzione di energia elettrica da fonti rinnovabili sul territorio della regione Molise e per il corretto inserimento degli impianti nel paesaggio – Attuazione dell'atto consiliare n. 117 del 10 luglio 2006, recante: "Piano energetico ambientale – regionale". Campobasso

Regione Molise (2009a) Legge Regionale Molise 22 del 7 agosto 2009 - Nuova disciplina degli insediamenti degli impianti di produzione di energia elettrica da fonti rinnovabili nel territorio della Regione Molise. Campobasso

Regione Molise (2009b) Delibera Consiglio Regionale Molise n. 1074 del 16 novembre 2009 – Nuove Linee Guida. Campobasso

Regione Molise (2010a) Delibera Giunta Regionale Molise n. 857 del 25 ottobre 2010- Modifica alle linee guida per lo svolgimento del procedimento unico di cui all'art. 12, comma 3 del D.lgs 387/03, in materia di produzione di energia elettrica da fonti rinnovabili. Campobasso

Regione Molise (2010b) Legge Regionale Molise n. 23 del 23 dicembre 2010 - Modifiche ed Integrazioni alla L.R. n. 22 del 7 agosto 2009. Campobasso

Regione Molise (2011) Delibera Giunta Regionale Molise n. 621 del 4 agosto 2011 - Linee guida per lo svolgimento del procedimento unico di cui all'art. 12 del D. Lgs. n. 387/2003 per l'autorizzazione alla costruzione ed all'esercizio di impianti di produzione di energia elettrica da fonti rinnovabili sul territorio della Regione Molise. Campobasso

Towards the Definition of the Urban Saving Energy Model (UrbanSEM)

Rocco Papa, Carmela Gargiulo and Floriana Zucaro

Abstract European cities are essential actors for transition to a low carbon society on a 2050 horizon. Urban activities account for 80 % of energy consumption in Europe as well as most GHG emissions. The need for a new paradigm based on energy efficiency and saving thus represents both a challenge and an opportunity to local authorities who have to deal with the complexity of urban systems and energy issues. In light of this realization, the Smart Energy Master project conducted by the Department of Civil, Architectural and Environmental Engineering at the University of Naples Federico II aims to develop a model of governance for local energy saving and efficiency. One of the results of this research project is the Urban Saving Energy Model aimed at integrating the different subsystems in which a city can be structured with energy consumption at a neighborhood scale. This paper describes the model in question and some of the results achieved by applying the UrbanSEM to three Naples neighborhoods.

1 Introduction

Evolution of urban areas and changes in land use have occurred partly thanks to the development of new forms of energy and new technologies. In historical periods of varying durations, pre-industrial, industrial and postindustrial settlements (Sjoberg 1960; Biocca 1987; Mega 2005; Fistola and Mazzeo 2009) have marked the transition from an economy chiefly based on agriculture to an industrial economy, characterized by technological innovation and new forms of work organization,

R. Papa (✉) · C. Gargiulo · F. Zucaro
DICEA, University of Naples Federico II, Naples, Italy
e-mail: rpapa@unina.it

C. Gargiulo
e-mail: gargiulo@unina.it

F. Zucaro
e-mail: floriana.zucaro@unina.it

R. Papa and R. Fistola (eds.), *Smart Energy in the Smart City*,
Green Energy and Technology, DOI 10.1007/978-3-319-31157-9_9

hence of urban space. Such changes have resulted in considerable urban development and demographic growth.

At the beginning of the 19th century there were about 20 large urban conglomerations (those with over 100,000 inhabitants). Starting from the 20th century, the phenomenon of urbanization has progressively led to an ever greater increase in the population residing in urban areas, reaching "an invisible but momentous milestone" (UNPF 2007) in 2008 when, for the first time in humankind's history, the urban population exceeded its rural counterpart (UNPD 2011).

According to current trends in population growth, predicted scenarios estimate that by around 2030 five billion people will live in urban areas, i.e. about 48 % of the world's population.

Although world urban growth envisioned in the next few decades will be driven chiefly by developing countries, where urbanization is occurring with much higher intensity and speed than in the industrialized nations, the sustainability of all cities will be challenged. In developing countries, a top priority is how to manage the future dynamics of land use change sustainably so as not to aggravate already existing environmental, economic and social disequilibria. At the same time, in cities in industrialized countries the use of resources needs to be limited, aiming at efficiency and especially at saving: "the prime imperative of the 21st century is the need to reduce energy use for reasons of its associated costs and emissions impacts" (Roaf 2010). The reduction in energy consumption in the urban context thus constitutes a crucial issue in European and international policy for the years to come in order to ensure a low-carbon future.

In light of the above considerations, it appears increasingly clear that it is necessary to adopt an approach which is more aware of area governance issues, able to combine the complexity of urban systems with the multidimensionality of energy issues.

The sectoral nature which has up to now characterized the energy challenge in urban areas, especially favoring the building sector, is in contrast with the systemic approach on which the UrbanSEM is based, an interpretative model operating at the neighborhood scale, which takes due account of the physical characteristics of man-made and natural spaces, of mobility, and distribution of activities taking place in cities and the behavior of its inhabitants.

This contribution, which describes the results of setting up this kind of model, is structured in four parts which reproduce the development of the whole research work: an updated review of the literature on energy consumption on an urban scale; definition of the variables on which the interpretative model is constructed; illustration of the techniques of statistical analysis used; and interpretations of the results attained.

2 State of the Art

The crisis of the no-limits growth model, starting from the 1970s, has forced public decision-makers and academics to tackle issues which now play a major role in all development policies, such as restricting consumption and industrial decarbonization.

The consequent need to reconcile the principles of sustainability with rapid urbanization has led the scientific debate, up to now chiefly at the theoretical level, to tackling the energy question by virtue of energy use being one of the principal components of the city, of its activities and its physical and functional organization. As cities account for 70 % of global energy use and for 40–50 % of GHG emissions worldwide, many scholars have begun to inquire how the organization of settlements, both in terms of spaces and activities, affects energy demand and production (Stone 1973; Keyes 1983; Laconte et al. 1982; Alberti 1999) as well as "flows of energy and resources", in accordance with ecosystemic theories and the urban metabolism (Lynch 1961; Forman 1995; Turner 1990).

Many studies have highlighted the relations between activities, urban structure and energy consumption. Among such studies, those of Owens have shed light on the interactions between energy and cities, as far as considering them a single integrated system. Elements such as the morphology, density, functional and economic structure of urban areas affect both energy demand and supply, chiefly in relation to distribution. Owens was, in practice, the first to raise questions on the reciprocal influence between urban structure and energy consumption, maintaining that all the components of an urban system affect energy requirements.

With a view to resolving such issues, scientific research has on the one hand focused on investigating the most efficient type of urban development in energy terms, and on the other it has set up many models to identify urban characteristics and elements that affect energy consumption, albeit chiefly in terms of the building sector. Though such energy models may be classified according to different criteria (Hoffman and Wood 1976; Nakata 2004), they are generally structured into three main categories (Swan and Ugursal 2009; Zhao 2012):

- *Top-Down* (Xu and Masui 2009; Lu et al. 2010): macroeconomic models based on the formulation of statistical relations between economic conditions, energy consumption and the adoption of technology, by elaborating empirical data collected in time.
- *Bottom*-Up (Kanudia and Loulou 1998; Gielen and Chen 2001; Cormio et al. 2003; Lin et al. 2010; Phdungsilp 2010; Zhang et al. 2011; Feng and Zhang 2012): models based on statistical techniques and engineering techniques. Statistical techniques to determine energy consumption for end uses are based on the study of behavior by analyzing energy bills and on sociological surveys. Instead, engineering techniques are based on the detailed study of the physical characteristics of the building and enable analysis of effects stemming from the use of new technologies.
- *Hybrid models* (Hadley and Short 2001; Mirasgedis et al. 2007; Turton 2008; Bohringer and Rutherford 2009; Liu et al. 2011): models which use the energy requirements of typical constructions and adapt them to estimate a city's consumption.

In particular, the study of bottom-up models, which are able to relate many input variables besides identifying variables linked to energy consumption, has shown that the analysis of energy consumption in urban areas is, in reality, a still neglected aspect within the scientific debate. Applications of bottom-up models to whole neighborhoods or to parts of the urban area are still few and far between. By the same token, setting up variables for different urban components remains a consolidated approach only at the theoretical level rather than in terms of application. Although many scholars claim to study this relationship with a multidimensional 1 to n approach, almost always such contributions are reduced to investigating the biunivocal relationship between each component of the urban system and energy consumption. Further, increasing the degree of uncertainty of the results is the fact that the urban characteristics in question are considered independent of one another. In other words, hardly ever are the relationships linking the various urban elements taken into consideration. However, such relationships may be considered decidedly important for understanding and managing energy consumption in such a highly complex system.

3 UrbanSEM: An Interpretative Model

The SEM research project is co-funded by the 2007–2013 National Operative Program (PON), Research and Competitiveness, Smart Cities and Communities, "Integrated Action for Sustainable Development—Energy Efficiency and Low Carbon Technologies" and enjoys a broad partnership which integrates universities, firms, research institutes and public administration.

One of the research products is the Urban Saving Energy Model (UrbanSEM), an interpretative model constituting the starting-point for setting up a decision support tool for public administration, professionals, land technicians, engineers and architects and for those interested in finding efficient and effective planning solutions for energy saving at the urban scale.

Within the panorama of studies tackling the subject of energy saving, UrbanSEM may be considered innovative in at least two aspects: in the approach used and in the area scale of application, besides its aim to seek to quantify the effects of the characteristics of urban elements upon energy consumption. Indeed, unlike almost all research that privileges the study of specific sector contexts, such as energy production plants, transport systems or the energy performance of buildings, UrbanSEM "looks" at the urban system in its entirety, starting from interpretation of the various elements (spaces, activities, etc.) of which it consists and the relationships between them, with a view to identifying and quantifying energy consumption arising from their urban inter-relationship. Moreover, most of the scientific attention is focused on scales of detail like that in construction, while UrbanSEM works at the neighborhood scale. Underlying the research is the consideration that in-depth reasoning at the neighborhood scale may facilitate the difficult interpretation of characteristics and complex relations between the urban system and energy consumption.

UrbanSEM was constructed on the basis of the following diverse subsystems into which a city which has the prime aim of saving energy may be structured:

- urban/environmental subsystem: so as to define the morphological and geometric characteristics of the urban fabric (physical characteristics) and climatic (environmental) characteristics which significantly affect energy consumption, such as building density, aspect ratio (height of the buildings on the road front/road width), compactness factor (building surface area/volume);
- residential subsystem: with a view to identifying the residential characteristics that significantly affect energy consumption, such as height, covered surface area, construction date;
- socio-economic subsystem: so as to define the main factors characterizing the behavior of the population, such as age, education, employment and income, which most affect energy consumption;
- highly humanized subsystem: so as to identify types and characteristics of non-residential buildings for services, business, public administration, etc., which have broad margins for restricting energy consumption, such as the estimated energy intensity of such buildings;
- the mobility subsystem: so as to define levels of accessibility to urban areas through both endowment in terms of public and private transport networks and of land use intensity, factors which have a significant effect on energy consumption, such as road network density and the distance of public transport rail stops from the city center;
- subsystem of open constructed spaces: so as to identify the relationship between the energy consumption of buildings and the physical and geometric characteristics of adjacent spaces (roads, squares, etc.), both in terms of accessibility to natural illumination and control of thermal radiation.

Extensive research that has attempted to measure the incidence of parameters of urban form and geometry upon energy consumption, has supplied mainly qualitative rather than quantitative results as a point of reference. Our research work sought to quantify the "statistical weight" of each urban variable or groups of variables in energy consumption, thereby allowing identification of the physical elements on which to be able to intervene significantly for the purpose of energy saving.

To set up the interpretative model, four main working phases were implemented:

1. definition of the reference variables of the urban system (for each urban subsystem);
2. acquisition of data referring to the variables of the urban system and energy consumption;
3. definition of the measuring techniques of variables;
4. choice of the techniques of statistical analysis to interpret the relations between urban variables and energy consumption.

The model's area of application and experimentation includes the neighborhoods of Chiaia, Vomero and Arenella in the central area of the city of Naples, whose different morphological and settlement characteristics represent a significant test for setting up a model which can be easily replicated in varied urban contexts. Indeed, the overall area, both in size (circa 10 km^2) and in demographic size, is comparable to the size of a small-medium city.

4 Environmental Urbanistic Results

Thanks to the study of the scientific literature of reference and the above models which, albeit at the building scale, have explored the relationship between urban components and energy consumption, we were able to identify the reference variables of the urban system. Of all these urban characteristics, this work focuses on results obtained for those concerning environmental and residential urban subsystems.

The 18 physical, environmental and construction variables on which the interpretative model is based represent the most significant and most widespread characteristics in studies hitherto conducted, whose incidence on energy consumption can only be assessed qualitatively, in accordance with the fact that both the interactions and the effects of the urban form and geometry on energy consumption continue to be little studied and controversial (Alberti 1999). In practice, we are dealing with a set of variables based "on the literature review and not on empirical findings or field work" (Papa et al. 2014).

Thus a set of variables was defined, structured into three macrocategories, namely physical, environmental and building, through which to study the main characteristics of urban morphology and form. Physical variables describe the geometry of the urban fabric, environmental variables concern climatic and morphological characteristics, and building variables describe the physical characteristics of buildings (Table 1).

Table 1 Set of variables considered

Physical characteristics	Environmental characteristics	Building characteristics
Construction density	Green plot ratio	Construction period
Building height	Green area radius of effect	Number of stories
Aspect ratio	Green ratio index	Glass surface areas
Cover ratio	Gradient	Covered surface area
Compactness average		Surface area of dwellings occupied by residents
		No. of dwellings
		No. of rooms
		Masonry buildings for residential use
		Reinforced concrete buildings for residential use

The data required to calculate urbanistic and environmental variables were obtained by using statistical (data from the 2001 ISTAT census) and cartographic sources (via elaborations in a GIS environment of maps for the three neighborhoods in question). The data were then georeferenced, i.e. associated to census sections so as to produce a series of thematic maps for the variables and significant processed data.

The data on energy consumption concern the year 2011 and refer to individual users, identified with cadastral data, both with regard to consumption of electricity and that of methane gas. Further, the data refer to sales (€) and kWh and m^3 consumed annually and were classified into two macrocategories split into various categories:

- Individual users: these include domestic residential users;
- Non-individual users: these include all non-residential users, such as non-commercial organizations, those who have a VAT number, large users and non-residential domestic users.

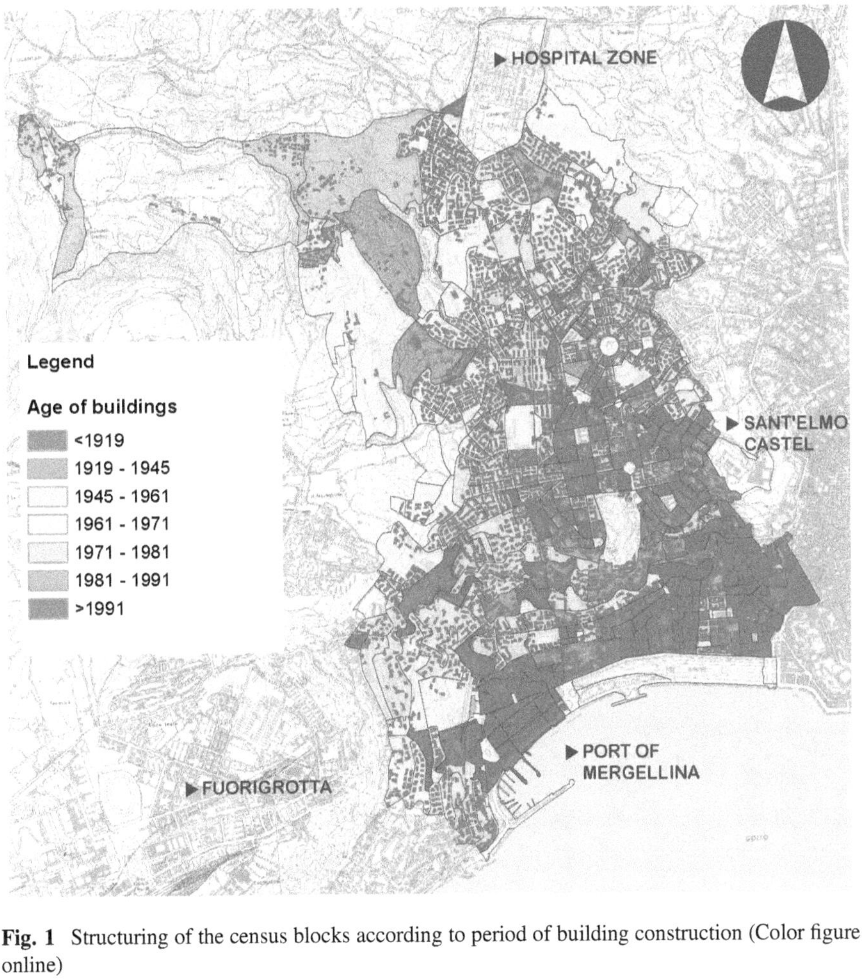

Fig. 1 Structuring of the census blocks according to period of building construction (Color figure online)

Geolocalization of consumption, i.e. the association of such data to census blocks, was performed by creating one geodatabase for electrical energy and another for gas.

For the interpretation of the relationships between urban variables and energy consumption, use was made of several techniques of multidimensional data analysis which allow the degree of correlation between the variables in question to be determined. In practice, thanks to such techniques, it is possible to determine, by virtue of the relationships established among all the variables involved, the main factors responsible for energy consumption.

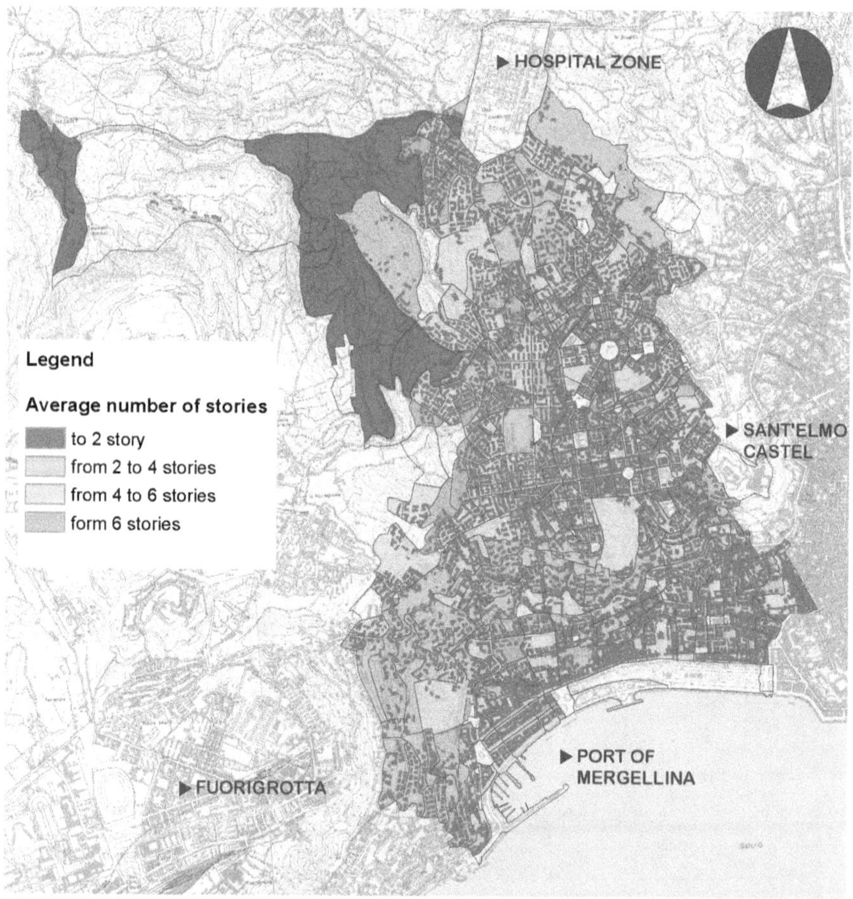

Fig. 2 Structuring of the census blocks according to the mean number of stories of residential buildings (Color figure online)

5 Measurement of Variables

The work involved in setting up a model able to detect the physical, environmental and building elements (and their reciprocal relationships) which have significant impacts on levels of energy consumption, at the neighborhood scale, as mentioned above, was unable to refer, in many respects, to a theoretical and operationally consolidated framework.

Against this background, it was necessary to choose and apply techniques for measuring the urbanistic and environmental variables which might allow us to reach not only qualitative but also quantitative results, to test a technique that might also be used in other urban contexts and to interpret not only the elements but also the relations between them which have most impact on energy consumption.

For each of the variables considered, different classes of values were thus defined, identified so as to make the differences emerge between census blocks to which they were associated, by virtue of the fact that such differences may prove significant in terms of energy consumption. In other words, the significant intervals of the values assigned to variables were identified in order that the relationships among the urbanistic and environmental variables and those concerning energy consumption might be interpreted on the basis of actually representative values of the model's area of application (Figs. 1 and 2). While this operation for most of the variables in question was facilitated in that, as already stressed above, almost all the research occurs at the building scale, for others it required setting up a complex scientific and technical procedure with a view to plotting such variables against energy consumption.

For the two variables, building density and radius of influence of green areas,[1] which on the basis of the scientific literature of reference have a major impact upon urban energy consumption (Yamamoto et al. 2004; Doherty et al. 2009; Zhang et al. 2011; Howard et al. 2012), a twofold objective was reached: on the one hand, measuring them, respectively, against the external surface temperature of buildings and air temperature, with a view to ascertaining whether and to what extent these variables affect temperature variations in urban areas; on the other, defining the reference values which might then be significant, once measured against energy consumption. The decision to link both variables to temperature is also connected to the fact that some scholars such as Oke showed that urban morphology, especially density, impact upon the urban heat island (UHI) phenomenon, indirectly linked to the issue of energy consumption, which creates a warmer microclimate within urban areas than in the surrounding periphery and rural zones, and which thus entails an increase in energy consumption especially in the summer months.

To be able to reproduce the microclimate behavior (external temperature, relative humidity, wind speed and direction) and environmental (size and type of green areas) and urbanistic characteristics (building density, height, covered area) in the

[1]By radius of influence of green areas we mean the maximum distance beyond which the cooling effect is no longer experienced.

three neighborhoods we used ENVImet environmental and microclimatic simulation software developed by Professor Michael Bruse of the German university Johannes Gutenberg Mainz. Although ENVImet is the most suitable and reliable software available (Roset and Vidmar 2013), it is unable to simulate elements such as gradient. Thus, not being able to identify the sample areas within the three neighborhoods, insofar as the latter are morphologically very different, it was thought appropriate to perform modeling of the study area, partly with a view to providing a methodology that might be applicable in any context.

In practice, the census blocks were represented by square modules with a specific density concerning the coverage ratio and height of the buildings (Fig. 3), at the center of which were positioned green areas to simulate the cooling effect. The advantage of this type of configuration is that it allows simple management

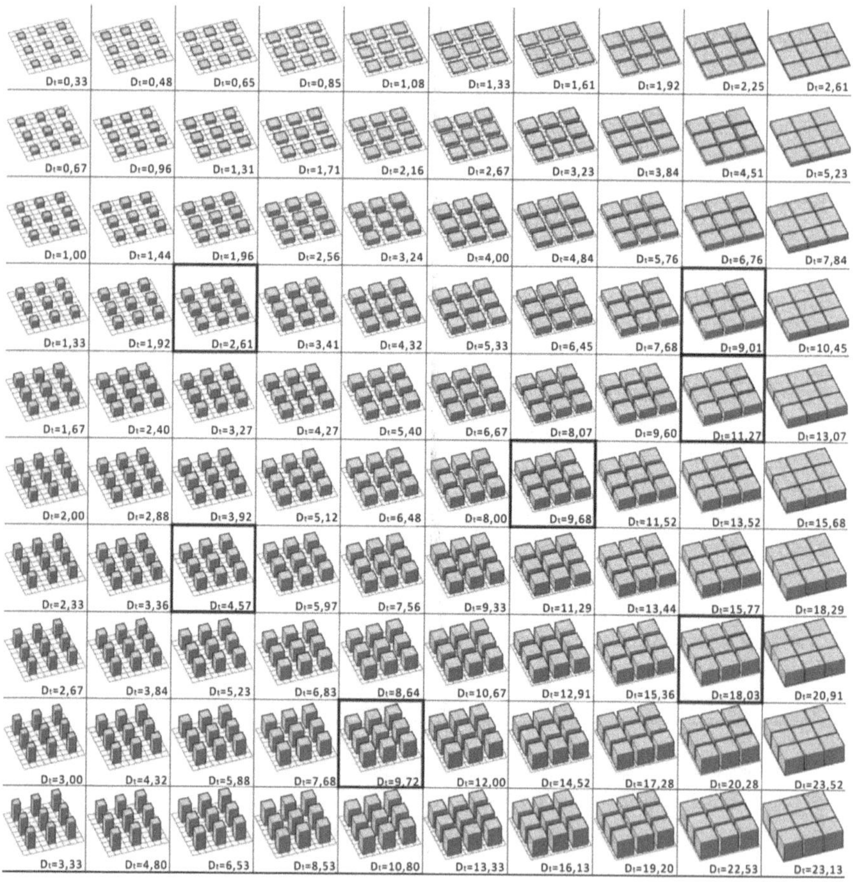

Fig. 3 Modular schemes to simulate building densities, in which the rows show the building heights and the columns the cover ratios (Color figure online)

both of the density values, given that they are ratios between surface areas, and simulation in ENVImet which operates on a square grid.

Unlike the green areas, for which simulations were only carried out on the hottest day in the last ten years, for building density the worst climatic conditions in summer and in winter were simulated so as to be able to study the effect of this variable on electricity consumption (summer air conditioning) and gas (winter heating).

The output supplied by the software, i.e. mean external temperature of the facades, was plotted against the building density to obtain significant ranges of values for energy consumption. These results show that between the building density and the variation in external surface temperature there is an inverse correlation. They refer to five classes for summer solar radiation and three classes for winter (Table 2).

For green areas, on the basis of differences between the average air temperature recorded at five-meter intervals starting from the green area and the temperature inside the green area itself (ΔT), the radiuses of influence were defined for each of the three cases simulated.

It emerged that surface areas below about 5000 m^2 do not lead to major reductions in urban temperatures and hence their presence contributes neither to mitigating UHI phenomena nor to reducing energy consumption for summer air conditioning (Fig. 4). These results were calculated in accordance with Chen et al. (2012): the maximum distance beyond which the cooling effect of green areas is no longer felt, i.e. the radius of influence, corresponds to the point in which the tangent to the curve is horizontal and hence the first derivative of the cubic polynomial function becomes zero (Fig. 5).

The subsequent phase of interpreting the relationships between the environmental variables and consumption of electricity and gas was carried out by using several techniques of multidimensional statistical data analysis.

Application of multivariate analysis to research fields like area analysis has great validity, thanks to its ability to process and summarize considerable quantities of information and data linked by complex interdependencies (Benzecri 1973; Diday et al. 1982; Lebart et al. 1995), a characteristic of territorial phenomena, which are otherwise difficult to govern. The purpose of this set of statistical methods is thus in line with that of the present research, aiming to know and define the latent physical

Table 2 Building density classes

Building density classes (summer solar radiation)	Building density classes (winter solar radiation)
Class 1: 0.0025–4.101	Class 1: 0.0025–4.174
Class 2: 4.102–7.033	Class 2: 4.175–8.477
Class 3: 7.034–9.898	Class 3: 8.478–19.486
Class 4: 9.899–12.451	
Class 5: 12.452–19.486	

Case 1	Radius of influence	ΔT
Small green area: 900 m^2 High density: 13.28 m^3/m^2 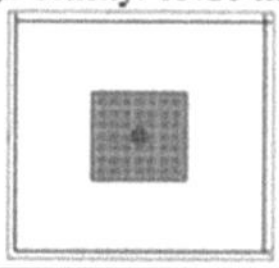	48 m	0.2° C
Case 2	**Radius of influence**	**ΔT**
Medium green area: 4900 m^2 Average density: 8.48 m^3/m^2 	122 m	0.8° C
Case 3	**Radius of influence**	**ΔT**
Large green area: 32400 m^2 Low density: 2.08 m^3/m^2 	145 m	1.2° C

Fig. 4 Simulation results by radius of influence of *green* areas (Color figure online)

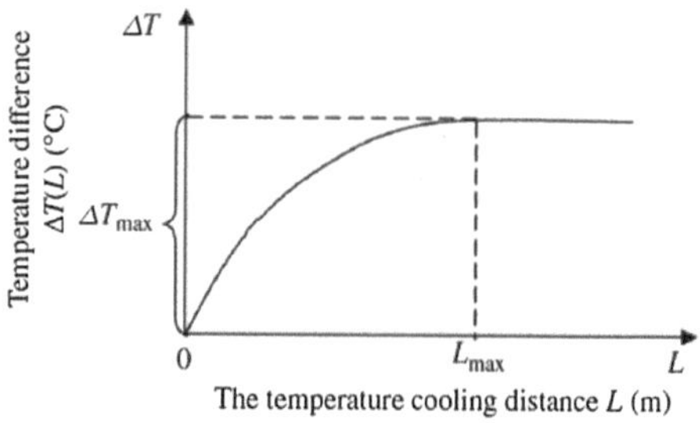

Fig. 5 Model of the effect of cooling of green areas according to Chen et al. (2012) in which L_{max} is the radius of influence of *green* areas (Color figure online)

structure of the urban system in question, so as to interpret its relationship with energy consumption.

The strengths of these techniques may in practice be attributed to two main aspects: on the one hand, the ability to reduce large quantities of data to a small number of variables, ensuring a minimum detectable loss of information and, on the other, the possibility to apply them even in the absence of set hypotheses concerning relations between the variables which describe the phenomenon in question (Gargiulo and Papa 1995).

The multidimensional statistical analysis techniques used in this research correspond both to factorial methods and to automatic classification methods. In particular, the statistical analysis methods chosen were Principal Component Analysis (PCA) and cluster analysis (Cluster). PCA allows a reduction in the number of variables which describe the units (or individuals) in question and representation of the latter's characteristics with a limited number of new variables, the principal components, breaking down the phenomenon according to structural axes of decreasing importance. The principal components are thus variables that cannot be observed directly which allow scores to be assigned to single individuals. Such scores are expressed by the coordinates of points on the factorial axes, the latter understood as latent structures of the phenomenon in question.

Cluster analysis allows the units in question to be grouped into classes according to (dis)similarity criteria (similarity or dissimilarity are complementary concepts, both applicable in the approach to cluster analysis), i.e., to determine a certain number of classes so that the observations are as homogeneous as possible within the classes (internal cohesion) and as heterogeneous as possible between the different classes (external division).

In particular, PCA was conducted on the physical, constructional and environmental variables, which constitute the active variables within the analysis, to then project on them those relative to electricity and gas consumption, which instead represent the illustrative variables. Indeed, energy consumption is not considered a component which structures the urban system, but represents the phenomenon in question. Hence the operative procedure adopted, which identifies the principal components within the variables characterizing the physical space, allows the determinants of consumption to be captured, in terms of physical, constructional and environmental variables. The results of a first analysis detected the need to verify both the individuals (blocks) considered and the variables relative to energy consumption.

As regards the census blocks, analysis showed the existence of a class consisting of only open built spaces. The single blocks belonging to the class are potentially misleading for the analysis since they modify the structure of the other variables considered, changing their variability. Such blocks (56 census blocks) were thus excluded from the analysis insofar as they were deemed "anomalous".

In the same way, the analysis carried out did not supply significant results, that is, no significant relationships emerged between the physical, constructional and environmental variables and electricity and gas consumption. The absence of a strong relationship between urbanistic characteristics and consumption made it worth setting up a new variable, both for electrical energy and gas, which would take account of surplus consumption. This choice stems from the fact that this first statistical analysis made us hypothesize the existence of a minimum consumption independent of the urbanistic characteristics considered and that, to be able to identify which physical, constructional and environmental components actually affect energy consumption, it is worth assessing the quantity of surplus electricity and gas consumption.

To calculate these new variables, termed Δ_{el} and Δ_{gas}, we proceeded as described below:

1. identification of the reference variable: both for electricity and for gas consumption a reference variable was identified: the number of rooms for electrical energy (the total in each census block), the built volume for gas (the total in each census block);
2. calculation of unit consumption per room and per built volume: for each census block we calculated total consumption, both of electrical energy and gas, according to the different types of users present. Such total consumption was then measured, respectively, against the total number of rooms and the total built volume to obtain the unit consumption.

$$C_{tot}/n^{\circ}\ rooms_{tot} = C_{room} \quad \text{(electrical energy)}$$
$$C_{tot}/\text{built volume}_{tot} = C_{mc} \quad \text{(gas)}$$

3. determination of fixed consumption: to determine fixed consumption reference was made to per capita income. The five lowest and most common values were considered. Among these we chose the lowest unit consumption per room and per m^3 (C_{room} and C_{mc});
4. calculation of Δ_{el} and Δ_{gas}: having defined fixed consumption, Δ was calculated, i.e., the part surplus to minimum independent consumption of electricity and gas:

$$\Delta_{el} = C_{tot} - (C_{room} * n^{\circ}\ rooms_{tot})$$
$$\Delta_{gas} = C_{tot} - (C_{mc} * \text{built volume}_{tot})$$

In setting up this procedure it is worth pointing out that:

- as the variable of reference, electricity consumption was considered, given that most of such consumption (~75 %) comes from electrical appliances. This shows that the presence of electrical appliances may be considered a "homogeneous" amount of electricity consumption, insofar as appliances such as refrigerators, washing machines or computers are now standard household endowment. What may be more variable, insofar as it is related to urban geometry, as well as habits and lifestyles, is the amount relative to lighting: the quantity of artificial light required is a function of various factors such as the proximity of buildings, exposure, size of accommodation unit, etc. Thus, for the purposes of determining surplus electrical energy consumption it was felt appropriate to refer to a variable able to express the strong variability of electricity consumption, such as the number of rooms;
- the built volume was chosen as the reference variable for gas consumption given that this consumption depends substantially on winter heating and hence the building volume;

- the correlation between rooms and habitations is very high (about 98 %); this means that considering one or the other variable for the purpose of calculating Δ does not involve any variation in the results;
- the correlation between total consumption of energy and residential consumption is 85 %; this means that considering consumption only with reference to residential consumption changes the rationale and the results only relatively.

6 Principal Components

In light of what was described in the section above, we proceeded to perform statistical analyses once again, eliminating 56 blocks with respect to the open spaces described in the previous section. With the same criterion the census block for the Cardarelli hospital was also excluded, insofar as the presence of this type of activity could give rise to a distorted interpretation of the results of the cluster, both as regards electricity and gas consumption.

In the case in question, the system proved weakly structured, in that, to reach an optimal threshold of the whole variance explained, it would be appropriate to consider the first five components (about 77 %). In particular, the first component explains 29.4 % of total variance, the second 21.6 %, the third 12.7 %, the fourth 7.5 % and the fifth 5.6 % (Fig. 6).

The first component, which explains 29 % of variance of the system of variables taken into consideration, may be termed the "axis of sprawl" (residential). This

1	7.0590	29.41	29.41	**
2	5.1946	21.64	51.06	**
3	3.0548	12.73	63.78	************************************
4	1.8113	7.55	71.33	*********************
5	1.3537	5.64	76.97	****************
6	1.0526	4.39	81.36	************
7	0.9945	4.14	85.50	************
8	0.7923	3.30	88.80	*********
9	0.7495	3.12	91.93	*********
10	0.6303	2.63	94.55	********
11	0.3725	1.55	96.10	*****
12	0.2483	1.03	97.14	***
13	0.1990	0.83	97.97	***
14	0.1473	0.61	98.58	**
15	0.1065	0.44	99.03	**
16	0.0779	0.32	99.35	*
17	0.0464	0.19	99.54	*
18	0.0375	0.16	99.70	*
19	0.0272	0.11	99.81	*
20	0.0208	0.09	99.90	*
21	0.0182	0.08	99.98	*
22	0.0025	0.01	99.99	*
23	0.0022	0.01	99.99	*
24	0.0013	0.01	100.00	*

Fig. 6 Eigenvalues, percentages of cumulative variance explained and a histogram of the physical, building and environmental variables of PCA

component depends first and foremost on volume, on the index of residential use, on height and on density, almost always referring to residential use. Upon analyzing the correlations between each variable and this component, it is observed that most (14) are negatively correlated with the first component, while the rest are inversely correlated (10). Through a rotation of its axis, this component may be considered as that for a compact layout, with tall, densely packed residential buildings. Indeed, it is no coincidence that the variable for the presence of green areas is one of the first variables.

With reference to energy consumption, Δ_{el} and Δ_{gas} are positively correlated with this component, although the score for such variables relative to consumption is modest ($\simeq$0.45). It may nevertheless be stated that low values of building density negatively affect energy consumption, in accordance with the relevant literature: low values of urban density have higher energy consumption. The most interesting aspect is that this influence is not so "strong" as broadly maintained by all those scholars who identify density as the key factor for the energy issue in urban areas.

The second component, which explains about 22 % of total variance of the whole system, may be briefly defined as the "axis of the size of residential stock". Indeed, this component depends strongly on building variables such as the number of dwellings and rooms, residential volume, the surface area of dwellings and the height of residential buildings. By contrast, it is inversely correlated with variables related to green areas, albeit with fairly insignificant values. In practice, this component is seen to describe the quantity of buildings.

With reference to energy consumption, both Δ_{el} and Δ_{gas} are positively correlated with this component, although those with a higher score concern energy. Thus it may be briefly stated that the more built-up an area is, the more energy is consumed.

The third component, which explains about 13 % of total variance, may be called "the green index", insofar as the variables to which it is positively correlated, and which have very high scores, are those relative to the presence green areas. By contrast, the last negatively correlated variable, with a significant score (−0.67) is the coverage index. This component therefore also provides an indication on the relationship between full spaces and voids, i.e., between built spaces and those not built-up, precisely because the variables relative to green areas and to built-up surface areas are in contrast. With reference to energy consumption, Δ_{el} and Δ_{gas} do not score highly, which suggests that the presence and characteristics of green areas do not affect consumption as much as is maintained in the literature.

Despite considering the percentages of variance explained by the fourth and fifth components negligible, they cannot but be included in this phase of interpretation of the results, insofar as they provide additional information on the structure of the urban system in question, supporting the first two principal components. The fourth component, which explains about 7 % of the variance of the system of variables in question, may be termed the axis of construction age. Indeed, this component depends first and foremost on the presence of masonry buildings and hence the characteristic which determines this axis is the construction period of the buildings. Finally, as regards the fifth component, which explains about 7 % of variance, it must be stressed that this component depends on the radius of influence of green areas.

The results of PCA enabled us to achieve a twofold objective: identify the interrelationships between physical, environmental and constructional variables which structure the urban system in question, and ascertain whether such relationships affect electricity and gas consumption. Indeed, the strength of this type of statistical analysis, like the innovative aspect of the whole work, is represented by the capacity to arrange the variables considered so as not to examine the single parts but, rather, the overall structure, highlighting its characteristic relationships.

The spatial configuration, the quantity of construction and green area characteristics are the main factors in the area in question. That said, their influence on energy consumption is important only for the second component.

7 The Clusters

The results of PCA describe with sufficient clarity the significant characteristics of the three neighborhoods in question, yet without supplying any information concerning the elements which may link groups of apparently heterogeneous blocks. Application of cluster analysis thus allowed such clusters to be identified as homogeneous in quality and characteristics. Cluster analysis was performed on the principal components so as to be able to operate on independent variables.

To identify the classes we chose an algorithm to group the statistical units so that all those that make up a given group have the same homogeneous characteristics and that each unit belongs to one and only one group. Following what was described in Sect. 4 we chose a hierarchical solution method which uses the maximum distance method.

The results are reported in two tables, with one referring to qualitative variables and the other to quantitative variables, ordered in both cases according to the test value. In practice, to characterize a class, the statistical software makes a comparison between the average of each variable of the class in question and the general average, and evaluates the difference, taking account of the variance of the variable, within the class. For this purpose, Student's t test was performed, i.e. a parametric statistical test which carries out the comparison for each variable, with a 5 % error. The variables reported a test value >2 are, on average, more present in the class compared with the general. The variables reported a test value less than −2 are, on average, less present in the class. The four resulting groups are represented in a GIS environment, with each cluster being assigned to a census block (Figs. 7 and 8).

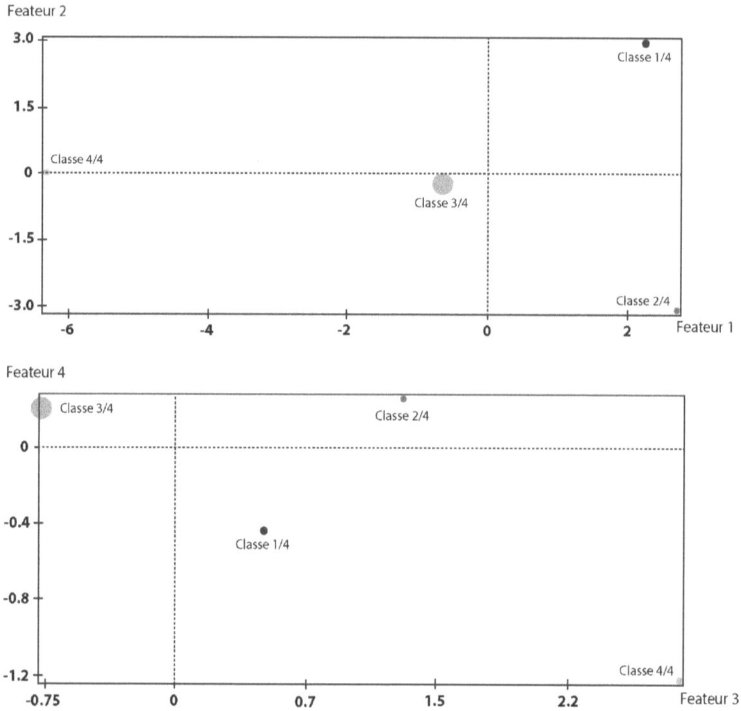

Fig. 7 Representation of the four clusters with respect to the principal components, taken two at a time

7.1 *Cluster 1*

The first group, which contains about 19 % of the statistical units, is characterized by buildings constructed chiefly in the decade 1961–1971 and by variables regarding the residential stock, such as number of rooms and dwellings, surface area of dwellings, and residential volume. This cluster, in the reference system consisting of the first two principal components, the sprawl axis as the abscissa axis and residential stock size as the ordinate axis, is positioned in the upper part of the first quadrant, by virtue of the high residential stock and the low density values. All the variables which may be attributed to urban indexes are indeed present in the lower part of the table of the cluster's characteristics (residential density, index of utilization, etc.).

This cluster contains the census blocks from low-quality postwar constructions, areas developed in non-organic fashion without a precise planning design behind them. Unsurprisingly, belonging to this class are almost all the census blocks in the Arenella neighborhood, built up more recently than the two other neighborhoods in question, including areas like Rione Alto near the hospitals, the Policlinico and Via Domenico Fontana.

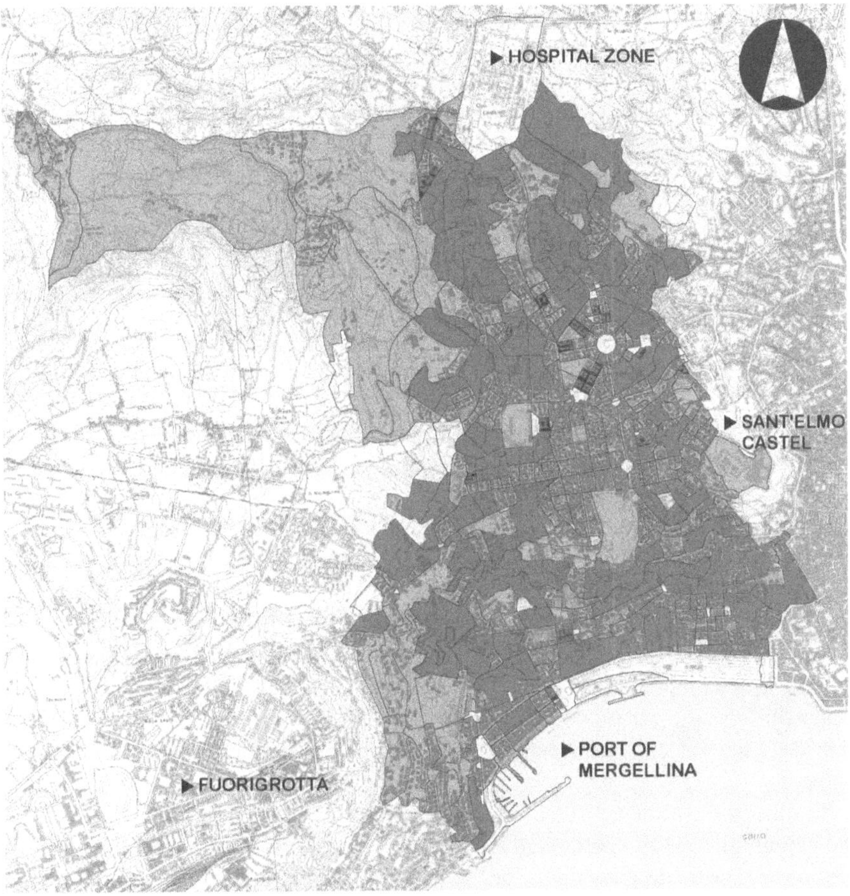

Fig. 8 Structuring of the census sections according to the four clusters identified: unregulated expansion (*dark green*), valuable fabric (*light green*), planned fabric (*orange*) and compact fabric (*red*) (Color figure online)

All these are areas with an unplanned urban structure, in which construction began as a consequence of the building saturation in the nearby neighborhood of the Vomero: "the hill of the Vomero had reached building congestion without any balanced relation between dwellings and services. Expansion of this neighborhood lacked an organic road system and had no reference to detailed plans" (Comune di Napoli 2004).

Continuing the description of the cluster with reference to energy consumption, especially Δ_{el} and Δ_{gas}, the latter two values, together with almost all the other energy variables, strongly characterize the cluster. These are therefore census sections with higher than average consumption of electric and gas energy compared with the whole study area. This state of consumption may be accounted for by the low construction quality, both in terms of building materials and methods, as well as construction techniques, found in this cluster.

It is worth pointing out that this class also includes, albeit constituting a very small fraction, census blocks from a construction fabric predating 1919 and a planned town structure, like those of the Chiaia neighborhood, by virtue of volume size. Overall, this cluster may be defined as an "unregulated expansion".

7.2 Cluster 2

The second group contains about 14 % of the statistical units. The aspect peculiar to this group consists in the presence of three variables concerning the green areas that are distinctive to this class. These are census blocks sharply distinguished by wide open (unbuilt) spaces. Indeed, as the average IGnR value is close to 1, these are areas with lower settlement densities and volumes than the other types of areas identified. Thus, just as for the third principal component, the green index, a clear separation occurred between the variables related to such environmental characteristics and those concerning the quantity of the built surface area, also in this group the variables which supply information on the presence of green areas are contrasted with those relative to the form and quantity of the built environment. Indeed, within the reference system consisting of the first two principal components, this group is positioned in the lower margins of the fourth quadrant, precisely because it has larger, unbuilt open spaces, hence with lower densities than the other groups.

In other words, the census blocks belonging to this group are situated chiefly in the neighborhoods of Chiaia and Arenella, and especially in the Posillipo area, in the former case, and Camaldoli, in the latter. These are areas in which the intense and not always controlled building process which is the hallmark of those belonging to the first class described above, is replaced by high-value residential building which chiefly began in the 1950s, with low buildings (about three floors) and a number of buildings such as to ensure a high real estate quality.

The markedly "green" nature which distinguishes this cluster would suggest a positive impact on energy consumption, as maintained by many scholars. Indeed, Δ_{el} is located in the lower part of the table of results, which indicates that the blocks of this cluster have lower consumption of electrical energy than anywhere else in the whole system in question. This result may be explained in the morphological and geometric characteristics in these areas: a sparsely built-up, low-density area, with buildings sufficiently far apart to allow the passage of natural sunlight and hence the possibility to limit the use of artificial lighting, partly thanks to the gradient, given that the areas lie between 150 and 250 m asl, and to the significant presence of green areas which contributes to limit the use of summer-time air conditioning. At the same time, the high green surface area ensures that the census blocks in this cluster have higher average gas consumption than those in the whole study area. Therefore, green areas do not contribute to limit energy consumption in general. Rather, they are an element to the benefit of electrical energy consumption, whilst accentuating that of gas due to the greater need to resort to winter heating. Overall, this class may called that of “the valuable fabric”.

7.3 Cluster 3

The third group contains most of the statistical units (about 61 %). This group may be termed “neutral” for analytical purposes insofar as, with respect to the first two principal components, i.e. those which have the most information content, it is closest to the origin of the axes. This class in practice represents the average of the urban system in question, characterized by tall compact buildings (six stories on average), built chiefly before 1930, and by high coverage indexes.

The census blocks belonging to this class appear to be representative of the compact fabric, planned to be built following a unitary design. In practice, these are foundation neighborhoods, i.e. neighborhoods which were designed and planned with a chess-board reticulate layout in mind, as in the cases of Piazza Vanvitelli and Medaglie d’Oro. It is no coincidence that this cluster comprises most of the census blocks of the Vomero neighborhood and also various blocks in the Chiaia neighborhood, like that of Viale Gramsci.

Besides historic buildings, this class includes those built during the postwar economic boom, from 1946 to 1960, a period in which urban planning was replaced by uncontrolled urban growth, which from the central area of the Vomero neighborhood expanded toward new areas within the Arenella neighborhood.

Continuing the description of the cluster with reference to energy consumption, especially Δ_{el} and $\Delta_{gas,}$ it has lower consumption than the average for the whole urban system in question. The correspondence between compact layouts and low energy consumption, amply discussed and maintained in the literature, is thus fully confirmed within this class, which may be called the class of the “planned urban layout”.

7.4 Cluster 4

The fourth group contains about 7 % of statistical units. The cluster is characterized by variables concerning density and the index of residential use, as well as by building variables such as height and glass surfaces. In the reference system consisting of the first two principal components, the cluster is positioned in the third quadrant, with low residential stock but high density. Indeed, these are census blocks consisting of a single building situated, in turn, within a single plot, particular building products, on average eight stories tall and spacious, chiefly erected prior to 1930 and to a much lesser extent between 1946 and 1960.

Continuing the description of the cluster, with reference to energy consumption, especially Δ_{el} and Δ_{gas}, there is lower than average electricity consumption with respect to the whole urban system in question and higher average gas consumption. These results may be explained by the fact that as we are dealing primarily with historic buildings with many glass surfaces and obsolete construction techniques, heat dispersion may be significant, which accounts for higher average gas consumption. Instead, their positioning within compact building fabrics may explain lower electricity consumption, as in cluster 3. Overall, the class may be called the class of "compact fabric".

Cluster analysis allows us to identify substantially three macrotypes of fabric, with specific physical, environmental, construction and energy characteristics:

- an urban fabric with buildings constructed after the 1960s, at least four stories tall and lower residential densities than the average of the whole reference system. Such characteristics result in higher electricity and gas consumption (class 1);
- a fabric with a low building density, substantial presence of green areas, buildings about three stories high and position mostly on the hill. Such characteristics result in higher average consumption of gas and lower electricity consumption (class 2);
- a fabric with a high building density, compact buildings at least six stories high, belonging chiefly to historical periods of construction. Such characteristics result in lower than average electricity and gas consumption (classes 3 and 4).

8 Conclusions

Given the quality and number of international research groups engaged in defining energy-saving approaches, methods and instruments we are still some way from establishing a consolidated framework to enable reliable results to be shared and, above all, effective solutions to be set up to limit energy consumption at an urban scale. Technical approaches and the knowledge framework set up in this work show that objectives maximizing energy efficiency and saving may be achieved only if they are treated in an integrated paradigm which first of all exploits the capacity to manage and implement urban changes to achieve the fundamental objectives of energy sustainability.

This study focused on city-energy relationships, starting from the interpretation of relations among the various physical and environmental elements which make up the urban system and may constitute a cue for reflection and stimulus for the scientific community both due to the systemic approach adopted and the geographical scale applied (an urban neighborhood).

On the basis of the results obtained both from PCA, which allowed us to identify the "latent structure" in energy terms of the neighborhoods in question, and from cluster analysis, which was used to interpret the relationships between significant urbanistic characteristics and energy consumption, we may formulate one or more significant hypotheses on defining several useful elements for setting up future strategies, solutions and interventions to promote energy efficiency and saving at the urban scale.

Very briefly, it may be stated that variables such as building density, volume and height are physical characteristics that affect energy consumption, although their "weight" does not appear as decisive or evident as is often maintained by the scientific literature of reference. The results of both types of statistical analysis carried out seem to suggest that attributing to such parameters the role of discriminants to classify part of the urban area as energy-guzzling or not may be hard to prove and in some cases mistaken.

This consideration may also be extended to the year of building construction, thus calling on readers to reflect on the benefits expected from many interventions aimed at improving the energy performance of building shells, which could prove of little significance compared to ambitious energy saving objectives established for such measures. In contrast, the presence of green areas seems to represent a positive element in restricting consumption of electrical energy, accentuating that of gas, in urban fabrics with low building densities. Correspondence between compact fabrics and lower consumption of both electrical energy and gas, extensively discussed and maintained in the literature is fully confirmed in the urban system concerned.

Our preliminary results appear to suggest that most energy consumption within some neighborhoods does not depend solely on the variables in question. By the same token, building densities, as well as the presence of green areas and height of buildings, affect energy consumption in different ways, making it impossible to identify an "ideal" (sustainable) urban form which is able to maximize energy efficiency and which still remains theoretical (cfr. Doherty et al. 2009). These results force us to reflect on the validity of the ambitious and, in certain cases, shared objective of identifying only one model which links urban form and energy consumption. There seems to emerge the need to reformulate this objective which is based on the hypothesis that there is not one but many models of consumption according to the different physical, environmental and building characteristics of the various urban contexts.

The decision to adopt an integrated holistic approach rather than a sectoral approach and to focus on neighborhoods rather than buildings allowed us to explore the issue of the city-energy relationship, at the same time taking many characteristics of urban space into account. This approach confirmed the complexity of relationships between such characteristics and energy consumption, and the scarce effectiveness of a sectoral approach whose results do not seem to agree with those described.

If, as occurred in the course of our research, together with physico-environmental characteristics, economic and social factors are also considered, the relationship between urban space and energy consumption becomes more complex. In this sense it may be stated that one of the most significant findings of this study is confirmation that the search for a single solution able to interpret the relationship between city and energy appears as hard as ever, as already stated in 2009 by Doherty et al.

The complex multidimensional nature of the city-energy relationship provides a number of interesting cues for future research developments. First of all, it would be desirable to verify the applicability of the interpretative model, which in this work was tested in a medium-size urban context, to a whole city, including a set of variables such as those which describe the functional specialization, climatic characteristics and behavior of residents in choosing transportation modes. A further research subject could involve the study of the role of green areas in energy consumption in small urban contexts.

References

Alberti M (1999) Urban patterns and environmental performance: what do we know? J Plann Edu Res 19(2)

Benzecri JP (1973) L'Analyse des données. Dunod, Parigi

Biocca D (1987) L'evoluzione dei sistemi urbani e il futuro delle città. Studi Storici 2

Bohringer C, Rutherford TF (2009) Integrated assessment of energy policies: decomposing top-down and bottom-up. J Econ Dyn Control 33

Chen XZ, Su YX, Li D, Huang GQ, Chen WQ, Chen SS (2012) Study on the cooling effects of urban parks on surrounding environments using Landsat TM data: A case study in Guangzhou, southern China. Int J Remote Sens, 33

Comune di Napoli (2004) Piano Regolatore Generale della città di Napoli

Cormio C, Dicorato M, Minoia A, Trovato M (2003) A regional energy planning methodology including renewable energy sources and environmental constraints. Renew Sustain Energy Rev 7

Diday E, Lemaire J, Pouget J, Testu F (1982) Elements d'analyse des données. Dunod, Parigi

Doherty M, Nakanishi H, Bai X, Meyers J (2009) Relationships between form, morphology, density and energy in urban environments. GEA Background Paper

Feng YY, Zhang LX (2012) Scenario analysis of urban energy saving and carbon emission reduction policies: a case study of Beijing. Resour Sci 34

Fistola R, Mazzeo G (2009) Evoluzione e Morfogenesi Urbana. In: Papa R (ed) Il governo delle trasformazioni urbane e territoriali. Milano, FrancoAngeli/Urbanistica

Forman RT (1995) Land mosaics: the ecology of landscapes and regions. Cambridge University Press, Cambridge

Gargiulo C, Papa R (1995) Strumenti di supporto alle decisioni per il governo delle trasformazioni metropolitane: una proposta di metodo. In: Gargiulo C (ed) La città come sistema a rete – innovazione, trasformazione, governo, Di.PI.S.T. Collana di Studi di Urbanistica, F. Giannini & Figli

Gielen D, Chen CH (2001) The CO_2 emission reduction benefits of Chinese energy policies and environmental policies: a case study for Shanghai, period 1995–2020. Ecol Econ 39

Hadley SW, Short W (2001) Electricity sector analysis in the clean energy futures study. Energy Policy 29

Hoffman KC, Wood DO (1976) Energy system modeling and forecasting. Annu rev energy 1

Howard B, Parshall L, Thompson J, Hammer S, Dickinson J, Modi V (2012) USA Spatial distribution of urban building energy consumption by end use. Energy Build 45

Kanudia A, Loulou R (1998) Robust responses to climate change via stochastic MARKAL: the case of Québec. Eur J Oper Res 106

Keyes DL (1983) Reducing travel & fuel use trough urban planning. In: Burchell RW, Listokin D (ed) Energy land use. Rutgers University Press, New Brunswik

Laconte P, Gibson JE, Rapoport A (1982) Human & energy factors in urban planning: a systems approach. Springer, New York

Lebart L, Morineau A, Piron M (1995) Statistique exploratoire multidimensionnelle. Dunod, Parigi
Lin JY, Cao B, Cui SH, Wang W, Bai XM (2010) Evaluating the effectiveness of urban energy conservation and GHG mitigation measures: the case of Xiamen city, China. Energy Policy 38
Liu GY, Yang ZF, Chen B, Zhang LX (2011) Energy-based multiple spatial scale analysis framework combining resources and emissions impact for urban ecological economic evaluation, Entropy 13
Lu CY, Zhang XL, He JK (2010) A CGE analysis to study the impacts of energy investment on economic growth and carbon dioxide emission: a case of Shaanxi Province in western China. Energy 35
Lynch K (1961) Good city form. The MIT press, Cambridge
Mega VP (2005) Sustainable development, energy and the city: a civilization of concepts and actions. Springer, USA
Mirasgedis S, Sarafidis Y, Georgopoulou E, Kotroni V, Lagouvardos K, Lalas DP (2007) Modeling framework for estimating impacts of climate change on electricity demand at regional level: case of Greece. Energy Convers Manage 48(15)
Nakata T (2004) Energy-economic models and the environment. Prog Energy Combust Sci 30
Papa R, Gargiulo C, Zucaro F (2014) Climate Change and Energy Sustainability. Which Innovations in European Strategies and Plans. TeMA Journal of land-use, mobility and environment, special issue
Phdungsilp A (2010) Integrated energy and carbon modeling with a decision support system: policy scenarios for low-carbon city development in Bangkok. Energy Policy 38
Roaf S (2010) The sustainability of high density. In: Ng E (ed) Designing high-density cities for social and environmental sustainability. Earthscan, Londra
Roset J, Vidmar J (2013) Evaluation of simulation tools for assessment of urban form based on physical performance, Environmental Physics For Architects
Sjoberg G (1960) The preindustrial city. New York
Stone PA (1973) The structure, size and costs of urban settlements. Cambridge University Press, UK
Swan LG, Ugursal VI (2009) Modelling of end-use energy consumption in the residential sector: a review of modelling techniques. Renew Sustain Energy Rev 13(8)
Turner BL (1990) The earth as transformed by human action. University of Cambridge, New York
Turton H (2008) ECLIPSE: an integrated energy-economy model for climate policy and scenario analysis. Energy 33
UN Population Fund (2007) State of the world population 2007. Unleashing the potential of urban growth. New York
UN Population Division (2011) World population prospects—the 2010 revision. New York
Xu Y, Masui T (2009) Local air pollutant emission reduction and ancillary carbon benefits of SO_2 control policies: application of AIM/CGE model to China. Eur J Oper Res 198
Yamamoto T, DehghaniSanij H, Rasiah V (2004) Assessment of evapotranspiration estimation models for use in semi-arid environments. Agricult Water Manage 64
Zhang L, Yang Z, Liang J, Cai Y (2011) Spatial variation and distribution of urban energy consumptions from cities in China. Energies 4(1)
Zhao F (2012) Agent-based modeling of commercial building stocks for energy policy and demand response analysis. Georgia Institute of Technology

Planning for the Conservation of Historic Districts in Sardinia, Italy

Strategic and Energy Efficiency-Related Issues, and an Ontological Approach Concerning a Small Town

Sabrina Lai, Federica Leone and Corrado Zoppi

Abstract A comparison between Sardinian strategic plans (SPs) and implementation plans of historic centers (IPHCs) shows that a general lack of coordination and integration among these municipal planning instruments and a sort of a communicative short circuit are taking place. On the one hand, SPs tend to neglect the importance and the intrinsic value of cultural heritage within historic districts, and, consequently, to undervalue the systemic and general potential of interventions (often limited to punctual and fragmented restorations of buildings) in the historic centers; and, on the other hand, IPHCs propose analyses of municipal historic settlement systems characterized by excessively philological and self-referential attitudes. This paper proposes a discussion on the definition and implementation of IPHCs with the general goal of orienting their conservative character, mainly based on the urban settlement system's restoration and restructuring, in order to generate conditions favorable to local economic and social development, following the strategic planning conceptual framework. Moreover, within the framework of the Regional Landscape Plan (RLP), and after providing the reader with a thorough presentation of some important technical issues related to IPHCs and a discussion on the semantics of the term "ontology," this paper discusses some key points concerning the ontology of the IPHCs procedure, that is the spatial analysis of the IPHCs and implied planning measures.

S. Lai (✉)
Dipartimento di Ingegneria Civile, Ambientale e Architettura, Università di Cagliari, Via Marengo 2, Cagliari, Italy
e-mail: sabrinalai@unica.it

F. Leone · C. Zoppi
Dipartimento di Ingegneria Civile, Ambientale e Architettura, University of Cagliari, Cagliari, Italy
e-mail: federicaleone@unica.it

C. Zoppi
e-mail: zoppi@unica.it

R. Papa and R. Fistola (eds.), *Smart Energy in the Smart City*,
Green Energy and Technology, DOI 10.1007/978-3-319-31157-9_10

1 Introduction

In the framework of regional and urban planning processes in Sardinia, in the context of the Regional Landscape Plan (RLP), established by the Decision of the Sardinian Regional Government (DSRG) no. 36/7 of 5 September 2006, the implementation plans of the historic centers (IPHCs) are planning tools that implement the Planning implementation code (PIC) of the RLP within the so-called "Areas characterized by historic settlements." For these areas, the PIC defines a set of prescriptive rules and planning criteria (articles nos. 51–53 of the part of the PIC related to "Cultural and historic spatial framework," as defined in articles nos. 47–59). More precisely, article no. 52 identifies the IPHC as a plan whose approval is contingent upon the cooperation between the Sardinian regional administration and a given municipality; this is a necessary precondition for a municipality to exert its ruling power over the local transformation processes related to the municipal spatial jurisdiction, which implies a considerable pressure on the local administrators in order to implement valuable and effective planning processes concerning the municipal historic centers.

With its approximately 24,000 km^2, Sardinia is the second-largest island in the Mediterranean Sea; its population density is rather low, as it counts less than 1.7 million inhabitants according to the 2011 National Census. The vast majority of Sardinians live close to the sea, and especially in the two densely urbanized areas around the two major cities (Cagliari and Sassari), while the inner areas of the region, hilly and mostly badly connected to the rest of the island, are sparsely populated. This uneven concentration of population parallels the island's unbalanced economic development, as in coastal areas the majority of economic activities takes place.

It is therefore not surprising, given this context, that the RLP, the first statutory landscape plan with regional dimensions produced in Italy under the new legislation, focused on the coastal zone because of the complexity of development conflicts arising from tourism (on which a large part of the economy of the island relies) and other development, and owing to the fact that thirteen out of the fourteen previous landscape plans covering coastal areas, which contained some restrictions on coastal development, had been quashed between 1998 and 2003.

Following approval of the plan in 2006, restrictions and prohibitions (on development of land and on certain changes in land uses) stemming from the plan are currently in force, in order to protect a part of the island considered economically strategic and environmentally sensitive. Restrictions and prohibitions are set out by the plan by means of a system of rules.

The planning activity of the regional administration of Sardinia has undergone a deep change after the approval of the RLP, which establishes the directions for nearly any future planning activity in Sardinia, and requires that actual sectoral and local plans, as well as plans for protected areas, be changed to comply with its directions.

As a consequence, in the planning processes of the IPHCs, heavily influenced by the control of the technical staff of the regional offices, a strong consistency and implied uniformity emerge and take two major forms: first, a strong attention to historical, typological and morphological characteristics in terms of the territorial analysis of historic urban settlement systems, which are identified by the RLP as "Centers of antique and primary development"; second, a strong prescriptive ruling framework characterized by a markedly conservative attitude.

Sardinian strategic plans (SPs) are studied and defined either by the municipal councils, or, less frequently, by groups of municipalities or province administrations, and funded by means of financial programs stated by: (i) paragraph 1.1 of the Decision of CIPE (The Interministerial Committee for Economic Programming of the Italian government) no. 2004/20, titled "Additional resources, premiality, extraordinary destinations and reserves," and by criteria and procedures established by the Interinstitutional Table for the "Reserve for urban areas" of FAS (the governmental Fund for Underdeveloped Areas) in November 2004, which states that a part of the financial resources of each Italian Region will be utilized to define and study municipal, metropolitan areas' or groups of municipalities' SPs with a resident population of at least 50,000 inhabitants (paragraph B.11, titled "Interventions concerning innovative planning/projecting activity and immaterial investment in urban areas"); (ii) Annex 2 of Note no. 125/GAB of 17 March 2005, titled "Modalities to activate resources," where funds for SPs are increased.

Following these measures, about forty SPs were defined, mostly at the municipal level; such SPs delineate the strategic framework of the ongoing projects of the local contexts in the medium and long run.

How these plans integrate and confront with municipal Masterplans and implementation plans of Masterplans is still an issue of debate and discussion, in theoretical and technical terms.

In the vast majority of cases, the SPs identify and prepare the implementation of strategies that consider the historic centers of municipalities fundamental for the definition of urban development policies aiming at generating, especially in the medium and long run, an effective improvement in quality of urban life. For instance, the SP of the city of Sassari implements a holistic approach and defines a "Direction D9: urban transformation/regeneration" where the interventions related to the historic center are integrated into a system of operations that addresses several important and interdependent issues, such as: hydraulic reclaim of the subsoil, urban refuse collection, urban retail sale organization, e.g. through natural commercial centers, energy saving and efficient use–oriented through appropriate plants (Comune di Sassari 2007: 153).

The SP of the town of Villacidro is rather different from that of Sassari, since it considers urban renewal of its historic center almost exclusively dependent on reuse and requalification of buildings, rather than an issue that should involve city planning in general and systemic terms. Some punctual interventions are planned as "catalyzing operations" implemented through hierarchically-ordered projects named "flag projects," "carrying projects" and "supporting projects." The catalyzing action mainly related to interventions concerning the historic center is named "socially-oriented building sites," which implies the implementation of projects based on the functional rehabilitation of historic buildings located either inside or outside the historic center. In this case, the rehabilitation of the historic center does not have a strategic character, but an ancillary role with respect to higher-order strategic goals (Comune di Villacidro 2008). Between the two extremes represented by the SPs of Sassari and Villacidro lay a range of intermediate situations, among which the SP of the towns of Settimo San Pietro, Sorso and Stintino.

A comparison between SPs and IPHCs shows, rather surprisingly, a general lack of coordination and integration among the processes of definition of these municipal planning instruments, and a sort of a communicative short circuit since, on the one hand, SPs tend to neglect the importance and the intrinsic value of historic and cultural resources of the centers of antique and primary development, and, consequently, to undervalue the systemic and general potential of interventions in the historic centers that are often limited to punctual and fragmented restoration of buildings; and, on the other hand, IPHCs propose analyses of municipal historic settlement systems characterized by excessive philological and self-referential attitudes, which do not take into account planning frameworks that may possibly found, in the medium and long run, realistic plan implementation processes, in terms of the logical framework of the objectives and financial feasibility.

In this very strict normative framework, recent energy efficiency-related opportunities concerning urban renewal in the historic centers are offered by the regional legislation[1] which implements into the Sardinian legislative system the so-called "Piano-casa" ("Housing Plan," HP), which makes it easier to obtain permits for the renovation or the change of use of the buildings within historic centers, provided that the projects improve substantially the energy efficiency-related characteristics of the buildings.

This paper proposes a discussion on the definition and implementation of IPHCs with the general goal of orienting their conservative character, mainly based on the urban settlement system's restoration and renovation, in order to generate conditions favorable to local economic and social development, following the strategic planning conceptual framework.

In the second section, we tentatively define a system of foundational elements for a "Strategic plan of a historic center" (SPHC), by considering some recent experiences as reference points. In the third section, we critically analyze some approved or adopted municipal IPHCs, with reference to these foundational elements. In the same section, we also discuss some possible theoretical and technical-practical paths to go beyond the dualism of the SPs' and IPHCs' planning approaches, and the key-role of the Sardinian legislation related to HP in mitigating the very strict rules concerning urban renewal in the historic centers on the basis of energy efficiency-related technologies.

Moreover, within the framework of the provisions of the RLP, and after providing the reader with a thorough presentation of some important technical issues

[1]Regional Laws no. 2009/4 and no. 2011/21.

related to IPHCs and a discussion on the semantics of the term "ontology" (fourth section), this paper discusses some key points concerning the ontology of the IPHCs procedure, that is the spatial analysis of the IPHCs and implied planning measures (fifth section). This discussion regards the IPHCs' definition procedure in the context of the municipal planning processes stated by the provisions of the PIC of the RLP, with the objective of proposing the ontology as an important contribution to the definition and implementation of this procedure (sixth section).

We develop the ontology on the basis of these normative standpoints, and implement its construction through Protégé, a software program developed by the Stanford Center for Biomedical Informatics Research of Stanford University and freely available at: http://protege.stanford.edu.

2 Foundational Elements of a Strategic Plan of a Historic Center

A critical analysis of the IPHCs of Sardinian municipalities, which reflect properly and precisely the planning guidelines of the regional administration, shows a lack of an explicit strategic vision that should characterize planning processes. To the contrary, IPHCs are mostly concerned with the analysis of the historic urban settlement system and of building typologies, and lay the grounds for projects that mainly consist of limited and conservative interventions. Therefore, there is no evidence of theoretical and practical connections between IPHCs and SPs, even though several SPs, established around just one year before the IPHCs, put in evidence a potential strong link between planning processes related to historic centers and local economic and social development.

Under this perspective, in this section, we propose two parallel analytical grids. The first aims at identifying the strategic potential of the IPHCs, on the basis of a detailed analysis that could substantially increase the strategic effectiveness of these plans, that is, implement their transformation into SPHC. The second grid indicates the strategic priorities related to the historic centers of the SPs of four Sardinian municipalities (Assemini, Cagliari, Elmas and Villacidro) in order to detect if, and to what extent, these priorities are recognized in their IPHCs.

As a reference point, we assume the GOPP (Goal-oriented project planning; Bussi 2004) strategic planning approach, which has recently been used in several planning and programming processes, implemented under the direction of the Sardinian regional administration. The GOPP methodology is based on a logical framework and was adopted in the first place for the definition of the so-called *Integrated development projects* in 2006, which were one of the main technical and financial instruments the Sardinian regional administration used in order to implement the investment policies of the 2000–2006 Regional operational program of the European structural funds. The GOPP approach makes it possible to address, in systemic and structured terms, economic, social and spatial planning and programming issues, and represents a theoretical and technical framework to implement social learning processes aimed at catalyzing local development policies, which imply active participation of the local communities.[2] The logical framework consists of a hierarchically-structured objective tree, which should be defined through an incremental and participatory process by the local communities. This process starts with a SWOT[3] analysis, which recognizes and classifies positive and negative current conditions that characterize the spatial, economic and social situation of the local context (e.g., the municipality) where a public policy is going to be projected and implemented. From the SWOT analysis a hierarchically-ordered problem tree is derived, where the lowest-level problems are identified as the causes of the highest-level problems, in a bottom-up causation chain. The problem tree generates the hierarchically-ordered objective tree that is a mirror copy of the problem tree, where objectives are future positive situations and can be represented as the overcoming of the (current) problems. The systemic and structured representation of the analysis of the local context through the SWOT analysis makes the definition of the problem and objective trees clear and straightforward, which is the logical framework of the public policy at stake. The objective tree and the planning operations, associated to each operational (lowest-level) objective, are named "matrix of the project" (Bussi 2004). Through the "Document integrating the guidelines concerning strategic planning" (Regione Autonoma della Sardegna 2005, pp. 6–7) the GOPP was assumed as the methodological reference point for SPs by the Office of Local public bodies, Financial affairs and Regional and urban planning of the Sardinian regional administration.

Effective strategic approaches to historic centers' strategic planning can be recognized in some recent experiences implemented by the municipalities of Reggio Emilia (Strategic plan for the qualitative enhancement of the historic center; Comune di Reggio Emilia 2005, 2011) and of Vicenza (Masterplan of the historic center of Vicenza; Fantin 2013). In both cases, a strategic approach is explicitly mentioned and implemented into the plans in order to study future scenarios for the historic centers, on the basis of a system of objectives that comes from the overcoming of a system of problems (negative current situations: a problem solving-based goal-oriented approach). A very similar logical framework can also be identified in the debate proposed in a monographic issue of Urbanistica Dossier related to LUDA (Large urban distressed areas).[4] In particular, Mueller et al. (2005) propose a GOPP methodology, the so-called CoSGOP (Collaborative strategic goal-oriented programming), to define strategies and programs to address urban

---

[2]See, for example, the materials available online in the institutional Internet site of the Sardinian regional administration at http://www.regionesardegna.it/ argomenti/programmazione/progettazioneintegrata/comepartecipare/presentazioneprogetti.html (accessed April 2014).

[3]SWOT is the acronym of strengths, weaknesses, opportunities and threats.

[4]The issue describes the experience of "LUDA Project—Improving the quality of life in large urban distressed areas," funded by the European Commission through the Fifth Framework Program—Energy, Environment and Sustainable Development, Key-action 4—City of Tomorrow and Cultural Heritage (Bentivegna 2005).

requalification programs towards cooperation between public and private stakeholders, based on the analysis of case studies related to the urban contexts of Bratislava, Dresda, Edinburgh, Florence, Lisbon and Valenciennes.

Starting from these methodological and conceptual premises, and in order to clarify the analytical framework of our discussion, it is fundamental to understand how the SPs of the four municipalities we consider in order to analyze their IPHCs deal with the issue of their historic centers' strategic planning. All these municipalities used the GOPP methodology to set up the logical framework of their SPs, and they founded their goal-oriented SPs on a context analysis ordered through a SWOT representation. The participatory issue is fundamental and gives the whole plan a bottom-up character which the identification of future urban scenarios is based upon. However, not all the analyzed SPs consider the municipal historic center as a primary issue. For instance, the SPs of the municipalities of Elmas and Assemini identify their respective historic centers as marginal urban areas, and treat them as such in defining the future municipal scenarios, by giving much more importance to other spatial contexts, such as the productive outskirts and the urban parks for open-space recreational activities. From this point of view, the projects related to the historic center are critical, since no strategic future vision concerning the urban historic settlement system can be recognized and integrated into the implementation process of local plans.

The analysis of the GOPP-based strategic approaches to the definition and implementation of plans for the historic centers provides the municipalities with sets of objectives that could be very useful to assess the strategic effectiveness of their IPHCs, and to identify suitable planning paths to improve the quality of life and to catalyze local economic and social development. These sets may eventually make more comprehensive and multifaceted the almost-monotonically conservative and philological character of the current IPHCs.

In our view, a tentative general set of objectives to define and implement SPHC could be the following, based on two general objectives, 1 and 2.

1. General objective: improving the quality of municipal life in the short run, which includes the following specific objectives:

   - promoting the urban system of the historic center and its relationships with the rest of the municipal area;
   - improving the quality of the historic center's built environment, which contributes substantially to the historic center's perceived features, which implies a particular attention to urban maintenance and renewal;
   - increasing the quality and potential of the historic center's public spaces in terms of aesthetic attractiveness, urban fabric and functionality;
   - organizing and increasing the quality of commercial and retail sale activities;
   - promoting the image of the historic center through marketing campaigns related to the local, regional, national and international tourist markets.

2. General objective: promoting local development in the medium and long run, which includes the following specific objectives:

   – making housing in the historic center more interesting and attractive;
   – implementing cooperative actions between the public and private sectors to generate a system of urban services qualitatively valuable and competitive in terms of capacity of responding to social demand, also by means of innovative tertiary activities;
   – improving accessibility, mobility efficiency and the situation of thru-traffic flows in the historic center, by encouraging the use of public transport, pedestrian and cycling paths, and discouraging the use of private transport;
   – implementing participatory practices to support planning processes.

With reference to the second grid of objectives, our discussion is related to the SPs of four Sardinian municipalities, Assemini, Cagliari, Elmas and Villacidro; for each, we consider only the strategic operations concerning their historic centers, which, theoretically, should be identified in their IPHCs as well. However, in the SPs of Elmas and Assemini there is no evidence of strategic operations specifically related to the historic centers, since they have a general spatial scope. However, these strategic operations can possibly have important impacts on the historic centers of the two municipalities. The municipalities of Cagliari and Villacidro identify site-specific policies related to their historic centers, regarded as key parts of the municipal areas in the strategic visions of the plans. Table 1 shows the second grid.

Table 1 Strategic operations of SPs

Municipality	Strategic operation
Assemini	Requalification urban projects: A1. reorganization and enhancement of existing collective services and areas A2. promotion of functional intermix A3. promotion of private and public collaboration Requalification building projects: A4. requalification and restoration of existing heritage in relation to environmentally friendly architecture principles
Cagliari	Characterization and orientation of historic centers neighborhoods, in terms of requalification, in order to increase their attractiveness and use: C1. definition of infrastructural actions in order to revitalize the historic neighborhoods Enhancement of all tourist components and rediscovery of commercial and artisan uses in relation to the municipal center and historic neighborhoods, which represent essential parts of a new supply system based on environment, culture, commerce, tourism and the local products, related to Sardinian tradition and cultural identity: C2. Development of natural shopping centers and small artistic centers, in particular in the historic neighborhoods

(continued)

Table 1 (continued)

Municipality	Strategic operation
Elmas	Requalification of the residential offer: E1. reorganization and enhancement of existing public areas E2. requalification of existing heritage E3. definition of urban planning policies aiming at addressing the needs of the current population (inhabitants and workers) and at promoting the increase in the endowment of public capital Reorganization of internal mobility: E4. usability improvement for disabled consumers E5. construction of pedestrian and cycling paths
Villacidro	Enhancement of the historic center as unique and recognizable system in order to improve usability with particular regard to both development of cultural tourism and increase in urban quality: V1. development of an integrated project concerning the overall relaunch of the historic center V2. integration of new cultural and social functions within the presently underused heritage V3. development incentive of high-quality tourist accommodations V4. development of a total or partial pedestrianization strategy in order to protect a public good and to enhance its peculiarities

The strategic operations are extracted from the SPs of Municipalities of Assemini (Comune di Assemini 2007), Cagliari (Comune di Cagliari 2008), Elmas (Comune di Elmas 2007) and Villacidro (Comune di Villacidro 2008)

As the above-mentioned analyses make it clear, in relation to the majority of examined SPs, the objectives and strategic operations implement the two general goals of the first grid that may possibly improve the living quality standard in the short run and promote local development in the medium and long run. As a result, the historic center acquires a strategic value, shared by local communities and administrators.

3 Analysis of Potentiality and Critical Aspects of IPHCs in Relation to Foundational Elements of a Strategic Plan of a Historic Center and the Key Role of the Energy Issue

The qualitative enhancement of historic centers represents a significant question in the national, regional and local planning contexts. In particular, the Sardinian regional administration identified in the renewal, improvement and promotion of historic centers a very important opportunity to support local, economic and social development. The regional administration in 1998 established Regional Law no. 98/29 titled "Protection and enhancement of the historic centers of Sardinia," and, in 2006, approved the RLP, which recognizes the strategic role of the "Centers of antique and primary development," which are defined through a cooperative planning activity by the regional administration and the municipalities, and are classified as landscape goods; as such, they are subject to a special protection regime under the provisions of the National Law enacted by decree no. 2004/42 on cultural and landscape goods.

On the one hand, this section of the paper aims at identifying potentialities and critical elements of the IPHCs of four Sardinian municipalities, such as Assemini, Cagliari, Elmas and Villacidro, in relation to the objectives defined in section two. A critical analysis is proposed in order to understand if the strategic operations defined in the SPs have been accepted by the IPHCs, lending a strategic value to these plans. From this conceptual viewpoint, it is important to provide reasons for the choice of these four municipalities and to provide some information on the IPHCs. First, we chose these four municipalities because they are the only Sardinian municipalities that developed both an SP and an IPHC in order to make the SPs/IPHCs comparative analysis possible. Second, only Elmas and Villacidro concluded the approval procedure in relation to art. 9 of the Regional Law no. 98/28, as stated by two Decisions of Directorate General for Urban Planning of the Sardinian regional administration, no. 4283 of 28 September 2012 and no. 2407 of 26 October 2010.

Second, the section concludes with a discussion concerning the key role of the energy issue, related to the Sardinian Legislation on the HP, in making easier and faster and, by doing so, more attractive for entrepreneurs, interventions on the buildings of the historic tissues of Sardinian municipalities.

3.1 Potentialities and Critical Elements of Four IPHCs

With reference to the first general objective "improving the quality of municipal life in the short run," which is taken into account much more than the second one, the four IPHCs address in particular the second specific objective ("improving the quality of the historic center's built environment, which contributes substantially to the historic centers' perceived features, which implies a particular attention to urban maintenance and renewal"). The four IPHCs, starting from context analyses that put in evidence a significant decay of historic centers, project interventions aimed at protecting the comprehensive characteristics of their centers of antique and primary development, paying particular attention to conservation of the historic identity of the built environment. On the other hand, not only do these operations focus on urban decay, but also they aim at revitalizing the urban historic contexts characterized by insufficient endowment of public services. The municipalities also address this issue through projects that pursue the fourth specific objective ("organizing and increasing the quality of commercial and retail sale activities"). Moreover, even though all IPHCs give provisions concerning enhancement and strengthening of public spaces in terms of aesthetic attractiveness, urban fabric and functionality, only the municipality of Cagliari emphasizes the strategic importance of this objective in terms of enhancing their attractiveness and functionality. Finally, it is certainly emblematic that the fifth specific objective ("promoting the image of the historic center through marketing campaigns related to the local, regional, national and international tourist markets") is almost totally neglected by the IPHCs. From this point of view, it has to be noticed that, in operational terms, the issue of the centers of antique and primary development is dealt with as an almost-exclusively local question, and, as such, as a challenge that does not deserve but a limited consideration. Only the municipality of Assemini underlines the importance of an increased awareness of the local communities, even though no planning policy is explicitly defined. The first specific objective ("promoting the urban system of the historic center and its relationships with the rest of the municipal area") is addressed, even though in general terms, only by the IPHC of Cagliari, which proposes an analysis of the synergic relationships between historic center and the rest of the city.

In regard to the second general objective "promoting local development in the medium and long run," the analyzed IPHCs do not show interest in strategic visions that go beyond a short-run horizon, with the exception of Cagliari, since they identify, as their only focal point, what indicated by the first specific objective related to the second general objective, that is "making housing in the historic center more interesting and attractive." On the other hand, the municipality of Cagliari promotes interventions and operations aimed at improving the quality of housing in the historic center by boosting not only the stability of the current resident population, but also the demand of new inhabitants, such as students, for houses. The other three specific objectives (namely "implementing cooperative actions between the public and private sectors to generate a system of urban services qualitatively valuable and competitive in terms of capacity of responding to social demand, also by means of innovative tertiary activities," "improving accessibility, mobility efficiency and the situation of thru-traffic flows in the historic center, by encouraging the use of public transport, pedestrian and cycling paths, and discouraging the use of private transport" and "implementing participatory practices to support planning processes") are almost completely ignored, with the exception of Cagliari. The municipality of Cagliari plans to strengthen and redevelop endowment of public services in the historic center, as well as pedestrian and cycling mobility. Finally, the IPHC of Cagliari is built on a participatory process, based upon a set of public debates, and shows a strategic vision related to the local economic and social development in the medium and long run, and, by doing so, this plan goes far beyond the provisions of the regional guidelines. In relation to the second grid, where strategic operations of the four IPHCs in relation to the SPs of the same municipalities are analyzed, the study emphasizes how, with the exception of Cagliari and Assemini, the IPHCs do not transpose integrally the objectives and strategic operations, defined by SPs, within their IPHCs. Moreover, Assemini, Elmas and Villacidro do not mention the existence of an SP for their territory. On the other hand, the municipality of Cagliari takes strategic operations of its SP into consideration, by analyzing the SP within a particular section of the IPHC's report.

In addition, Cagliari defines actions that address the strategic operations, identified by SP, lending a strategic value to IPHC, consistent with the SP.

The IPHC of Villacidro represents an emblematic case because only one of the four strategic operations is accepted and adopted in the IPHC. The position of Assemini municipality is intermediate. Indeed, although the IPHC accepts most of the SP's strategic objectives, with exception of one, it does not define any specific operations related to such objectives.

3.2 The Energy Issue

We have already put in evidence that lack of strategic visions characterizes the approach of Sardinian municipalities to the IPHCs, mainly because of the need to comply with the strict rules of the RLP that require strong consistency and implied uniformity in terms of historical, typological and morphological characteristics of proposed interventions related to existing buildings and urban tissues.

However, the regional legislation concerning interventions on the building stock, with particular reference to the housing stock, the so-called HP legislation, defines a general strategic approach which aims at supporting a comprehensive qualitative enhancement based on interventions concerning the improvement of energy efficiency and saving of the building.

The approach established by the HP legislation implicitly assumes that the RLP fails to implement effective processes to improve the urban quality of historic centers, and states that a strategy to make it happen can be defined through the energy issue. The Sardinian Parliament approved in 2009 Regional Law no. 4[5] that implements a simplified procedure to increase the volume of existing buildings into the technical and administrative rules concerning building activities.

This law states that interventions on existing buildings located in the historic centers of Sardinian municipalities implying an increase in the built volume by up to 20 percent are allowed, provided that the buildings are recent (newer than 50 years old) and not significant in terms of their monumentality (art. 2, paragraphs 1 and 7). A necessary condition for these interventions is that the building which results from the volumetric increase has to comply with the provisions of Law enacted by decree no. 2005/192, which transposes Directive no. 2002/91/EC on the energy performance of buildings into the Italian legislation. Moreover, paragraph 3 of art. 2 establishes that the allowed increase in volume of an existing building can reach up to 30 percent (50 percent more) if the entire resulting building either

[5]Regional Law no. 2009/4 on "Disposizioni straordinarie per il sostegno dell'economia mediante il rilancio del settore edilizio e per la promozione di interventi e programmi di valenza strategica per lo sviluppo" ["Extraordinary regulations to support the regional economy by means of a new development of the building sector, and to promote interventions and programs strategically connected to local development"].

complies with the provisions of Law enacted by decree no. 2005/192 or presents a decrease in energy demand greater than 15 percent.

The regional government has recently issued a new law proposal (no. 130/A, issued on October 23, 2014) which is consistent with Law no. 2009/4 in regard to energy saving requirements as necessary conditions to comply with for volumetric increases of existing buildings.[6]

Therefore, the energy issue matters in terms of implementation of strategic policies into planning approaches related to Sardinian historic centers. Planning policies based on strategic approaches related to energy saving and efficiency could effectively integrate the fragmented and philological provisions of the IPHCs and orient them towards strategic visions focused on local economic and social development and enhancement of the comprehensive urban quality.

4 Ontology and Ontologies

A generally-accepted meaning of the term "ontology" in contemporary theoretical debates in urban and regional planning is "discussion of the substance of an object," that is a discussion of the most important characteristics of its essence, especially in epistemological debates. For instance, Hillier (2010) points out that, according to some contemporary scholars, "ontology" indicates the paradigm of "relational ontology," that is a discussion of the substance of the relations between agents and structures (capital, social classes, agreed-upon semantics, etc.) which do not possess their own essence, formed only through their being in relation. Moreover, Hillier stresses, with reference to DeLanda (2006), that the reference point of planning practice should be the observation and analysis of the relations between the elements which constitute the empirical reality (e.g., agents and structures), whose existence does not depend on the fact that human beings perceive it. These relations generate emergences, that is unexpected phenomena, for those who are familiar with the single elements but who are not aware of their mutual relations as well: the ontology of agents, structures and relations is a "realistic ontology" of the scientific paradigm of the (planning) disciplinary paradigm, which is based on the empirical analysis of the relation. These emergences have an autonomous existence with respect to agents and structures. The realistic ontology (the ontology of relations) is an ontology of the reality (Hillier 2010), and a scholar's disciplinary role is to be aware of and to describe this reality, by identifying and analyzing its relational substance.

The "substantial" attribute of the term "ontology" leads to an effective comprehensive view of the contemporary debate concerning ontology and ontologies. In this context, ontology is not referred to, according to the meaning described so far, as one

[6]The law proposal states stricter rules than Law no. 2009/4 since it establishes that volumetric increases of existing buildings have to comply with the implementation codes of the IPHCs as well. This implies that no intervention would be allowed were the municipality's IPHC not approved.

or a set of conditions which define the substance of a concept (reality–agents, structures, relations) as much as the substance of its agreed-upon representation, that is its formal definition. Ontology is the identification of a concept, of a domain—in other words, the cognitive contents that a set of agents identify as the particular characteristics of a domain. Therefore, ontology is not connected to substance, that is the essence of an object, as much as to the agreed-upon available knowledge (scientific, technical, based on traditions and on common sense, etc.) concerning an object.

Formal ontologies are not connected to substance or to essence, but to the essence of representations, or definitions; that is, they propose an agreement on cognitive contents, rather than the substance analysis of an object. According to Smith (2003), ontologies are descriptions of domains of objects as closed data models whose nodes define concepts. These concepts are strategically identified and make sense only in the context of the universe they try to model. Moreover, Smith illustrates that, historically, the use of formal ontologies comes from the fact that several disciplines are experiencing a dramatic Tower of Babel syndrome which needs to be addressed somehow. Those who deal with complex systems of data and knowledge have peculiar and often idiosyncratic frameworks for representing information. The semantics used for the same term may vary, or the semantics for different terms could take the same meanings. Formal ontologies could make it easier to deal with this syndrome. Also according to Guarino (1998), as quoted by Pretorius (2004), a formal ontology is a projected representation which consists of a specific agreed-upon set of words which describe concepts belonging to a knowledge domain and a set of agreed-upon propositions concerning the meanings of these word as well. Pretorius agrees with Smith since, in his view, the concept of ontology originated in the field of artificial intelligence.

5 The Spatial Analysis of Historic Centers and Implied Planning Measures

The starting point for describing the PIC-based technical approach to the definition and implementation of an ontology of the spatial analysis and implied planning measures of IPHCs is represented by the provisions of art. 52 of the RLP, titled "Areas characterized by historic settlements. Prescriptive rules."

The article is articulated into four parts. First, it deals with the question of spatial analysis, which has to be carried out through: (i). a description of comprehensive urban characteristics of the municipal area, such as orography, water resources and hydrography, natural resources, settlement system(s), municipal walls and ramparts and urban landmarks; (ii). an analytic description of the historic centers' built environment and open spaces, which is implemented through detailed and accurate records related to each public and private block, buildings and built structures, private yards and public spaces (squares, parks, gardens, widening-road areas, etc.), with particular reference to the built environment's degradation and obsolescence state.

Second, a classification of the allowed interventions is proposed, which makes reference to the category definitions of art. 3, paragraph 1, of Law enacted by decree of the President of the Italian Republic no. 2001/380. The allowed interventions are limited to ordinary and extraordinary maintenance and conservative restoration and refurbishment.[7] In this way, the RLP states, on the one hand, that the spatial analysis should result in prescriptive statements which constitute the normative core of IPHCs, and, on the other hand, that these statements should comply with a general conservative approach, which is implemented through very restrictive rules related to allowed operations on the built and unbuilt environment.

Third, the most invasive and transformative interventions, that is radical renovation of the existing buildings and change of parts of the urban historic morphology (block shapes, streets, squares, etc.), are limited to the parts of the historic centers where spatial analysis puts in evidence that the original characteristics of buildings and urban morphology are dramatically and irreversibly altered.

Even in the case of radical interventions, art. 52 states that renovation projects have to focus on the general objective of defining transformations which will eventually generate conditions consistent with the not-irreversibly altered or well preserved parts of the historic urban settlements and morphology.

Fourth, the IPHCs should rule over the functional uses of the buildings belonging to the historic centers, by assuming, as a general normative point of reference, that the residential uses should be the most preferred, and, as a consequence, that houses should maintain their residential use status and the number of the existing residential units should possibly increase as a consequence of the IPCHs' operations. Moreover, municipalities are encouraged by the provisions of art. 52 to strengthen the availability of public services for residents so that historic centers may become more attractive for people and families who are deciding about the location of their houses. Alternatives to residential uses are identified as accommodation activities for tourists, such as hotels, residences and multi-building hotels, private and public service activities, medical services, craftmanship workshops and retail shops. Finally, art. 52 indicates that a careful consideration of the opportunity of maintaining unbuilt the areas which result from the demolition of ruins should be taken into account in the rules of IPHCs. Hence, the provisions of article 52 of the PIC of the RLP state that the IPHCs should be based on spatial analysis, whose interpretation of the urban settlements and morphology of historic centers is the point of reference for the conservative and transformative operations established by IPHCs. This is the conceptual framework of the ontology we propose in the following section.

[7]For the normative category definitions of these interventions see points "d," "e" and "f" of art. 3, paragraph 1, of Law enacted by decree of the President of the Italian Republic no. 2001/380.

6 An Ontology of the Spatial Analysis

As stated in the previous section, the spatial analysis for an IPHC comprises, among other, an analytic description of the historic centers' built environment and open spaces, to be implemented through detailed and accurate records concerning each spatial unit[8] and detailing both conditions and provisions for each element in that unit. In this section, we describe how an ontological approach can support this task.

The ontology of the domain "Description of the historic centers' built environment for an IPHCs" was developed according to the phases suggested by guidance documents and methodological reports produced by the Ordnance Survey, according to which the process whereby an ontology is built can be broken down into a series of steps, the first being the identification of the purpose of the ontology and of its scope. These two aspects are crucial for ensuring both that the ontology is correctly formalized and that it is useful, meaning that it contains only those concepts, relationships and constraints that are judged to be relevant, with regard to the possible ways in which the ontology can be used. With reference to the first point (purpose), the ontology here proposed aims to represent and describe the built environment of a given historic center in order to support the making of IPHCs. With reference to the second point (scope), this ontology aims at providing plan makers with a robust descriptive framework on which they can ground the selection of allowed building activities. Once purpose and scope have been established, the following step consists of the construction of a glossary ("knowledge glossary"), comprising two tables. The first table ("table of concepts") contains a list of core and secondary concepts together with their definition in natural language, and of the source of the definition. The second table ("table of relations") lists and defines, again in terms of natural language only, relationships between concepts pertaining to the domain. The definition of concepts and relationships was based on documentary sources only, which comprise national and regional pieces of legislation and technical documents. The use of such sources was an attempt to ensure that definitions are agreed upon by, and shared among, domain experts. Through a series of checks at internal level of definitions and relations, the two tables were enhanced and modified recursively.

The knowledge glossary, in spite of its being an explicit and shared (at least, potentially) specification of the entities that constitute the abstract and simplified model of the domain, is not a formal specification of the domain itself, up to this point. For the computational aspects, the software program Protégé was next used;

[8]Spatial units are identified through the first step of the required spatial analysis, that is upon analysis and description of the urban morphology, of typological, architectural and landscape features of the built environment in the historic district, also taking into account aspects such as blocks layout, streets' and paths' hierarchies, cadastral subdivisions and properties. The identification of spatial units (which can comprise more than only a single building) is of outmost importance because each spatial unit is treated as the minimum unit for projects and interventions, meaning that building activities should concern the spatial unit as a whole. For this reason, we refer to it as "minimum unit".

concepts were arranged in a hierarchical manner, that is they were organized and grouped into classes and subclasses on the basis of the relationship "is a." To state that a given element belongs to a subclass of a class is in fact tantamount to affirming that this element belongs to the class and therefore that it inherits its properties. In this way, key concepts identified in the previous phase were first organized in a taxonomy according to the following classes: "Activity," "Assessment," "Building," "Obsolescence," "Material," Minimum Unit," "Part," "Use" and then the corresponding definitions were inserted for each class and subclass.

Next, for each class and subclass, appropriate slots were defined and created. Slots can be used either to characterize the elements of a class by means of attributes of different types (for instance, string, integer, float, enumerated), or to describe the relationships between instances, which are defined as the elements belonging to a given class; in other words, features represent the finest level of granularity and form the basis of the hierarchy. As an example, the slots assigned to the class "Building" are: "Cadastral code," "Code," "Construction year," "Area sqm," "Height," "Volume cubm," "Storeys no." and "Zone." These are all descriptive attributes, all required and having single cardinality (meaning that only one value for each attribute is allowed); the type of the first, second and third is "string" (meaning that any alphanumeric string is allowed), while that of the fourth, fifth and sixth is "float"; the type of the seventh, accounting for the number of storeys, is integer, and finally the type of the eighth is enumerated, since an IPHC should concern only areas classed in the zoning scheme of the Masterplan as "A" (historic center zone) or "B" (residential completion zone) or "S" (essential services for a residential zone, such as for instance open-space leisure areas or parking spaces). As far as the other five ("Has assessment," "Has obsolescence," "Has part," "Has use," "Is contained in") are concerned, they make the relations between the class "Building" on the one hand and, respectively, the classes "Assessment," "Obsolescence," "Part," "Use" and "Minimum Unit" (or subclasses thereof) on the other hand explicit. Two of these five slots have single cardinality, that is, an instance of the class "Building" can have only one assessment value[9] and it can belong only to a given minimum unit, while the remaining three have multiple cardinality, since a building can serve more than one purpose (for instance, a part of it could be residential and a part commercial), it can be broken down into several components or parts (for instance, roof, walls, porch) and it can have (in our framework) three different values of obsolescence (a structural/physical obsolescence, a functional obsolescence, and an image obsolescence).

The construction of the ontology continues with the creation of instances and the filling-in of the values of the slots, and this is done by entering these values in

[9]By "assessment" we mean the final judgment on whether a single building in the historic fabric has retained traditional characters, which have to be preserved, or it has been altered either reversibly or permanently. For modern buildings, this judgment states whether they are consistent whit the historic fabric or not. The assessment is important because it entails strong directions towards allowed building activities and operations.

appropriate forms that prevent users from including values which are inconsistent with the ontological hierarchy previously defined.

Once the instances have been created and their slots have been filled in, the ontology is fully and formally defined, even though it can be continually adjusted and integrated; moreover, the ontology can be represented graphically as a graph tree in which classes, subclasses and instances are represented as nodes, and relations as arches, allowing users to navigate the whole hierarchy, or only a part. Graphs can be tailored to the user's needs, meaning that the user can choose whether to display all of the ontology, or only a part of it, by selecting the nodes to be represented or by filtering the relationships to be shown, which allows for a more effective and more understandable representation and exploration in case of complex ontologies.

7 Conclusions

This paper discusses the definition of planning processes concerning historic centers in terms of sustainable development. There are two parallel analyses proposed in the paper. One identifies the potential strategic values that convert IPHCs into SPHCs. The other concerns the relations between SPs and IPHCs of four Sardinian municipalities.

From this methodological viewpoint, the first above-mentioned analysis has emphasized six critical questions. The first issue regards the difficulty of small municipalities in defining strategic operations. Indeed, this situation is clearly traceable through a comparison between Cagliari, the Sardinian regional capital, and the other municipalities. In addition, the issue is particularly critical in relation to the second general objective that concerns the medium and long run. Under this perspective, the increased availability of resources in terms of skills and money could favor the development of a systemic vision that lends strategic value to SPs in the case of small municipalities.

The second issue concerns the conceptual gap between the strategic operations, defined by the four SPs, and strategies and actions, identified by IPHCs. With the exception of Cagliari and, partially, of Assemini, not only do the IPHCs ignore the existence of an SP, but also they neglect the most part of strategic operations, which puts in evidence a missing dialogue among the public departments at different scales of planning. The third issue concerns the lack of strategy of the current IPHCs, which only aim at fulfilling some practical questions targeted by various departments of the regional public administration, identified by the repletion of such task as filling in specific forms related to the architectural analysis of buildings, and the detailed description of blocks and parcels. Therefore, technical guidelines and handbooks were issued by the Sardinian regional administration related to IPHCs. Fourth, following this conceptual vision, historic centers represent single entities within municipal conurbations, which raises the fourth issue: except for Cagliari, all IPHCs neglect the identification of functional relations among the historic center, the rest of the urban tissue and the peripheral and periurban areas of a municipality. This problem originates from the lack of a systemic vision and strategic values that convert IPHCs into SPHCs. Fifth, in the light of our discussion, a theoretical and practical path that may possibly help overcome the almost-absolute lack of strategy of the IPHCs, and, at the same time, be a promising future development of our research, could be the radical redefinition of the actual IPHCs through a thorough and analytical implementation of the GOPP methodology, and the identification of the logical frameworks of the IPHCs.

Sixth is the energy issue. This issue could play as a paragon to implement strategic visions related to other strategic questions: public services, recreation, school, etc. A properly-designed premiality system, not necessarily consisting of volumetric increases, implemented in a fair and participated way, could prove very useful in increasing the effectiveness of the IPHCs.

Moreover, with the second analysis, this paper has attempted to build an ontological representation of the analysis of the built environment of historic centers as a means to support an IPHC, based on the interpretation of normative acts and documents. The ontology can be useful for four reasons. First, this approach provides all the participants involved in the making and implementation of the plan (be they private citizens, planners, public officers or decision makers) with a better understanding of the domain of interest, through an iterative learning process that can continually be refined; this learning process is, in principle, inclusive, because the construction of the glossary can be improved by integrating the definition of concepts, relations, and descriptive attributes, here carried out solely on the basis of documentary sources in a participative way. Such a collective conceptualization of the domain would also greatly improve the chances of sharing and reusing the ontology in the domain field.

Second, since the ontology here proposed is a domain ontology, therefore aimed at structuring, representing and communicating knowledge on a specific area of interest, the ontology can be updated, refined and reused in the given domain, and it can lay the bases for the development of task-dependent or application-oriented ontologies in the same domain, for instance focusing on administrative and procedural tasks.

Third, the ontology effectively supports, as desired, the analytic description of the historic centers' built environment and open spaces.

Fourth, the ontological approach here utilized can be readily exported; although the domain ontology here developed is grounded on the normative framework that regulates IPCHs in Italy, and more precisely in Sardinia, this ontology could be easily reused to describe in detail features and characteristics of historical settlements.

Finally, an important limitation to exportability of the ontology in other contexts lies in the fact that the definitions of concepts are based, at least to some extent, on Italian laws and regulations, technical documents and vocabularies; for this reason, the table of concepts was built in Italian in the first place, and afterwards translated into English, which can cause some issues of semantic precision in English.

Acknowledgments Sabrina Lai, Federica Leone and Corrado Zoppi have made substantial contributions to the paper's conception and design, background and concluding remarks (Sects. 1 and 7). Corrado Zoppi defined the question of the problematic dualism of strategic plans and IPHCs (Sect. 3.2 of Sect. 3 and Sect. 5). Federica Leone and Corrado Zoppi designed the discussion on the issue of the founding elements of a strategic plan of a historic center (Sect. 2). Federica Leone analyzed potentialities and critical points of the IPHCs with respect to these founding elements (Sects. 3 and 3.1). Sabrina Lai and Corrado Zoppi designed the discussion on differences between ontology and ontologies (Sect. 4). Sabrina Lai designed the conceptual framework and implementation of the formal ontology related to the spatial analysis of IPHCs (Sect. 6)

References

Bentivegna V (2005) Gli aspetti della governance urbana nelle LUDA [The aspects concerning urban governance in the LUDA]. Urbanistica Dossier 8(74):5–11

Bussi F (2004) Progettazione e valutazione di progetti con il Quadro Logico (Projecting and project assessment through the logical framework). Resource document. http://www.crotoneuropa.it/documenti/strumenti/1_Bussi_Progettazione_E_Valutazione_Di_Progetti_Con_Il_Quadro_Logico.pdf. Accessed 10 Mar 2015

Comune di Assemini (2007) Piano strategico (Strategic plan). Resource document. http://www.sardegnaterritorio.it/documenti/6_288_20100525114322.zip. Accessed 10 Mar 2015

Comune di Cagliari (2008) Piano strategico di Cagliari (Strategic plan of Cagliari). Resource document. http://www.sardegnaterritorio.it/documenti/6_288_20100525113848.zip. Accessed 10 Mar 2015

Comune di Elmas (2007) Piano strategico (Strategic plan). Resource document. http://www.sardegnaterritorio.it/documenti/6_288_20100525115755.zip. Accessed 10 Mar 2015

Comune di Reggio Emilia (2005) Piano strategico per la valorizzazione della Città storica di Reggio Emilia (Strategic plan for the enhancement of the historic City of Reggio Emilia). Resource document. http://www.municipio.re.it/download/cittaStorica/doc_indirizzi_piano_strategico.pdf. Accessed 10 Mar 2015

Comune di Reggio Emilia (2011) Piano strutturale comunale di Reggio Emilia, P3.2 Strategie e azioni per la Città storica (Municipal structural plan of Reggio Emilia, p.3.2 Strategies and actions for the historic City). Resource document. http://rigenerazione-strumenti.comune.re.it/strumenti-di-pianificazione/psc-piano-strutturale-comunale-vigente/, http://www.comune.re.it/download/pscre/1-PSC/Elaborati_tecnici/P3.2_Strategie_e_azioni_per_la_citta_storica.pdf. Accessed 10 March 2015

Comune di Sassari (2007) Piano strategico della Città di Sassari: Progetti—Allegato 3 [Strategic plan of the City of Sassari: Projects—Annex 3]. Resource document. http://www.sardegnaterritorio.it/documenti/6_83_20100526092101.zip. Accessed 10 Mar 2015

Comune di Villacidro (2008) Piano strategico del comune di Villacidro (Strategic plan of the municipality of Villacidro). Resource document. http://www.regione.sardegna.it/documenti/ps_Villacidro.zip. Accessed 10 Mar 2015

DeLanda M (2006) A new philosophy of society: assemblage theory and social complexity. Continuum, New York

Fantin M (2013) Il Masterplan del Centro storico di Vicenza [The Masterplan of Vicenza's historic district]. Urbanistica 150–151:1–34

Guarino N (1998) Formal ontology in information systems. In: Guarino N (ed) Formal ontology in information systems. In: Proceedings of FOIS'98. IOS Press, Amsterdam, pp 3–15

Hillier J (2010) Introduction to Part Two. In: Hillier J, Healey P (eds) The Ashgate research companion to planning theory. Ashgate Publishing Limited, Farnham, pp 235–250

Mueller B, Curwell S, Turner J (2005) Un modello per il miglioramento delle LUDA: lo sviluppo del collaborative strategic goal oriented programming (A model for improving the LUDA: the implementation of the collaborative strategic goal oriented programming). Urbanistica Dossier 8(74):14–19

Pretorius AJ (2004) Ontologies: introduction and overview. Resource document. http://www.starlab.vub.ac.be/teaching/Ontologies_Intr_Overv.pdf. Accessed 10 Mar 2015

Regione Autonoma della Sardegna (2005) Pianificazione Strategica. Documento integrativo delle linee guida in materia di pianificazione strategica di cui alla nota dell'Assessore Regionale degli Enti Locali, Finanze ed Urbanistica n. 125/GAB del 17 Mar 2005. Resource document. http://www.sardegnaterritorio.it/documenti/6_83_20061023163522.pdf. Accessed 10 Mar 2015

Smith B (2003) Ontology. In: Floridi L (ed) Blackwell guide to the philosophy of computing and information. Blackwell, Oxford, pp 155–166

Cities Dealing with Energy Issues and Climate-Related Impacts: Approaches, Strategies and Tools for a Sustainable Urban Development

Adriana Galderisi, Giuseppe Mazzeo and Fulvia Pinto

Abstract In respect to the numerous environmental challenges that cities have to face in the next future, namely to energy issues and to the increasing impacts of climate change, this contribution aims at providing innovative approaches, strategies and tools for improving cities' response to climate-related issues. According to this aim, the first chapter focuses on the two concepts of vulnerability and resilience, both of them concerned with how systems react to internal and external pressures, highlighting the complementarities between a vulnerability and a resilience-based approach for promoting a sustainable, safe and less energy consuming urban development; then, focusing on current strategies addressing mitigation and adaptation to climate change, the main relationships among current strategies and spatial planning will be explored; finally, an innovative tool for the evaluation of spatial plans at local scale and of built-up areas, capable to assess the overall sustainability of cities' development paths in the face of climate change, will be presented.

A. Galderisi (✉)
DICEA, Università di Napoli Federico II, Naples, Italy
e-mail: galderis@unina.it

G. Mazzeo
ISSM, Consiglio Nazionale delle Ricerche, Naples, Italy
e-mail: gimazzeo@unina.it

F. Pinto
DASTU, Politecnico di Milano, Milan, Italy
e-mail: fulvia.pinto@polimi.it

R. Papa and R. Fistola (eds.), *Smart Energy in the Smart City*,
Green Energy and Technology, DOI 10.1007/978-3-319-31157-9_11

1 Resilience and Vulnerability: Key-Concepts for a Sustainable Urban Development

1.1 Urban Development, Energy Consumptions and Climate Change

Nowadays, cities are struggling to cope with numerous and interwoven phenomena that significantly threaten their future development. Demographic, environmental and economic changes currently affect urban areas all around the world and some of these structural changes will probably increase in the next future.

Europe is one of the most urbanized regions in the world: at present, the 73 % of European population is living in urban areas and it "is expected to be over 80 % urban by 2050" (UN 2014).

The constant growth of urban population, combined with the patterns of urban development, is inducing more and more severe impacts on natural resources; moreover, cities are responsible for about the 69 % of all GHG emissions and, according to the demographic trends and to urban lifestyle, this percentage could significantly increase in the next future.

Shifting the focus towards the environmental challenges that cities have to deal with, it is worth emphasizing that available disaster statistics show an increase in the total number of natural disasters in recent decades and, above all, of the hydrological and meteorological events (floods, mudslides, droughts, changes in temperature and heat waves) in respect to the geophysical hazards. The reports of the Intergovernmental Panel on Climate Change (IPCC) clearly underline that the intensification of these events has to be interpreted as an outcome of the climatic variations that largely depend, in turn, on human activities. Moreover, it is widely recognized that the increasing degradation of natural environment, which is also a consequence of the urban development, significantly compromises the potential contribution of natural capital to climate change mitigation and adaptation.

Therefore, the complex interactions between human activities and nature let clearly arise the need for changing our perspective in the face of environmental challenges that cities have to deal with, recognizing that on the one hand, these challenges largely represent an outcome of current urban development pattern and lifestyle; on the other hand, that cities—due to the growing urban population as well as to the significant concentration of assets and economic activities—are also the most powerful engine for promoting economically, socially and environmentally sustainable societies.

Thus, the key question that nowadays is challenging scholars and practitioners is how to improve cities' capacities to deal with the more and more complex interactions among urban development patterns, urban lifestyles, climate change and related disasters in a context of growing uncertainty, reducing current vulnerabilities and moving towards more sustainable patterns? According to numerous institutional documents and to the vast scientific literature focused on this issue, the concept of resilience might represent a relevant scientific approach for understanding the interwoven systems of humans and nature and for dealing with current and emerging environmental challenges. Identified by the Hyogo Framework for Action 2005–2015 as one of the strategic goals for disaster reduction and for a sustainable development, resilience is nowadays a controversial term that, as remarked by Grünewald and Warner (2012), "seems to be going the way of sustainable development or governance, meaning all things to all people, and as a result, there is a risk that it will become an empty shell".

The most recent document on Sustainable Development—"The Future We Want"—emphasizes the close relationships among urban development processes, risks and climate change, underlining "the importance of considering disaster risk reduction, resilience and climate risks in urban planning" (UN 2012) in order to build up sustainable cities. In the same line, some scholars have recently stressed the need for analysing climate change in the wider framework of the Disaster Risk Reduction, placing the latter as "a subset of wider development and sustainability processes" (Kelman et al. 2015).

To achieve this aim, in the following we will deepen the two concepts of resilience and vulnerability, both of them concerned with how systems react to stress factors due to external and internal pressures and widely interpreted as core-concepts, both in disaster risk and in climate change community for driving cities' development towards sustainability goals.

1.2 Resilience and Vulnerability: A Debated Relationship

The concepts of vulnerability and resilience have significantly evolved starting from the Seventies, due to the contribution of different disciplines and the development of theoretical and operational approaches in different domains of practice.

The concept of vulnerability, to which large attention has been devoted in the disaster field and more recently in the field of urban adaptation to climate change, has significantly widened due to the contributions of numerous research domains (from structural to systems engineering; from social to economic science, etc.). Hence, its primary focus has shifted from the physical vulnerability of buildings and infrastructures to a multi-dimensional and dynamic perspective, comprising social, economic and environmental aspects (Birkmann et al. 2013). Furthermore, numerous European Research Projects (e.g. the FP6 Armonia Project, the FP7 Ensure and Move Projects, etc.) have been addressed to deepen the multidimensional concept of vulnerability, highlighting the variability of the different vulnerability facets over time and across different geographical scales (from local to global) (Menoni et al. 2012) and providing methodological and operational tools for assessing the different facets of vulnerability to heterogeneous hazard factors, including climate-related hazards.

Therefore, despite the lack of a standard procedure, vulnerability analysis and assessment are largely established as risk analysis tools as mirrored in both the EU strategy on climate adaptation and in the European Programme for Critical Infrastructure Protection (EPCIP).

Unfortunately, most of the recent strategies and action plans addressed to counterbalance climate change do not seem to benefit significantly from the results already achieved in the field of vulnerability analysis (Galderisi 2014). The latter is still largely considered as a technical and sectoral knowledge mainly bounded into the risk community, struggling to overcome the disciplinary boundaries and finding little or no place in urban planning culture.

In addition, the resilience concept has significantly widened in the last decades due to the contribution of different disciplines (from physics to psychology, from ecology to management science). Indeed, it has enlarged its focus from natural ecosystems to complex adaptive systems and its perspective from a stability-based concept, addressed to guarantee a timely and effective bounce-back of a system after a disturbance, to a multiple equilibrium perspective, addressed to enhance the capacity of complex adaptive systems to move towards new equilibrium states (bounce forward) in the face of internal and external pressures.

Following the Hyogo Framework for Action 2005–2015, a growing attention to the concept of Resilience has been paid by organizations and institutions at all levels (from global to local). Most of the official documents on Disaster Risk Reduction and on Climate Change Adaptations on a global, European and national level, provide an interpretation of Resilience as an overall desirable condition, an ultimate goal (Fekete et al. 2014) or, even, a "by-product" of many different actions, policies and actions, focusing on multiple objectives (Pizzo 2014).

Thus, despite the large attention nowadays devoted to the resilience concept also in the urban planning domain (Davoudi 2012), the numerous available definitions of resilience and of resilient city as well (Papa et al. 2015) and the many on-going European Research Projects focused on resilience (e.g. emBRACE, RAMSES, TURAS, etc.), so far Resilience seems widely to be interpreted as an umbrella concept, while an effective translation of this concept into operational terms is still missing. Most of the numerous institutional documents that, at different levels (international, national or local), envisage strategies and policies for climate change mitigation and adaptation, largely embrace the idea that vulnerability and resilience represent two sides of the same coin and refer to the vulnerability analysis and assessment as operational tools capable of providing effective measures for building up resilient cities (Fekete et al. 2014).

Nevertheless, this idea does not adequately reflect the heterogeneous schools of thought that have been so far developed: the relationship between these concepts is still a controversial matter, a consensus is far from being achieved and the different positions range from a flip-side approach, to the idea of two separate concepts, although partially overlapping, which may lead to different outcomes.

1.3 Vulnerability and Resilience as Complementary Concepts

Despite the heterogeneity of approaches and interpretations arising from the vast scientific literature on the relationships between vulnerability and resilience, nowadays numerous scholars, overcoming the flip-side approach, emphasize that these concepts, although different, should be considered as complementary (Turner 2010) in order to drive urban development towards sustainability and to frame effectively urban strategies to counterbalance climate-related issues.

To better understand such a position, first it is worth emphasizing that—even though the concept of vulnerability has significantly enlarged its focus—vulnerability analysis and assessment are generally focused on the fragilities of a given system and aim to support strategies and actions capable of improving the intrinsic features of the exposed elements and systems in the face of expected or likely events. On the opposite, the concept of resilience is mainly focused on the adaptive characteristics of a system, an individual or a community, in order to empower them to better cope with unknown problems, unexpected phenomena (Fekete et al. 2014), such as those arising from the complex and interconnected dynamics of urban development and climate change.

Moreover, as mentioned above, along its evolution path the focus of the resilience concept has progressively shifted from a bounce-back towards a bounce-forward perspective. Resilience is nowadays generally understood as the result of the "dynamic interplay" among "persistence, adaptability and transformability across multiple scales" (Folke et al. 2010), based on a continuous learning process that allows systems to resist and absorb disturbances and to move "towards a more desirable trajectory" (Davoudi et al. 2013).

On summing up, it is possible to envisage at least two main reasons for looking at vulnerability and resilience as complementary concepts:

- a vulnerability-based approach, focusing on fragilities of elements and systems in respect to specified stress factors, is crucial for allowing urban systems to better withstand recurrent or expected threats; meanwhile, a resilience-based approach, addressed to strengthen systems' capacities in the face of changing or unexpected circumstances, can be appropriate for allowing urban systems to better cope with the growing uncertainty related to climate related phenomena in the next future;
- a resilience-based approach clearly encloses the idea of transformability, which represents the capacity to create a fundamentally new system when ecological, political, social or economic conditions make the existing one untenable (Walker et al. 2004). Hence, it goes far beyond a vulnerability-based approach, mainly related to the optimization of the existing system, by emphasizing the capacity of living systems to continuously adapt or change, by "inventing new practices in front of novel problems" (Grøtan 2014).

Therefore, based on the interpretation of vulnerability and resilience as two interconnected and complementary approaches, it is possible to state that whereas a vulnerability based approach seems to be crucial for minimizing urban damage in the face of the expected climate-related impacts, a resilience-based approach is essential for moving cities towards more sustainable and less energy consuming urban development patterns capable, in the long term, of reducing the climate impacts on urban areas.

2 Climate Change Mitigation and Adaptation Strategies: The Role of Urban Planning

2.1 Cities and Climate Change

Climate change is a global process; it has many negative impacts that affect profoundly communities in Europe and around the world. The impacts of climate change are several and very different from each other; they include extreme weather events such as storms and hailstorms, floods, heat waves and droughts. These extreme weather events often result in injuries and deaths; they damage the building stock and infrastructure, cultural heritage, economy, trade, agriculture and ecological systems and cause large economic losses. Urban planning and territorial governance have an important role in preventing the effects of climate change.

In the Western World, in recent decades, the degradation of the natural environment, air pollution and climate change have increased the demand for socio-economic models with "green" and "sustainable" features. Many climate adaptation measures have been implemented at different scales. However, some questions remain open. How to implement the urban dynamics to make the city less vulnerable to climate change and less prone to producing pollution and greenhouse gases? How to implement coordination between mitigation and adaptation in a coherent and integrated approach? First, it is necessary to identify the factors that cause the urban communities' responsiveness to global environmental risks such as those related to climate change. The responsiveness also depends on the perceptions of risks and on their social acceptability. These elements differ according to the stakeholders, the represented interests and the territorial scale reference.

Local policies for sustainable development with actual effects are very rare. Some cities have achieved good results, for example, reducing CO_2 emissions by more than 20 % in ten years as Heidelberg, Vaxjo, Malmo, etc. These cities have built neighbourhoods where the energy supply is entirely ensured by renewable energy. These experiences respond to strong political will, in contexts of advanced decentralization, but most of the cities cannot coordinate the various policies in the area. Sustainable policies adopted in the cities have focused on four main lines of action: climate strategies, eco-construction, sustainable mobility and urban planning, the last one being the more difficult goal. The mitigation measures are defined according to social, political and economic conditions of each city.

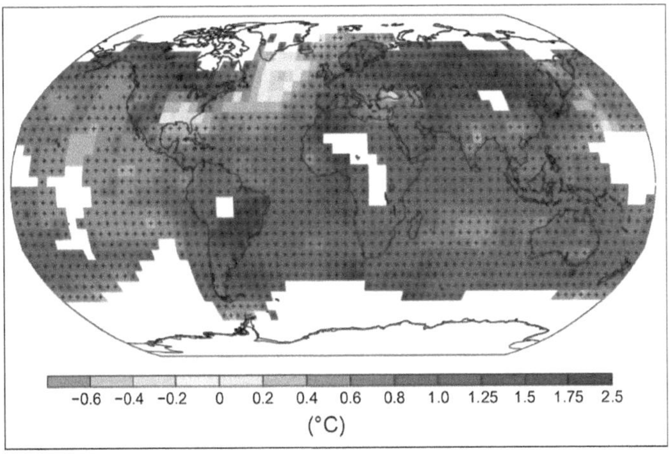

Fig. 1 Change on surface temperature from 1901 to 2012 (IPCC 2013)

The Fifth Assessment Report (AR5) of the Intergovernmental Panel on Climate Change (IPCC 2013) provides an update of knowledge on the scientific, technical and socioeconomic aspects of climate change. Since the 1950s many of the observed changes are unprecedented over decades to millennia. The atmosphere and ocean have got overheated, the amounts of snow and ice have diminished, sea level has raised and the concentrations of greenhouse gases have increased.

Climate change has visible effects in every part of the planet, but its effects are very strong in high-density areas due to human activity, which in recent centuries has settled in these localized areas, often by changing the morphology and original shape according to their needs (Fig. 1).

Cities produce most of the greenhouse gas emissions and they themselves are more exposed and vulnerable to the impacts of climate change. Urban planning has an important role, because it directly affects the environment, often by reducing agricultural areas with a consequent increase in land consumption, pollution and alteration of the natural environment.

> Cities are the places where most people in Europe will experience climate change impacts first; they accommodate around three quarters of the population, to share which is expected to increase to further (EEA 2012).

Currently, a critical specification of urban areas is the rigidity of its structure. City is often unable to adapt quickly to abnormal stress, such as climate change. Moreover, some features of the city—for example the low percentage of permeable surfaces or the morphology—further inhibit its ability to react. In this context, urban planning plays a key role to guide the city towards a more sustainable and resilient shape. Cities need operational and programming tools aimed at ensuring an adequate level of safety, managing risk prevention with a strategic and cross-disciplinary integration between different disciplines. It is necessary to create an efficient synergy among the available knowledge, able to change the answers according to the specific needs of the area, reaching, therefore, a flexible adaptation strategy.

Scientific knowledge must be the basis of urban policies, highlighting the vulnerability of the area to extreme events and pointing the way for a coherent, effective, efficient and resilient planning. The role of urban planning is essential within the challenge posed by climate change, both in terms of adaptation and mitigation. Planning has been often developing according to too many levels of the decision-making process, and there is no interaction among them, which is the key to solve these problems.

2.2 *Adaptation, Mitigation or "Adaptigation"?*

The vulnerability of cities to climate change is determined by many factors. The location certainly has an impact on the exposure to climate-related hazards, but the most important factors are related to socio-economic aspects and to the ability of administrators to develop appropriate strategies for adaptation and mitigation. In recent years, the concepts of adaptation and mitigation have become the basis of climate policy, identifying several strategic possibilities in order to manage these issues.

It is important to define the scope of the mitigation and adaptation: the first is a global goal that requires the national levels' intervention, which participates through international agreements; the second is an activity mainly developed at the local level, which, therefore, requires the local governments' commitment. Moreover, it is necessary to distinguish between adaptive and mitigating measures of the phenomenon (Pinto 2013).

The adaptation includes those actions that tend to minimize the effects of climate change, while mitigation strategies are aimed to reduce the intensity of the phenomena. Adaptation seeks to manage the consequences of climate change (Fig. 2).

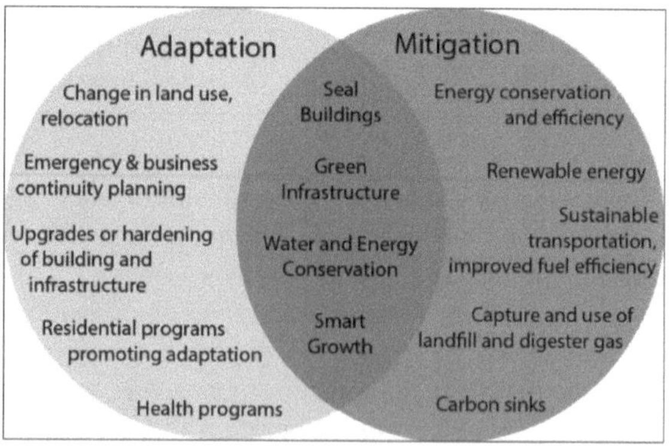

Fig. 2 Climate change mitigation and adaptation options (Winkelman 2012)

It does not intend to solve the causes of the problems, but try to make the coexistence with the phenomena tolerable, until subsequent deteriorations require further adjustments. The adaptability of the system is directly correlated to its resilience and inversely correlated to the sensitivity of the system related to the suffered perturbation or to changes in pre-existing conditions.

Consequently, improving the adaptation of the system means, above all, to decrease the vulnerability to climate change, increase the risk threshold, reduce the potential negative impacts and promote a more sustainable development. The capacity to adapt to changes considerably depends on the Countries, regions and socio-economic groups. The ability to adapt is a function of different factors such as economic power, technology, information, institutions and equity.

Mitigation strategies can be seen as the best option for reducing the risk of disasters related to extreme weather events exacerbated by global climate change.

The current challenge for the city is to implement policies and projects aimed at reducing emissions of urban areas and at increasing their resilience. The city will have to contribute to "mitigate" climate change and, at the same time, to "adapt" to climate change. It is essential that climate changes are tackled in an integrated way (Kress and Schibel 2007).

If climate change induces a vast range of hazard factors, it is necessary to prevent them by acting on the causes, namely by reducing emissions of greenhouse gases from human activities and stop or at least slow down the accumulation in the atmosphere, but it is also essential to act on their effects, by limiting the territorial and socio-economic vulnerability to climate change.

The two strategies are not alternative but complementary: a greater commitment to climate change mitigation determines smaller demands for adaptation and vice versa. At the local level, the management of the impacts of climate change requires more and more the definition of an integrated policy for the climate with a focus on the environmental, economic, urban and social aspects.

> Neither adaptation nor mitigation alone can avoid all climate change impacts; however, they can complement each other and together can significantly reduce the risks of climate change (IPCC 2007).

In the past, governments have focused their attention exclusively on the mitigation of climate change, while in recent years they tend to increase the integration of policies for adaptation and mitigation. According to Richard Langlais, good planning integrates both perspectives: adaptation and mitigation. He affirms that to separate the two perspectives in planning is becoming obsolete and it can also lead to contradictory measures (Langlais 2009).

According to the IPCC, climate change, adaptation is defined as "initiatives and measures to reduce the vulnerability of natural and human systems of the actual or expected climate change". Mitigation of climate change, on the other hand, means "the implementation of policies to reduce emissions of greenhouse gases and to increase the use of renewable energy resources and renewable fuels".

Langlais uses a simple example of a municipality planting a new forest along a local river. As he says, this measure should be seen as both adaptation and mitigation, that is: "adaptigation", by which he means a response to climate change that integrates a focus on adaptation with a focus on mitigation, to avoid conflicts and promote synergies. "Seeing this newly planted forest as both adaptation (it buffers the impacts of flooding) and mitigation (it absorbs carbon) is a good planning", writes Langlais. The most important thing is to keep "adaptigation" in mind: adaptation and mitigation may be the result of the same measurement.

2.3 *Towards an Integrated Urban Planning*

Planning is able to influence the way and the density in which the soil is used, but at the same time is able to connect the macro-scale (cities and districts) to the micro-scale (buildings) in a single view. The planner often acts within the limits of political and economic factors; nevertheless he plays a key role in the adoption of measures for the improvement of urban microclimate. In fact, the success of projects and mitigation plans also depends on the planner's ability to improve the awareness of likely future changes in the microclimate caused by the modification of the urban fabric. The planner should be able to understand and incorporate the principles and techniques that can be applied to pursue the goal of reducing the impact of territorial transformations on the urban climate.

It is essential to understand the two-way relationship between the built environment and the environmental thermal conditions: planners change the environment through the building and, on the other hand, the buildings suffer from environmental surrounding changes. The future buildings should be designed in the light of the impacts reduction about environmental thermal conditions and, at the same time, the buildings should control the adverse effects related to the changes in these conditions on the urban spaces users.

It is necessary to concentrate the efforts on the following three areas with synergies between adaptation and mitigation: urban planning, building design and decentralized energy production. That is why local governments have a central role in climate change adaptation. Many actions have been undertaken in this direction. After the adoption of the European climate change package in 2008, the European Commission proposed the Covenant of Mayors to support the efforts of local authorities in the implementation of policies in the sustainable energy field. To achieve and exceed EU targets for energy and climate, the signatories of the Covenant of Mayors are committed to develop an Action Plan for Sustainable Energy (SEAP). This document defines the energy policies that administration has taken in order to reach the EU target of reducing CO_2 emissions by 2020 (Pinto 2014).

Recently, the Institute for Environmental Protection and Research, ISPRA, has developed a strategy for local adaptation to reduce the risks posed by climate change, suggesting the development of the individual communities' resilience. The Institute has prepared some Guidelines for Climate Change Adaptation at Local level. The latter provide a practical and operational support to local governments that are interested in starting a process of adaptation and in developing plans for the Climate Change Adaptation at Local level (PAL).

Moreover, local governments can use urban planning to counterbalance the phenomena related to climate change. A systemic view of the city and a different set of local centres of activity near public transport nodes help to reduce the amount of land used for construction and tend to reduce travel and transport emissions.

However, responding to climate change and adaptation requires space within and around buildings. An average density of settlements, together with a differential use and green areas, tends to lead to a reduction of greenhouse gases and contribute to adaptation. The integration of functions—residential, productive, recreational, infrastructure—in urban areas leads to further benefits of adaptation. Innovative systems of air conditioning help to limit emissions. The orientation and organization of buildings and covered areas make it possible to replace the conventional air conditioning system with solar and district heating. The planning of green areas around buildings and green roofs to reduce the temperature leads to a substantial reduction of energy consumption.

On summing up, a combination of strategies for adaptation and mitigation can lead to optimal results in terms of both reducing vulnerability and increasing resilience, improving environmental quality and local economy (Fig. 3).

An integrated and holistic approach to mitigation and adaptation can be a great starting point to turn a threat into an opportunity for an urban development capable of achieving a better future not only in respect to the climate related challenges, but also in economic and social terms.

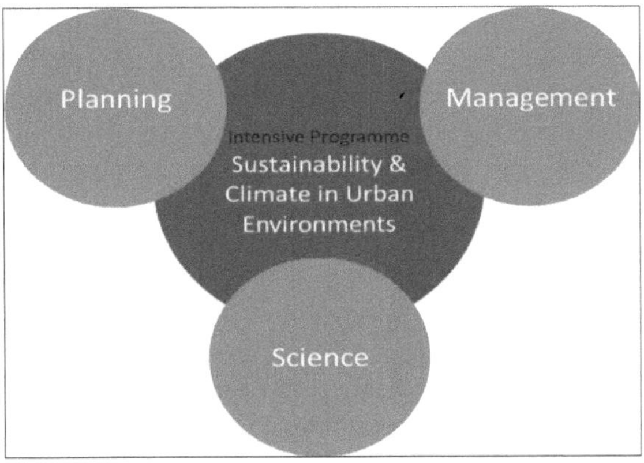

Fig. 3 Guidance for local government (Mc Cormick 2012)

3 Sustainability of the Planning Tools: Towards a Certification

Many opportunities can be considered to reduce the anthropic impact on the environment: new technologies for the production and the use of energy, but also the awareness of new behaviours and the construction of new regulative perspectives are the factors for acting on urban planning, on building project, on urban design and on a more sustainable mobility. "The new framework demands a renovated mission and a new tool-kit for urban planning focusing on increasingly complex and urgent issues" (Moccia 2013).

The current planning system is based on tools covering the municipal territory, implemented through local plans afterwards. While the former has a general, strategic framework, the second is strictly connected with the detail design. For this reason, the local level is the most suitable for an operative testing.

It is interesting to deepen the potentialities of the local plan towards a tool with advanced characters of sustainability. This sustainable local plan will foresee the minimum environmental impact of volumes and functions to be set up, will provide for the realization of public spaces with zero or almost zero impact, will promote the integration of technologies for the reduction of consumptions and for the energy production. The consequence is to deepen the possibility to measure expected and real results through the formulation of a system of certification that we can name "Urban Labelling". In addition, the same tool could be applied to existing and/or new construction/conversion of urban areas.

3.1 Evaluating the Local Plans

If we consider the indicators used to evaluate the sustainability of an urban system, we can highlight that general indicators are mainly used. In this field studies are evolving and it is not possible to define a potential end point. Urban Footprint or GHG emissions are the main samples and define the sustainable state of a city; they are used also to build national or international classifications of urban sustainability. These indicators are not used to evaluate urban plans.

Among urban plans, local plans are the most suitable for this trial; local plans, in fact, are tools that transform the general indications contained in the urban plan in a drawing with defined technical characters. They are the place where it is possible to test the urban design, to create the future form of an urban district, and to apply a real evaluation of sustainability (Fig. 4). In addition, the tool could be used to evaluate existing parts of cities, and to assign them a sustainability label. A local plan and an urban district, in fact, are formed by the same elements (buildings, green, mobility, …). In the first case, they are planned for a future achievement, in the second case they are physical realities.

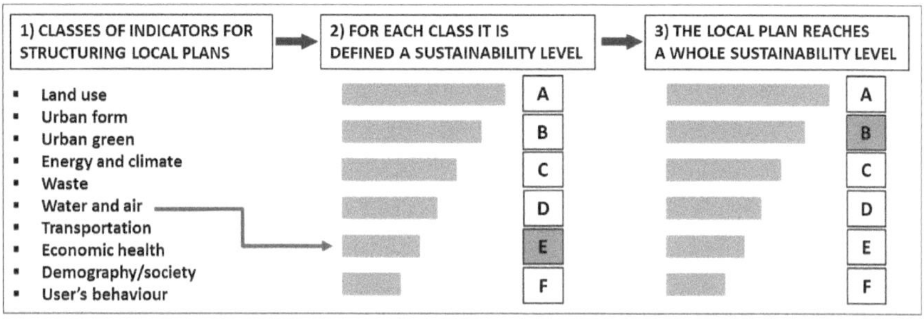

Fig. 4 Evaluation of a local plan as arrangement of single elements

It is necessary, however, to consider that a city is not a machine, namely a system where an input I is transformed in an output O by the use of a work.

The labelling of a mechanism (a television, a refrigerator, or a washing-machine) is assigned only on the base of technical characters. A city is different. It can be used in infinite ways and the changing behaviour of the citizens can change its performance. For this reason when we define a city as sustainable, we should consider the presence of the human variable and the necessity to confront with it.

In this way, the social characters of a city can be of great help. A high social control level eases the "enforcement" of more restrictive behaviour rules.

3.2 *Existing Urban Certification Systems*

The evolution towards a more risky urban environment can be summarized by such trends as warmer atmosphere, extreme weather events, growing of the greenhouses gas concentration, reduction of water quality, and so on.

The listed phenomena have an increasing impact on urban systems, both in terms of dimension, and of speed (Loarie et al. 2009). The city can be considered as a changeable space for physical and functional relationships, and the spatial influence of the sustainability development is uncertain. That is the reason why, it is necessary, above all, to define the appropriate evaluation scale, with the awareness that the right boundaries should be adopted case by case, but with a preference for the district level. In the latest years, cities have increasingly recognized their role for achieving sustainability goals, as shown by several documents (Charter of European Cities and Town Towards Sustainability, Aalborg, 1994; Leipzig Charter on Sustainable European Cities, 2007; Local Governments for Sustainability Preparing for Tomorrow Strategy 2010–2015, ICLEI; World Sustainable Capitals, c40.org, 2010), but it should be underlined the lack of a recognized assessment system.

Urban or local assessment systems are currently under development by several research groups (Cole 2011). The main cases are the LEED for Neighbourhood Development (LEED ND), proposed by US Green Building Council, and the two

CASBEE tools: CASBEE for Urban Development (2006) and CASBEE-City (2011). Other assessment systems are used. In Italy, for example, GBC system has developed a version for neighbourhood evaluation, similar to LEED ND. In Finland the development of Ekovikki, near Helsinki, was assessed by PIMWAG ecological criteria (Faninger-Lund and Lund 2000). Further indications can be found in Zhou et al. (2015), particularly about indicators system.

LEED ND starts from the consideration that land use and neighbourhood design patterns create a particular physical reality and force the behaviours with meaningful effects on the environmental performance of a place. A neighbourhood planned for the car use, and another planned for pedestrian and for the public transport use have two different impacts on the environment.

This is noticeable everywhere, because the entire world is affected by the negative impacts of the urbanized space on the wide spaces. "Environmentally responsible buildings and infrastructure are an important component of any green neighbourhood, further reducing greenhouse gas emissions by decreasing energy consumption. Green buildings and infrastructure also lessen negative consequences for water resources, air quality, and natural resource consumption" (LEED 2013).

The sustainability of a ND is defined by the score reached by the case-study. LEED 2009 for Neighbourhood Development Certification Levels have 100 base points plus 10 added points (6 for Innovation and Design Process and 4 for Regional Priority Credit). The score detects 4 potential level of sustainability: Certified (40–49 points); Silver (50–59 points); Gold (60–79 points); and Platinum (80 points and above). The score is the arrangement of five categories of indicators:

- smart location and linkage (27 maximum points);
- neighbourhood pattern and design (44 maximum points);
- green infrastructure and buildings (29 maximum points);
- innovation and design process (6 maximum points);
- regional priority credit (4 maximum points).

The CASBEE City (Comprehensive Assessment System for Built Environment Efficiency) tool is specifically designed for city assessment. It helps local governments and other stakeholders to identify the environmental, social and economic characteristics of the city and to quantify the effectiveness of the urban-wide policies. CASBEE-City is based on hypotheses of environmental efficiency and it produces an evaluation of a city based on two aspects: (1) the environmental load imposed by the city on the wider space outside its boundary, and (2) the quality of life (environmental, social, economic) inside the city.

A city characterized by low environmental load and high quality presents a high Built Environment Efficiency (BEE) value and is regarded as a sustainable city within the CASBEE framework (CASBEE 2012).

The two examples above presented are addressed to the building of environmental assessment systems for urban districts. Notable differences are present: while LEED analyses a specific neighbourhood and certifies the sustainability level on the basis of its physical and technological characteristics, CASBEE builds an index (BEE) that connects urban quality and urban load, extending in this way the analysis of the existing connections among city and surrounding area. Finally, neither of the two tools seems to take into account the previous planning phase and the sustainability of the urban/local plans.

3.3 Certification for Local Plans and Neighbourhoods: A Proposal

The experimental nature of these examples offers the possibility of widening the topic, namely the building of a new tool, an Urban Labelling addressed to a sustainability analysis. The tool could be applied on: (1) the action urban plans, namely "operative plans" or "local plans", and (2) the urban districts, as part of the city and as recognizable parts of an urban structure. For specific characters, the first is an assessment of a governance-planning tool, while the second is an assessment of the physical shape of an urban sector (Mazzeo 2014).

Urban Labelling is an open procedure for detecting the sustainability level of an urban structure, existing or planned. The environmental certification carries out the evaluation by using a set of values going from the environmental neutrality (Z/NZN —Zero/Near Zero Neighbourhood) to the highest environmental unsustainability.

Urban Labelling improves the traditional planning systems with an evaluation tool aimed to define the urban sustainability both quantitatively and qualitatively, and placing it on a default grading scale (A–F, for example). The formalization will be based, as said, on measurable indicators (formed by numerical data, if possible, but also by qualitative factors) from which a clear and shared assessment can derive. This leads to measure the sustainability of a plan or an urban structure (a built environment).

The formulation of this rating system is based on the establishment of classes of indicators finalized to the identification of the sustainability level that, in its turn, is considered as a combination of vulnerability and resilience criteria, as defined in Sect. 1.3, also if this paper does not assign the classes of indicators to one or to both the criteria. The detected classes are:

- A1: indicators of land use (soil quality and consumption);
- A2: indicators of urban form (population density, mass density, public spaces);
- A3: indicators of urban green (quantity, biodiversity, urban farms);
- B1: indicators of energy and climate (carbon intensity, energy intensity, renewable and clear energy, production and storage);
- C1: indicators of waste generation and treatment;
- C2: indicators of water and air quality;
- C3: indicators of transportation (facilities, accessibility, modal structure);
- D1: indicators of economic health;
- D2: indicators of demographic and social health;
- D3: indicators of user's behaviour (education, monitoring).

The classes from A1 to C3 refer to the physical and functional aspects; those from D1 to D3 refer to the characteristics of urban space use and to the anthropic behaviours, once the area will be transformed. The classes from A1 to C3 enclose the indicators formed by volumes, areas, height of buildings, densities, type of activity, sustainability indexes of buildings, green and permeable surfaces, origin and type of materials, colours, production of energy, recycled water, flexibility of the spaces and their adaptability to change, production of physical and not physical pollutants. For the most part, they are defined in quantitative terms. The classes from D1 to D3 enclose quantitative and qualitative indicators related to the level of social welfare and to the use of the urban spaces by citizens and users, derived from the knowledge on the use of the city, from the process of activities performing and from the time-related changes in behaviour, changes that can be read in terms of persistence or change (EEA 2013).

For each of these indicators minimum and maximum thresholds have to be identified, using regulatory, experimental or research values internationally recognized (EPA—Malaga City Council 2012; Zhou et al. 2015) and characterizing them in relation to the sustainable use of the space. For their characteristics the evaluation of plans uses the first set of indicators (physical and functional indexes), while the evaluation of urban sectors uses both sets of indicators (physical/functional and welfare, use and behaviour indicators).

Another point to underline is that a plan can be applied normally to a territorial system in which the state of disaster is only an extreme scenario. Nevertheless, it is important to improve the basic levels of the plan response increasing their resilience and reducing their vulnerability.

The attainment of a label level occurs through a process involving several steps (Fig. 5). Once established the object to analyse (an action plan or an urban sector), it is possible to use the indicators belonging to the two classes (physical/functional or welfare, use and behavioural). For each of these the minimum and maximum variance thresholds are to defined, by associating them with a sustainable quality (high, medium, low and/or other degrees).

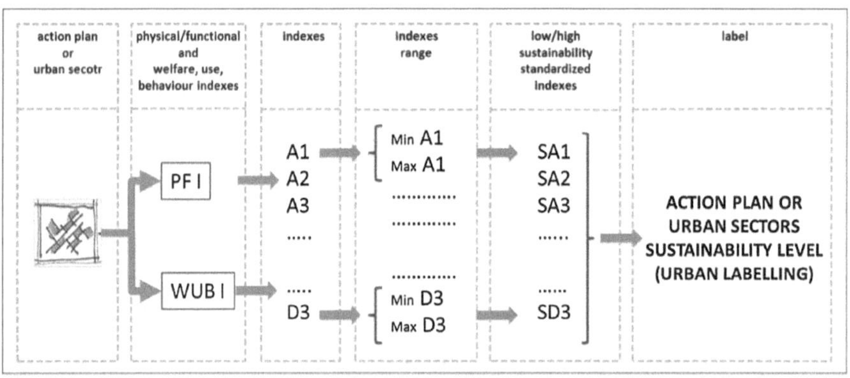

Fig. 5 Procedure for the emission of the sustainable label

This is a basic passage in order to homogenize the sustainability scale: for example, if we consider the building density and the soil sealing, we can say that low sustainability is associated with low densities and with high soil sealing.

The shift from data (I) to sustainability indexes (SI) may be implemented with the choice of a standardization method. At this point a homogeneous system of indicators in a default scale has been established, forming the bases for evaluating the sustainability level of plans or urban sectors.

In this way we reach the aim to define a certification system for plans and urban sectors able to define levels of consumption for the resources and levels of impact for the environment. This certification system should be able also to give each urban sector a synthetic index by defining its sustainability level, an index that will be extended to the city when all the urban sectors will be assessed. The pursuit of this aim could be favoured by introducing the urban labelling inside the procedures of plan making and approval.

Another expected result is the spreading deriving from the use of such a methodology, nearly a dissemination effect, especially if it will become the base for promoting incentive policies increasing the overall sustainability of planning tools and their implementation.

4 Conclusion

This contribution has been addressed to provide innovative approaches, strategies and tools for improving cities' response to the numerous environmental challenges threatening their future and, above all, to climate-related issues.

To this aim, we have firstly focused on the concepts of vulnerability and resilience, highlighting, on the one hand, the crucial role played by these concepts for facing the challenge of climate change and addressing as well the wider goal of a sustainable urban development; on the other hand, the potential complementarities among a vulnerability and a resilience-based approach in promoting an integrated approach to climate issues. In detail, whereas a vulnerability based approach seems to be crucial for minimizing urban damage in the face of the expected climate-related impacts, a resilience-based approach is essential for moving cities towards more sustainable and less energy consuming urban development patterns capable, in the long term, to reduce the climate impacts on urban areas.

According to this perspective, in the second part, current approaches to mitigation and adaptation strategies have been briefly discussed, emphasizing that so far they have been often considered as separate issues. On the opposite, the need for embracing an "adaptigation" perspective has been emphasized, focusing on the key role played by urban planning in pursuing both mitigation and adaption issues. In detail, the city will have to contribute to "mitigate" climate change and, at the same time, to "adapt" to climate change. Adaptation and mitigation can complement each other and together can significantly reduce the risks of climate change.

Finally, in the last paragraph, a first attempt to provide an evaluation tool capable of guiding planners along the path aimed at renovating existing or building new urban neighbourhoods, counterbalancing climate issues and addressing meanwhile the broader issue of a sustainable urban development, has been provided.

Acknowledgments Adriana Galderisi, Giuseppe Mazzeo and Fulvia Pinto have made substantial contributions to the paper's conception and design, background and concluding remarks (abstract and Sect. 4). Adriana Galderisi analyzed resilience and vulnerability as key-concepts for a sustainable urban development (Sect. 1). Fulvia Pinto analyzed the role of urban planning in the climate change mitigation and adaptation strategies (Sect. 2). Giuseppe Mazzeo deepened the building of a certification tool for the sustainability of local plans and urban sectors (Sect. 3).

References

Birkmann J, Cardona OD, Carreño ML, Barbat AH, Pelling M, Schneiderbauer S, Kienberger S, Keiler M, Alexander D, Zeil P, Welle T (2013) Framing vulnerability, risk and societal responses: the MOVE framework. Nat Hazards 67:193–211. http://dx.doi.org/10.1007/s11069-013-0558-5

CASBEE (2012) CASBEE for cities, technical manual, Japan sustainable building consortium. http://www.ibec.or.jp/CASBEE/english/download/CASBEE_City_manual_2012(E).pdf. Accessed 20 Apr 2015

Cole RJ (2011) Environmental issues past, present & future: changing priorities & responsibilities for building design. Proc. SB11 Helsinki

Davoudi S (2012) Resilience: a bridging concept or a dead end? Plann Theor Pract 13(2):299–307. http://dx.doi.org/10.1080/14649357.2012.677124

Davoudi S, Brooks E, Mehmood A (2013) Evolutionary resilience and strategies for climate adaptation. Plann Pract Res 28(3):307–322. http://dx.doi.org/10.1080/02697459.2013.787695

EEA (2012) Climate change, impacts and vulnerability in Europe, Technical report 12/2012. European Environment Agency, Copenhagen

EEA (2013) Achieving energy efficiency through behaviour change: what does it take? Technical Report 5/2013. European Environment Agency, Copenhagen

EPA—Malaga City Council (2012) Sustainable urban models, Malaga: CAT-MED project, http://www.catmed.eu/. Accessed 3 Feb 2015

Faninger-Lund H, Lund P (2000) Toward sustainable cities: case Ekoviikki in Helsinki and its solar project. In: Proceedings of ISES-Europe conference EUROSUN-2000, Copenhagen, Denmark, June 19–23. www.ptp.irb.hr/upload/mape/kuca/06_Heidrun_Faninger-Lund_TOWARD_SUSTAINABLE_CITIES_CASE_EKOVI.pdf. Accessed 3 May 2015

Fekete A, Hufschmidt G, Kruse S (2014) Benefits and challenges of resilience and vulnerability for disaster risk management. Int J Disaster Risk Sci 5:3–20. http://dx.doi.org/10.1007/s13753-014-0008-3

Folke C, Carpenter SR, Walker B, Scheffer M, Chapin T, Rockstrom J (2010) Resilience thinking: integrating resilience, adaptability and transformability. Ecol Soc 15(4):20. http://www.ecologyandsociety.org/vol15/iss4/art20/

Galderisi A (2014) Climate change adaptation. Challenges and opportunities for smart urban growth. TeMA J Land Use Mobility Environ 1:43–67. http://dx.doi.org/10.6092/1970-9870/2265

Grøtan TO (2014) Hunting high and low for resilience: sensitization from the contextual shadows of compliance. In: Steenbergen RDJM, van Gelder PHAJM, Miraglia S, Vrouwenvelder ACWM (eds) Safety, reliability and risk analysis: beyond the horizon. Taylor & Francis Group, London, pp 327–335

Grünewald F, Warner J (2012) Resilience: buzzword or useful concept? http://www.urd.org/RESILIENCE-buzz-word-or-useful?artpage=2-5#outil_sommaire_0
IPCC (2007) Climate change 2007: the physical science basis http://www.ipcc.ch/report/ar4/wg1
IPCC (2013) Climate change 2013: the physical science basis http://www.ipcc.ch/report/ar5/wg1
Kelman I, Gaillard JC, Mercer J (2015) Climate change's role in disaster risk reduction's future: beyond vulnerability and resilience. Int J Disaster Risk Sci 6:21–27. http://dx.doi.org/10.1007/s13753-015-0038-5
Kress A, Schibel KL (2007) Adattamento e Mitigazione. Un approccio integrato di politica del clima, Alleanza per il clima. http://www.climatealliance.it
Langlais R (2009) Adaptigation. J Nordregio, Lund University, http://www.nordregio.se
LEED (2013) LEED 2009 for neighborhood development, congress for the new urbanism, natural resources defense council, and the U.S. Green Building Council, Washington, update 2013, DC
Loarie SR, Duffy PH, Hamilton H, Asner GP, Field CB, Ackerly DD (2009) Velocity Clim Change Nat 462:1052–1055. http://dx.doi.org/10.1038/nature08649
Mazzeo G (2014) Labelling urbano come programma di lavoro sulla città. Eyesreg 4(1):27–30
Mc Cormick M (2012) Extreme climate risk. State resources overview. Office of planning and research, http://opr.ca.gov. Accessed 5 Apr 2015
Menoni S, Molinari D, Parker D, Ballio F, Tapsell S (2012) Assessing multifaceted vulnerability and resilience in order to design risk-mitigation strategies. Nat Hazards 64:2057–208. http://dx.doi.org/10.1007/s11069-012-0134-4
Moccia FD (2013) Ecological restoration methodology. In: Moccia FD, Palestino MF (eds) Planning stormwater resilient urban open spaces. Clean Edizioni, Napoli, pp 12–17
Papa R, Galderisi A, Vigo Majello MC, Saretta E (2015) Smart and resilient cities. A systemic approach for developing cross-sectoral strategies in the face of climate change. TeMA J Land Use Mobility Environ 1:19–49. http://dx.doi.org/10.6092/1970-9870/2883
Pinto F (2013) Isola di calore e resilienza urbana: strategie di mitigazione e di adattamento della città ai cambiamenti climatici. In: Sbetti F, Rossi F, Talia M, Trillo C (eds) Il governo della città nella contemporaneità. La città come motore di sviluppo. Urbanistica Dossier, vol 4. INU Edizioni, Roma
Pinto F (2014) Urban planning and climate change: adaptation and mitigation strategies. TeMA J Land Use Mobility Environ [S.l.]:826–840. doi:http://dx.doi.org/10.6092/1970-9870/2547
Pizzo B (2014) Problematizing resilience: Implications for planning theory and practice. Cities 43:133–140. http://dx.doi.org/10.1016/j.cities.2014.11.015
Turner BL (2010) Vulnerability and resilience: coalescing or paralleling approaches for sustainability science? Glob Environ Change 20:570–576. http://dx.doi.org/10.1016/j.gloenvcha.2010.07.003
UN (2012) The future we want, . Accessed 15 Mar 2015
UN (2014) World urbanization prospect. Population division world urbanization prospects. The 2014 revision, highlights. Department of Economic and Social Affairs. http://esa.un.org/unpd/wup/Highlights/WUP2014-Highlights.pdf. Accessed 8 Jan 2015
Walker B, Holling CS, Carpenter SR, Kinzig A (2004) Resilience, adaptability and transformability in social–ecological systems. Ecol Soc 9(2):5. http://www.ecologyandsociety.org/vol9/iss2/art5/. Accessed 15 Feb 2015
Winkelman S (2012) What does climate resilience look like? Center for clean air policy, www.cca.org. Accessed 13 Mar 2015
Zhou N, He G, Williams C, Fridley D (2015) ELITE cities: a low-carbon eco-city evaluation tool for China. Ecological Ind 48:448–456. http://dx.doi.org/10.1016/j.ecolind.2014.09.018

Evolved Frameworks for the Integrated Development of Territorial Services

Mauro Francini, Annunziata Palermo and Maria Francesca Viapiana

Abstract In recent years, through the adoption of strategic planning processes, the main objective of cities and territories is that of becoming smart, it its different declinations, associating original, flexible ad intelligent technological processes. This work describes two researches that, starting from the common basic assumptions, have defined and created evolved frameworks that can generate "smart services" in support of integrated territorial development. The first study focuses on the definition of a usable platform in the 'cloud' environment with which integrated "smart services" can be developed. These smart services do not only promote the territory, thus creating value for the public administration as well as facilitating and stimulating the local economy; they also contribute to qualifying some specific management aspects such as those referring, for example, to the service of smart energy, or rather the optimisation, monitoring and remote regulation of production/consumption of energy even by means of service chains that involve manufacturers and service provider. The general objective of the second research, instead, consists of the necessity of providing services based on access to a geo-referenced database that is interoperable and predisposed to evolve towards new development scenarios, containing cartographic information associated with various information on info-mobility and road safety.

M. Francini (✉) · A. Palermo · M.F. Viapiana
DINCI, University of Calabria, Cosenza, Italy
e-mail: francini@unical.it

A. Palermo
e-mail: annunziata.palermo@unical.it

M.F. Viapiana
e-mail: mf.viapiana@unical.it

R. Papa and R. Fistola (eds.), *Smart Energy in the Smart City*,
Green Energy and Technology, DOI 10.1007/978-3-319-31157-9_12

1 "Smart Services" for Promotion Integrated Territorial Development

1.1 Preliminary Considerations

The starting vision of the current research project consists of the necessity of promoting the territorial image by means of differentiated immaterial use supported by original, flexible, and intelligent (smart) technologies, which are capable of improving the state of the art both in terms of scientific results and in terms of effective and innovative territorial reverberations.

In order to pursue the vision defined in the elaboration of the conceptual phase, it was necessary not only to guarantee the most recent national and European regulations in terms of a standardised representation of territorial information and its full interoperability between clients, producers and end-users, but also to verify the coherence of the research intents with that provided for by the actual European programme policies.

It was impossible not to take into consideration the indications of Horizon 2020 which, in terms of new challenges for the EU, provides for a specific focus on the themes dealt with.

In particular, the Europe 2020 (Digital Agenda) strategy advocates for the advancement of research in terms of European market economy improvement over the next decade, on the basis of priority sectors which are closely connected and which are mutually strengthening: intelligent growth, through the development of an economy based on knowledge and innovation; on sustainable growth, through the promotion of an efficient economy under the profile of the use of resources and competitively; and on inclusive growth, through the promotion of an economy with a high employment rate which favours social and territorial cohesion.

Moreover, the research vision is reflected in that indicated in the European Union Plan Strategic Energy Technology (SET), "Plan for the development of low carbon technologies", with the aim of defining guidelines and energy policy objectives, with particular reference to the development and sharing of information on technological innovation. Furthermore, the programme has launched different European Union initiatives favouring the diffusion of a new sustainable urban-territorial model: Smart Cities, over the entire continent.

Specifically, the starting vision focuses attention on the realisation of an advanced framework, based on a usable platform in a "cloud" environment, in which "Smart" services can be developed which can create value for public administrations, as well as agevolating and stimulating the local economy (Francini et al. 2012).

In brief, the vision is characterised by two preeminent aspects: the creation of an open and interoperable platform which enables the efficient development of applications and services for the citizen/tourist, for Local Authorities and for SMEs responsible for territorial development; a set of applications for the supply of "Smart Services" regarding the following thematic contexts.

Innovation of the territorial tourism service to guarantee both tourists and residents easy access to services which meet their requirements, permitting them to interact with and integrate information (Smart Tourism).

Rationalisation of the local territorial public transport system, capitalising on cost savings induced by a homogenous management of the service whilst contributing to an increase in the perceived quality of the service (Smart Mobility).

Optimisation, monitoring and remote regulation of energy production/consumption services, also through service chains which involve manufacturers and service providers (Smart Energy).

The specific objectives inherent to the thematic area Smart Tourism refer to different aspects, above all the necessity to promote the improvement of territorial tourist services and an integrated use of the services by means of: value services for tourists and operators; a "territory database" (user generated), supported by indicisation models and univocal identification; interoperable and interactive; "many-to-one" territorial marketing; an integrated atmospheric modelling system for weather forecasts in points which are localised and of particular interest or along routes to attractive destinations; integration with smart mobility services.

The specific objective inherent to the Smart Mobility thematic area refers to the necessity of sustaining territorial attractiveness through the rationalisation and improvement of transport services, supplying the territory with advanced Simulation-Optimization instruments to support transport and fleet management models which consent an eco-sustainable mobility system, which is both safer and has lower costs. The key element is in the convergence on a sole model of different components of transport and of services/logistic platforms, as well as different models of requirement and consumption for an integrated and optimised management of all the chain participants. The model, therefore appeals to new generation transport methods (EV) and relative standards of use.

Finally, since the Smart Energy thematic area concentrates on operative models to individuate and optimise renewable energy resource use, as well as technical/functional models for optimisation activities of energy use and monitoring/regulation of consumption, the specific objectives of the thematic area are substantiated in: diffusion of low environmental impact alternative energies (also in sites of great natural, archaeological value), supporting the creation of spin-off applications that are currently poorly diffused; energy efficiency (reduction of emissions and costs) through the intelligent use of resources (gas, power, local generation); "smart" management of the electrical grid, through a streamlined model of a virtual power plant which aggregates and manages production distributed allowing market economies, management models, innovative consumption contributing to network balancing; integrated management and optimisation of the use of energy from renewable sources, by means of efficient planning models and their integration with models of territorial consumption and socio-behavioural elements to maximise asset value; scheduling building consumption and generation by means of distributed and scalable algorithm which can regulate, in an integrated and dynamic manner, all the devices in a building (smart appliances, solar panels, heat pumps, etc.) considering—both during the planning and control phase—the uncertainty connected due to weather conditions and to the behaviour of the building inhabitants/visitors.

1.2 Synthesis of the Research Project

The research project methodology being examined emphasises the particular and in-depth use of technology to propose concrete evolutions in business models, operating on tangible value drivers: address real requirements both in terms of cost efficiency (transformation and cost efficiency mobility), and in terms of business enablers (growth and enhancement of tourism and culture); enable new services and market operators; introduce growth factors to existing markets, whilst contributing to the total economic efficiency of the chain; support the growth of distinctive territorial competences with value which can find a positive economic response in terms of continuous improvement of the business and technological arrangement of the system.

The results of the work, therefore, are of a technological nature, linked to the development of a cloud-based platform and of *business contexts*, which are connected and integrated, but also of a socio-economic character, thus, with the aim of defining an infrastructural and applicative model/framework which finds an effective response with the different territorial realities, it is necessary to verify the basic assumptions implementing the model defined on the sample territory, as well as on a sample of users (Francini et al. 2012).

The phase following the structuring of the model regards the analysis, on a sample territory, of some elements which are useful to guarantee the identification of further eventual clusters enabling value models to integrate to the model defined in the first instance, offering the best combination of the energetic and tourist elements and referred to the mobility system with the potential of the environmental, natural and cultural resources of the same territory.

Such analyses regard, specifically, the checking of the existence of some suitable instruments, methods and information use strategies to be integrated, therefore reference is made to: innovative modes of availability and valorisation of the tourism service; characterising components of the tourist service and construction of interrelations (analysis of products by means of criteria of differentiability of advantages); techniques for the optimised information management regarding consistency and conservation of building heritage (urban and rural); detailed in-depth analysis of the areas that are at risk from landslides and/or floods for all the archaeological and environmental interest sites; modelling of the state of substance, study of non-destructive diagnostic techniques and use of innovative materials; innovative and effective strategies of surveying, cataloging, and diffusing tourist information; methodologies for the integration of tourist type informative contents with optimisation models/methods of the territorial service; creation of an integrated atmospheric modelling system; technologies and methods for the evolved and automatic collection of survey information from elements (active or static) from the network; simulation/optimization models and methods for problems regarding vehicle routing & dispatching, multimodal transport, demand responsive transport, real time fleet management problems; predictives for road traffic and real time calculations of journey times; for Single Integrated Ticketing; for the optimised management of the rolling stock life cycle (Level of Repair Analysis and back-up level); applications useful to increase road safety and facilitate traffic management in critical conditions.

1.3 Territorial Context of Experimentation

The sample territory for the analysis is situated on the Calabrian Tyrrhenian coast, in the province of Vibo Valentia.

The resources characterising the reference territorial context are formed by the presence of worthy historical-architectural elements, by important enogastronomic and popular traditions, as well as by a notable naturalistic heritage. To all this a consistent and high quality accommodation service is added. Therefore, it is not just by chance that the Costa degli Dei, of which Tropea is the most significant expression, represents one of the few places in Calabria in which tourist services, even if concentrated on summer holiday tourism, manage to cover almost six months per year, involving the presence of around 1.5 million visitors per year, mainly from Europe.

Specifically, the existing problems are differentiated according to territorial characterisation, or rather those connected those connected to the coast and those connected to the inland areas, which find a contact point in the missing dialogue which over time has not allowed the coastal areas to widen their tourist flows nor the inland areas to evolve.

The potentialities to be qualified, instead, refer to the territorial vicinity of the areas with vocations which are prevalently different which, only if integrated, can evolve in terms of widening and integrating tourist services through the potentialising of mature elements, such as infrastructures near the waterfront, and the requalification of important elements present in more inland areas.

Therefore, in such a context, the high and sufficiently continuous presence (in relation to Calabria as a whole) of a multifaceted and qualified tourist user, along with a mature productive and social fabric, which together form an excellent sample on which to test the project idea in its entirety.

In fact, the same Municipalities, given the propensity for tourism and the capacity to attract a high number of visitors at specific times of the year, record irregular energy consumption, characterised by peaks which are delimited and are distant from typical (average) consumption and difficult to predict, since they are strongly dependent on factors which are external to the territory. It is precisely since the issues of grid balancing in these municipalities are of particular importance in set periods of the year and the growth and exploitation of local renewable generation becomes a fundamental weapon in addressing consumption peaks in the most sustainable manner for the territory, that it represents the ideal context for the experimentation of Smart Energy services provided for successive to the experimentations already started and successively described in reference to the Smart Tourism service.

1.4 Achievements and Prospects of Development

The described control analysis on the sample territory has regarded the knowledge of cultural assets and of all the elements that characterise the entirety of the services dedicated to tourism, as well as the different variations of the same including that which refers to natural heritage.

With such an aim, in order to increase the potential of technological development to increase interactivity in tourist/cultural experiences and extend the rapport with the end-user and the adoption of tracing instruments, geo-referencing and territorial geo-localisation, as well as enhancing the perceived experience with sophisticated multimedia solutions and advanced interfaces with immediate consequences in the valorisation of resources and quality of life of the collectivity, it was necessary to precisely identify the historical/artistic/landscape points of interest, as well as the accesses useful for accessibility to sites via hiking and other types of itineraries (public transport, cars, train, etc.).

The objective of the aforementioned analysis was substantiated in the necessity of verifying which elements were present on the territory and which would allow the citizen, both visitor and resident, to further increase innovative information services, which can be supplied by a web portal via multimedia points present on the territory (totem point) or by means of commonly used devices such as PCs, laptops, smartphones, PDAs and tablets in an intuitive and personalized way, according to the users' own preferences.

Such services contribute to the valorisation of the cultural and environmental heritage of the territory proposing, moreover, solutions for the hiring and parking of ecological modes of transport, thematic maps (e.g. places of worship and panoramas, historical buildings, natural parks and also agrotourisms, restaurants, B&Bs, hotels, and offices), in which the objects shown can be selected and specific information can be supplied via a multi-channel approach (e.g. audio/video and textual information, photographs, weather, addresses and telephone contacts). Furthermore, the user can conduct specific research thanks to an ontological research engine.

Within such an activity an effort was made, to favour the user, to simplify and summarise the plurality of the available informative resources through convergence and ubiquity, not only to access knowledge in a different manner, but also to realise a type of social interaction of digital media through the internet. Therefore, the efficiency of the information available on line is not only dependent on a robust and fast infrastructure, which represents the "physical capital", but also on the real time availability of a semantic platform for the sharing of knowledge which represents the true "social and intellectual capital".

In this regard, it became necessary to develop web services and intelligent interfaces allowing users to access such environments in an interactive mode, by means of semantic and ontological searches.

The results, subject to the sample users, draw attention to the destination context, to the user profile and the media type, thanks to new technologies such as real-time audio/video, story-telling and gaming.

This work, therefore, has combined key elements which, together with the personalised transmission capacities offer a new form of multimedia interactive involvement which goes beyond the state of the art and radically changes they way of "consuming" cultural information.

In particular, it was agreed to extend the boundaries of streaming technology to that which is defined as an "immersive interactive experience" (IIE), where not only the content, but also the context (or rather the topic which joins together a set of pertinent contents) of the user can be changed in an interactive manner by means of active links, which are dynamically created starting from the current content.

The active links are visualised on the screen in such a way that the user can recall them at any moment enhancing their use of the information. This leads to a revisitation in media distribution in that the users no longer represent passive users of online information, but they gain an active role that is personalised and independent, reaching a level of experience sharing that was not previously possible (Francini et al. 2012).

A similar structuring of the territorial data (Fig. 1) in terms of indications inherent to the material use of the aforementioned assets, moreover, favours the external and internal accessibility to the reference context allowing, for example, the optimisation of the connections between the improvement of the energy and environmental balances of the territories and the management of the naturalistic and socio-cultural resources according to principles of equity and sustainability, which still have to be investigated in terms of territorial experimentation.

It therefore appeared that the informative base thus created, with the aim of rendering the product integrated and integrable, can be extended with related information, which is particularly useful to local public administrations in order to plan large scale interventions on collective technical domains, such as the cultural and naturalistic heritage, urban planning, prevention of calamitous events linked to hydrogeological instability, etc.

Further development possibilities concern the definition of an "operative model" (sales processes, support, operations) for the services of each thematic area and on the identification of a "commercialisation model".

This aspect, results as being particularly important with the aim of rendering the research results differentiated, not only from a technological viewpoint, but also regarding the profile of the services supported, which appeal to the in-depth use of the IT component in order to facilitate new methods of distribution,

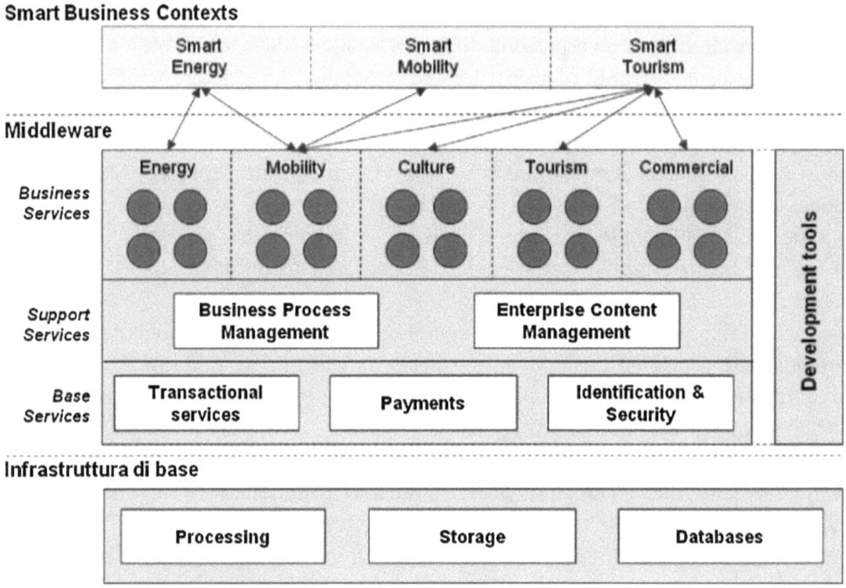

Fig. 1 Basic architecture of the system

commercialisation and use of the territory and the services connected to it, as well as to introduce new operators and encourage socio-economic development.

The extension of cloud, in this perspective, provides a frame to the entire system in that it consents the introduction of advanced Shared Services models which integrate benefits of cost efficiency, flexibility and time to market, of the cloud models, with business model and characterising/local aspect management differentiation elements.

The studies inherent to the definition of the aforementioned "operative model" of the services connected to each thematic area today refer only to the prefiguration of preliminary base concepts.

Instead, with regards to the Smart Energy thematic area, for example, the distinctive element consists in the evolution of a Utility Operator model, enabling new business models and market operators that capitalise on the aggregation of distributed local production and on territorial assets, or who exploit the potential of consumption regulation by addressing more efficient ways of balancing energy.

The peculiarities of the model can be summarised as follows: models/production planning instruments that integrate traditional techniques with local consumption properties and environmental elements enabling the definition of scenarios which identify optimal production in terms of valorisation of natural resources and impact on the distribution grid (therefore integrating the potential of real production); demand-response regulation models/instruments of consumption in relation to energy efficiency and production-consumption balancing at a micro-grid level in order to best exploit the local production potential; enablement of new models of market operators as "Service Aggregators"; Virtual Power Plant integrated models with an aggregation capacity of distributed production, monitoring and management of the same with the aim of creating real-time energy trading models and optimisation of plant maintenance processes.

2 Info-Mobility and Urban Functions for the Evolution of New Scenarios Smart

2.1 Preliminary Considerations

The present research was conducted within the National Operational Programme for research and competitiveness for the convergence regions PON M2M—"Mobile to Mobility Information systems and telecommunications for road safety".

The research programme, highlighting the lack of an analysis and intervention methodology which takes into account all the aspects that influence, or are influenced by accident rates (infrastructure, flow, landscape-environment, anthropy, etc.), aims to provide services based on access to a geo-referenced database, which is interoperable and predisposed to evolve towards new development scenarios, containing cartographic information associated with information on info-mobility and road-safety. The object of the research consists of the desire to define a renewed relationship between spatial planning and transports, which focuses on road-safety in the face of studies, which have mainly regarded the impact of transport in terms of land use and vice versa.

Specifically, studies on transport-spatial interaction have been produced in North-American countries and were prevalently conducted in urban contexts or on a metropolitan scale. The possibility of extending the aforementioned research to a regional scale with empirical approaches and by means of models which are capable of reproducing the interaction between levels of accessibility and the location of residential and economic activities (Coppola and Nuzzolo 2006), has been investigated.

Besides this experience, numerous models of microsimulation of spatial and transport use are applied (Wegener and Spiekermann 1996; Landis and Zhang 1998; Salomon et al. 2002), as well as complex models which, based on the theory of cellular dynamics (CA), consent the reproduction of the interactions which arise between spatial and transport systems (Batty 1997; Ferrand 2000).

Starting from these assumptions, the work develops concentrating on the study of the literature pertaining two cardinal concepts, namely road-safety and quality, in order to better understand the extent to which factors of an urban or architectural type can influence not only the articulation of mobility within specific contexts, but also safety, paying particular attention to the influence of the morphological-urban conformation of the reference context on determined components regarding mobility (traffic, public movement, public safety, movements of residents), as well as the importance of connected factors, for example, the lack of services and hygienic-sanitary conditions of buildings, public streets and spaces, the citizens' quality of life and the orientation of choices.

The starting objective consists of the need to select global urban, environmental and landscape indicators, or rather factors which are believed to be relevant for the evaluation of and the successive planning of choices within the analysis contexts.

This is conducted through the acquisition of specific data that is useful for the discovery of their critical and strong points, also with the support of interviews/specific surveys.

An objective which is essential in order to create a database of values and indices capable of providing guidelines which useful for: highlighting the parameters which mainly influence road safety and the articulation of mobility, even in proximity to the analysed urban contexts and, therefore the analysis of the selected data in order to generate reflections and possible solutions, highlighting the parameters which mainly influence the citizens' quality of life and their choices, and the analysis of data in order to generate technically sustainable solutions.

Some pertinent national and international normatives were analysed to support the aforementioned reflections, in order to compare the Italian dynamics with other European countries.

The relationship between road safety and the urban component concerning only the regulations of some Northern European countries. The hierarchical classification of roads occur in two phases: the first phase considers the functional category of the road (as in Italy), based on the geometrical properties of the infrastructure, on the type of traffic components and on the relative role played in the spatial context; the second phase takes into account the allowed speed assigned in relation to the presence of particular urban elements (schools, shopping centres, residences, etc.).

Through such settings, it is possible to correlate the functional classification of the road to the urban context in which it is found, thus defining the allowed relative speed.

European Union guidance correlates more frequently the urban component of transport policy, in line with the concept of sustainable urban transport.

On the basis of results obtained through different initiatives and experiments, some fundamental principles were established within the Sustainable Urban Transport Plan, which constitutes the main instrument for the management and control of sustainable actions on mobility; these include: guaranteeing accessibility for different types of users (residents, commuters, pensioners, disabled people); guaranteeing the safety of citizens both in terms of mobility and of health; containing air pollution and increasing the use of clean energy sources, and contributing to the increase of urban quality.

Following a study of the abovementioned literature, the parameters useful in order to widen the transport-spatial interaction were defined. They generally coincide with the different physical properties of the road infrastructure and functional of the accessibility of an area of interest, or rather the presence of urban elements which "incentivise" life quality, which also belong to a "periurban" field. These, in turn, contribute to a variation of the princes of lend and, therefore, the definition of policies and of spatial design. However, most importantly they can play an important role in order to guarantee appropriate road safety standards, in that they are generators of dangerousness points if associated with, for example, a greater attraction of vehicular and pedestrian flow or a particular concentration of weak road users.

To summarise, after having identified the urban uses involved, considering a specific experimentation field, some of them were located, with relative dimensional analysis and characterisation of the access modes (compared to the main viability) in typological and functional terms (services—schools, healthcare structures, sports centres; tertiary/productive sector—wholesale, shopping centres, manufacturing establishments; touristic sector—hotels, resorts, historical—cultural heritage), with the aim of creating a preliminary mapping of the potential risk factors useful for the insertion of specific data within a more extensive reference database (DB) (Palermo and Francini 2013; Francini et al. 2014).

The studies conducted come under the following operative objectives (OR) and the relative activities of the national operative plan:

- OR 2 Study of factors and creation of risk chain—Activity 2.2 "Contextualisation of risk factors and scenarios in real road settings";
- OR 4 Construction of multi-profile database—Activity 4.1 "User profiles and associate functional requisites" and Activity 4.2 "Structure of the system and the DB";
- OR 9 Construction of the test site—Activity 9.1 "Acquisition of the reference test site" and Activity 9.4 "Construction of GIS model and data input";
- OR 10 Experimentation and testing of prototype—Activity 10.1 "Validation of the Mobile prototype" and Activity 10.2 "Verification of efficiency/usability of the web platform for professional users".

2.2 Synthesis of the Research Project

In terms of *contextualisation of risk factors and scenarios in real road settings* many studies of the literature were analysed in order to define the factors and risk indicators, including that from the national Observatory on Italian autonomies for the coordination of communication and making the road network safe—born from an initiative of ANCI (*Associazione Nazionale Comuni italiani*) and UPI (*Unione delle Province d'Italia*) with the support of the Ministry of Infrastructures and Transport.

From the study, it emerges that road accidents seem to depend neither on the demographic dimension, nor on territorial density. Nor do they depend on the rate of motorisation of the cities. Instead, the correlation between the "resident population", "territorial density" and "rate of motorisation" from the study emerges to be approximately zero, even with worthy clarifications.

The specific urban references originate from such general elements which allowed the identification of the potential risk factors representing the information to be inserted as a priority in the reference DB, as well as creating a hierarchy of viability, an analysis of the characteristics pertaining to dimensions and the conservative-functional state and the typological classification of road intersections (roundabouts, crossroads, intersections with traffic lights, etc.), with a dimensional and functional analysis of them (presence and properties of horizontal and vertical road signs, presence of critical structural elements, presence of elements of disturbance, etc.), or rather: the location of urban uses, with relative dimensional analysis of the access modes (compared to main viability) in typological and functional terms: *residence, services* (green public spaces, equipped green public spaces, schools, hospitals, private clinics, car parks, sports centres, etc.), *tertiary* (retail, wholesale, banks, post offices, chemist's, public offices, etc.); *productive sector* (agricultural, artisanal and industrial activities); *tourist sector* (receiving and tourist structures); *historical-cultural heritage* (important and/or protected buildings, castles, historical buildings, archaeological areas, etc.); *urban fittings.*

The collection of this information, coinciding with the mapping of potential risk factors, further synthesised in a data collection table, represented the starting base for the structuring of a Spatial Information System suitable for the needs of the M2M project and characterised by a high level of flexibility and openness towards the possible access and integration with existing databases, both geo-referenced and non geo-referenced, with main reference to the competences of the involved territorial bodies (public administrations).

In order to define appropriate *user profiles and associated functional requisites*, and therefore to render the system one which can be integrated and interoperated. The conducted studies highlight the need to refer to what are known today as new instruments of participation; these instruments, or rather social networks, result as being particularly useful in an active participation logic of the users as they are essentially based on *gaming* logic. More specifically, reference is made to a particular type of social network: *geo-social networks* or *location-based social networks*, which arose from the use of services offered by GIS systems within the social networks. The user, due to these systems, can provide and share information regarding their own geographical position with other network users. In the current case, it results as being useful for the definition of factors, indicators and standards of quality, as well as the control of performance requisites of specific urban functions. Furthermore, it is necessary to highlight how all this is possible, in *real time*, thanks to the use of mobile devices such as smartphones or tablets, equipped with GPS.

Particular attention was paid to the definition of some of the system functional requisites, connected to the delineation of a specific "user profile" potentially characterised by eventual categories of sub-users, for whom dedicated access with suitable privilege levels to specific DB areas must be made available.

All this is essential in order to ensure the supply and management of the service, whilst meeting a high level of personalisation of the system, as well as notable levels of loyalty and maximum satisfaction in its use.

The M2M project operated a first distinction between *professional users, non professional users* and *commercial users*.

With reference to this classification, attention was focused, within the "professional user" type, identifying a specific sub-category, that relative to local authority technicians (primarily public administrations), that fine themselves managing spatial information directly or indirectly connected to road safety on a daily basis.

For this user-profile, by means of face-to-face investigations, supported by semi-structured questionnaires, it resulted necessary to make dedicated access with suitable levels of privilege available to DB areas for processing sensitive data and that with direct impact on the info-mobility platform management system, in order to determine, codify (both quantitatively and qualitatively) and represent the interconnections between road accidents and the location of particular urban functions.

Concerning the work conducted in relation to *the DB system and structure*, an analysis was conducted that was pertinent to the identification of possible database architectures for the creation of the prototype, examining the hardware and software aspects and evaluating the advantages and limits in terms of cost, complexity and performance of the possible solutions identified.

In this regard, we highlight that from the bibliographic study conducted of the possible DB architectures, *two-tier client-server* and *multi-level client-server* emerged.

The *client-server* architectures are organised in tiers, each tier corresponds to a node or group of calculation nodes on which the system is distributed. Each of these functions as the server for the preceding level and as the client for the successive level.

Starting from such suppositions, the specific activities conducted regarded the design of a system for the archiving of data and for their successive extraction in order to be visualised.

In order to acquire the aforementioned information referring to the context chosen as the location of the test site, two specific data survey forms were structures, which present elements which were successively reflected on in terms both of vertical and horizontal integration as well as the streamlining of redundant information pertaining to the acquisition of data.

The survey forms created refer to two different types: *general form,* relating to the entire reference spatial area for the test site which in turn can be divided into investigation sub-areas, to facilitate data acquisition; *detailed form*, pertaining to each property subject of interest, in which it is possible to better specify valuable information.

In reference to the *construction of the GIS model and relative dataloading*, with a view to integrating the collected information and above all in order to use the arranged cartographic data in terms of road accident prevention, as well as acquiring data, two preeminent directions were followed: control of correct data transfer, without losing information, but means of the arranging of an adequate implementation methodology useful for developing the programme properties, as well as its specific functions; explication of the interest variables to be inserted within the DB, in correlation with the defined risk factors and indicators.

Precisely in reference to the construction of the GIS model and the relative loading of data, work was conducted in order to insert data found for the town of Crotone in the reference GIS model, following studies conducted in terms of comparison of the alignment method and the manual method, from which it emerged that the alignment method resulted as being more appropriate as it also permitted the improvement of a series of information compared to that obtained using the graphic method (e.g. known point coordinates).

The method of alignment is characterised by the following phases: loading of aerial photographs (this step occurs exploiting georeferenced aerial photographs in ecw format pertaining to a flight made in 2000, with Gauss—Boaga, Monte Mario Italy 2 reference geographical coordinates); assigning of a system of Gauss-Boaga, Monte Mario Italy 2 geographical coordinates, to the dwg tables, setting the coordinates for some known points (such a step resulted as being necessary prior to data processing, since at the basis of loading data on a GIS system it is necessary to set a system of uniform geographical coordinates); overlapping of aerial photos with the dwg files; transformation of the product in shape file; product verification. The manual method initially thought of and described as an exclusive solution for the use of Quantum GIS 1.8.0 open source software. From a comparison with the analytical method, it resulted as being inappropriate, above all in terms of time necessary to reproduce the tables, which had already been processed and presented in dwg format, substantially conducting a new structuring of the same (Francini et al. 2014).

2.3 Territorial Context of Experimentation

The test site reference area is situated in the province of Crotone and refers to a section of the SS 106 highway.

In order to analyse sufficiently, the municipal parameter was used as the preliminary parameter of spatial reading and of the relative urban uses, with specific reference to the Towns of Crotone, Cutro and Isola Capo Rizzuto, which are crossed by the road section being examined.

In the specific case of the Town of Crotone, in which the successive verification of theoretical assumptions was concentrated, situated on the east coast of Calabria, with a surface area of 179.83 km^2, at 8 m.a.s.l., bordering with: Cutro, Isola Capo Rizzuto, Rocca di Neto, Scandale, and Strongoli. These towns are linked to Crotone by an intense road network based on two main road axis, highway number "Ionica" and highway n. 107 "silana-crotonese". All the other arteries branch off from these two axis.

Cutro, in particular, is connected to the municipal territory of Crotone by highway n. 106 and municipal road n. 44. The following are present in the borough: a provincial road, n. 43, and a highway, n. 109. Instead, Isola Capo Rizzuto is linked to Crotone by provincial road n. 50. Two other provincial roads are present in the borough (n. 48 and n. 45, which is connected to highway n. 106), as well as numerous municipal roads, the most important of which is n. 60.

Hereafter follows a summary of some captured data for the town of Crotone, which were also preliminarily synthesised by means of cartographic processing and in the aforementioned survey form.

For reasons of simplicity, the data analysis and restitution of the relative results, the municipal territory was divided into two parts: Crotone 1 (to the north); Crotone 2 (to the south).

With regards to urban uses, from the conducted analysis it emerges that the area identified as Crotone 1 is affected by limited elements representing services, the tertiary sector, the productive sector, the tourism sector and historical-cultural heritage.

From the analysis conducted in the area identified as Crotone 2, it emerges that, in reference to urban uses within the *services* category, numerous schools are present, the San Giovanni Dio hospital with more 300 beds, as well as four private clinics and several sports centres, all located and catalogued in detail (Fig. 2).

2.4 Achievements and Prospects of Development

In order to pursue the research objective, at this point, as already stated, since the studies on this subject belong to the theme of urban and spatial planning associated

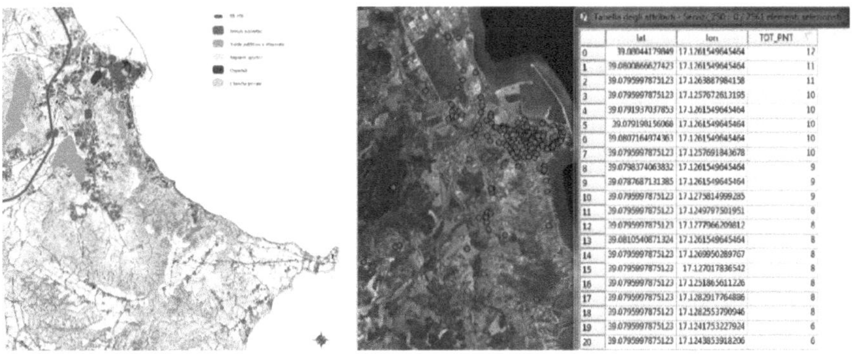

Fig. 2 Screenshot services category municipal territory of Crotone (1), Calabria. Original scale 1:10,000

to that of road safety, starting from the formulation of the initial hypothesis according to which urban uses can generate points of dangerousness, in that they are connected, for example, to a greater attraction of vehicular and pedestrian flow and/or a particular concentration of weak road users (children, pensioners, disabled people), the following categories of urban uses were defined which were useful for the general aims of the programme:

- services;
- tertiary/manufacturing sector;
- tourism sector.

The initial subdivision presented other analysis categories, in terms of location of uses, as previously indicated, with relative dimensional analysis and characterisation of access modes (compared to the main viability) in incorporated typological and functional terms; subdivision which was synthesised in a last analysis of the three aforementioned reference categories.

Successive to data loading, the interrogation criteria, which are summarised in the following table, were defined through location controls.

The interrogation criteria refer to:

- *relative distance* (variable on variation of urban use in connection to its level of dangerousness in terms of proximity to the investigated section and to the possible exploitation of the same use by weak users);
- *distance from the notable point* (which sums the previously chosen interrogation buffer value, 250 m, to the relative distance);
- *concentration level* (variable, in terms of range, for each use category, based on a first control interrogation on the effective presence of a maximum value of concentration of the urban use category in the reference buffer) (Table 1).

In relation to the definition of the starting ranges, the minimum assigned value is equal to 1 in that it is intrinsic to the definition of dangerousness, therefore a starting value of 0 would have no significance.

Following the normalisation of values and the definition of the reference range of the Urban Dangerousness Index (UDI) (as indicated above in the analytical representation), the values in percentages of γ coherent with the general formulation of the M2M project were identified.

In terms of validation of the Mobile prototype, as well as the successive verification of the efficiency and usability of the web platform for professional users, it therefore arises that each urban use characterises, in terms of relevance of dangerousness, a similar weight as the differentiation criteria are connected, both to the relative distances and the effective concentration level of some categories compared to others that, according to the definition in literature of each single urban use, will result as generally having an almost constant proportional relationship, even referring to a different analysis site (Francini et al. 2014).

Table 1 Characterisation factors of the urban dangerousness index

Urban land uses	Relative distance (m)			Distance from the notable point (m)			Concentration level		
	d1	d2	d3	D1	D2	D3	Low	Medium	High
Services	125	250	500	375	500	750	1–2	3–4	>= 5
Tertiary/manufacturing sector	250	500	1000	500	750	1250	1–9	10–19	>= 20
Tourism sector	200	400	800	450	650	1050	1–5	6–10	>= 10
yi							0.25	0.5	1
βi				1	0.	0.25			
IPU = Σyi * βi	IPU			γ			IPT = IP + γ * IP		
		0 < IPU < 0.3	Low	10					
		0.3 < IPU < 0.6	Medium	20					
		0.6 < IPU < 1	High	30					

To this purpose, it should be remembered that the destination field can be deduced from multilevel OSM, applications, which are adaptable on a global level (Crotone is only a reference adopted for the test site).

References

Batty M (1997) Cellular automata and urban form: a primer. J Am Planning Assoc 63:264–274

Coppola P, Nuzzolo A (2006) Accessibilità extraurbana e localizzazione delle attività socioeconomiche. XXVII Conferenza italiana di Scienze Regionali "Impresa, mercato, lealtà territoriale". Pisa

Ferrand N (2000) Multi-reactive agents paradigm for spatial modelling. In: Fotheringham AS, Wegener M (eds) Spatial models and GIS: new potential and new models. Taylor & Francis, London, pp 176–184

Francini M, Colucci M, Palermo A, Viapiana MF (2012) Intangible fruitions for the territorial promotion. TEMA J Land Use Mobility Environ Mobility Competitiveness 5:133–144

Francini M, Palermo A, Viapiana MF (2014) Urban land uses and smart mobility. TEMA J Land Use Mobility Environ 441–452 (INPUT 2014—Special iussue)

Landis JD, Zhang M (1998) The second generation of the California urban futures model. Part 1: Model logic and theory. Environ Plan 25:657–666

Palermo A, Francini M (2013) The impact of urban uses in road safety. XX Conferenza Internazionale Vivere e camminare in città - La sicurezza degli utenti deboli della strada. Brescia

Salomon I, Waddell P Wegener M (2002) Sustainable life styles? Microsimulation of household formation, housing choice and travel behaviour. In: Black WR, Nijkamp P (eds) Social change and sustainable transport. Indiana University Press, Indiana, pp 125–131

Wegener M, Spiekermann K (1996) The potential of microsimulation for urban models. In Clarke G (ed) Microsimulation for urban and regional policy analysis. European research in regional science 6. Pion, London, pp 146–163

Energy and Climate Change Polices in Europe: Overview and Selected Examples from a Spatial Planning Perspective

Rocco Papa, Carmela Gargiulo, Floriana Zucaro, Mario Cristiano, Gennaro Angiello and Gerardo Carpentieri

Abstract The challenges imposed by the changing climate and the energy-driven developments are very complex and need to be addressed from the global to the local scale. In the last decades, this issue has attracted the attention of policy makers at all levels of government, attempting to adopt an integrated and adaptive energy and climate strategy. This paper reviews and analyzes the main efforts that have been made in Europe to secure a transition toward a low-carbon and energy—efficiency society from a spatial planning perspective. To this aim, the paper presents an in-depth analysis of selected climate and energy policy documents elaborated at three different levels of governance: the European Union, the national and the local level. At the European level, our analysis shows that very limited attention has given to spatial planning as a strategy to reduce or ameliorate the impacts of the changing climate. At the national level, while mitigation policies are more inclined towards techno-centric solutions, adaptation policies partly recognize the anticipatory role of spatial planning to play in promoting robust adaptation measures. At the local level, where most of the causes and the effects of the changing climate are manifested, technological options are often well integrated with spatial planning. However, even at the local scale, energy and climate policies focus mainly on individual sectors or urban functions rather than systemic changes.

R. Papa (✉) · C. Gargiulo · F. Zucaro · M. Cristiano · G. Angiello · G. Carpentieri
DICEA, University of Naples Federico II, Naples, Italy
e-mail: rpapa@unina.it

C. Gargiulo
e-mail: gargiulo@unina.it

F. Zucaro
e-mail: floriana.zucaro@unina.it

M. Cristiano
e-mail: mario.cristiano@unina.it

G. Angiello
e-mail: gennaro.angiello@unina.it

G. Carpentieri
e-mail: gerardo.carpentieri@unina.it

R. Papa and R. Fistola (eds.), *Smart Energy in the Smart City*,
Green Energy and Technology, DOI 10.1007/978-3-319-31157-9_13

1 Introduction and Context

The relationships between climate change and energy are very complex, raising serious, high-level policy challenges that need to be addressed from the global to the local scale (Pasimeni et al. 2014). In the last decades, this issue has attracted the attention of policy makers at all levels of government, attempting to adopt an integrated and adaptive energy and climate strategy (Brunnengräber 2013; Lundqvist and Biel 2013; Goldthau 2014).

At the global level, the European Union has assumed a clear leadership role in international climate change and energy governance by promoting both internal and external ambitious policies (Oberthür and Roche Kelly 2008). At the national level, Member States have made, over the past two decades, significant advances in developing strategies to combat climate change and to reduce energy consumption. Some have implemented appropriate legislation to achieve these goals while others are in the process (Casado-Asensio and Steurer 2014). Policies at the national level set the legal framework within which local authorities implement practical strategies that could help reduce or ameliorate anticipated negative effects of climate change while contributing to the achievements of the energy targets set up at the national and global level (Zanon and Verones 2013).

Local-level polices, however, are not solely the outcome of a multi-scale, state-led processes and, in the recent years, several research (e.g. Juhola and Westerhoff 2011; Ostrom 2012) has outlined the importance of non-state actors in processes of decision-making on energy and climate change at the local level. Especially where national government-led policymaking have been limited, the participation in EU-funded research programmes or in NGO networks has represented the principle driver for the implementation of local policies. This particular form of governance is generally referred as network governance (Gustavsson et al. 2009; Juhola and Westerhoff 2011). It poses emphasis on the role of markets and networks in decision-making and on the fact that governments are increasingly dependent on the co-operation of various policy actors outside their hierarchy (Gustavsson et al. 2009; Juhola and Westerhoff 2011).

Besides being a multi-scale issue, the energy-climate challenge cuts across sectors and societal domains. The implications of this challenge for sectoral domains have been debated in recent literature. This includes the implications of the energy-climate challenge for, among others, the forest (Lindner et al. 2010), the agriculture (Kurukulasuriya and Rosenthal 2003), the tourism (Scott et al. 2012), the finance (Labat and White 2007) and the spatial planning sectoral domain (Wilson and Piper 2010).

As the energy-climate challenge has a clear urban dimension, it is not surprising that a growing body of literature has analyzed the critical role of spatial planning in developing mitigation measures to reduce energy consumption and GHG emission and adaptation measures to ameliorate the effect of climate change (Laukkonen et al. 2009; Grazii and van den Bergh 2008; Biesbroek et al. 2009; Measham et al. 2011; Hurlimann and March 2012; Zanon and Verones 2013; Papa et al. 2014). Drawn on the broader literature on urban sustainability, a first research strand has addressed the potential contribution offered by spatial organization and related physical urban planning to a reduction of GHG emissions and energy consumption, especially in the transport and residential sector (Grazii and van den Bergh 2008; Zanon and Verones 2013; Papa et al. 2014). A second research stand has addressed the anticipatory role of spatial planning to play in promoting robust adaptation measures in order to reduce the risks and capitalize on the opportunities associated with global climate change (Measham et al. 2011; Hurlimann and March 2012). Finally a third research stands has addressed the possible synergies and trade-offs between mitigation and adaptation policies by exploring the role that spatial planning can play in developing effective mitigation and adaptation options in an integrated manner (Laukkonen et al. 2009; Biesbroek et al. 2009).

We aim to contribute to this debate from a different angle that is to review and analyze the main efforts that have been made in Europe to secure a transition toward a low-carbon and energy—efficiency society from a spatial planning perspective. To this aim, we present an in-depth analysis of selected climate and energy policy documents elaborated at three different level of governance: i) the European Union strategy level where the European Council sets the overall strategies with predominantly strategic declarations and roadmaps; ii) the national strategic decision-making level aimed to implement the European Union policies into national legislation and iii) the local level where national legislation are being implemented in the cities and regions.

In particular, in Sect. 2, we present the strategy of the EU in the fields of energy and climate change. This section covers the main EU polices including the European Energy and Climate Change Package, the European Strategic Energy Technology Plan, the European Road maps to a low carbon economy, the Energy Efficiency Plan for 2011 and finishes with a brief consideration of the implication of the EU policies for the spatial planning domain. In Sect. 3, we present an overview of the national climate change and energy policies adopted by EU member States, followed by an analysis of two examples in Denmark and Italy. In Sect. 4, we present two relevant case studies of European cities that have recently adopted integrated energy and climate change plans aimed to meet the challenges imposed by the changing climate and the regime of resource scarcity. The work ends with a discussion on the examples analyzed and with some suggestions on future research directions, offering some insights to this end.

2 EU Policies for Energy and Climate Change

Since the early 70s the environmental and energy issues have begun to play a central role in European policies, even if at the time the action of the European Community was not yet mature enough to establish precise commitments and obligations within legislation, which had implications in the field of territorial transformations. The documents developed until the mid-90s (the European Act of 1987, the Maastricht Treaty of 1992, the European Energy Charter of 1994 and the Treaty of Amsterdam in 1997), for example, addressed the energy issue almost exclusively in terms of security of supply and investment, rather than as an environmentally and climate treat (European Communities 1987, 1992, 1997).

Although the EU has pledged to take a lead in reducing greenhouse gas emissions, being the first to adopt on a voluntarily basis the objective of stabilizing CO_2 emissions at 1990 levels by 2000, it didn't identify climate protection as one of the priority action areas until 2001 in the VI Environmental Action Plan "Environment 2010: Our Future, Our Choice" (European Commission 2001), in order to "stabilize the atmospheric concentration of greenhouse gases at a level that will not cause unnatural variations of the climate terrestrial." In this document spatial planning, seen as an indispensable tool for the development and use of land, has an important role, which influences the nature of places and how they can function, "in the complex play of forces and pressures which give rise to environmental problems the role of planning and land management is fundamental: they cover a wide range of decisions, usually made at local or regional level, determining the character and intensity of land uses and activities which may often have a major impact on environmental conditions." The plan calls for the need to promote the integration of environment and climate policies in regional planning and to encourage the dissemination and exchange of experiences of sustainable planning, especially at local and regional levels. For this reason, among the many proposed actions it included communication initiatives aimed at directing attention to planning and the environment, and to support the cohesion policy Community assigning a greater role to European cities to achieve the goals of sustainable development.

Numerous documents and European directives that relate to the next low-carbon future especially promote energy efficiency, renewable energy sources and mitigation of greenhouse gases (Fig. 1), without, however, in most cases, making explicit reference to the strong contribution that actions of territorial transformation may make instead. This is the case of the Energy Services Directive (ESD 2006/32), which set a 9 % estimated energy savings target by 2016 and imposed in each MS to identify its own saving national target (with an intermediate 2010 target) and to prepare three National Energy Efficiency Action Plans (NEEAPs), 2007, 2011 and 2014 to illustrate the strategies and advancements. Both in the ESD and the national plans we can, in fact, note the absence of measures and actions related the design and use of territories and if it can be understood in the case of the Directive, since this mainly concerns the market and the supply of energy, it is a big point of weakness in the other case. The ineffective coordination between the national and

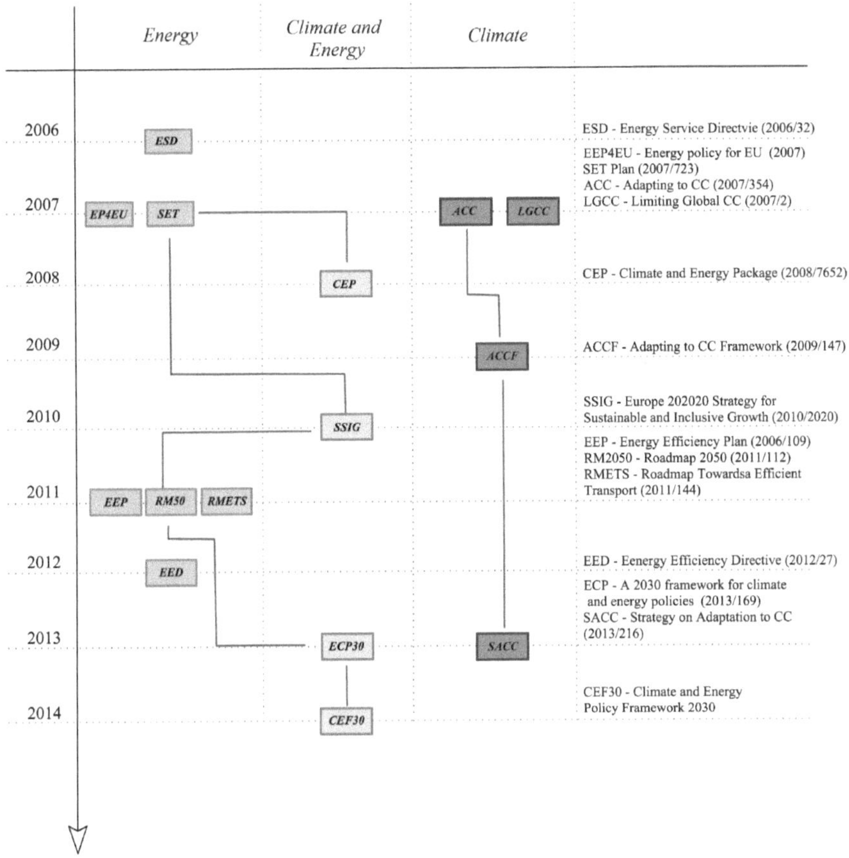

Fig. 1 EU greenhouse gas emissions for the major activities in the period 1990–2050 (100 % = 1990) (Color figure online)

the regional and local authorities, in terms of actions to be implemented and subjects to be involved, can hinder the achievement of energy goals identified at the national level and can increase the uncertainty in which engineers, planners and administrators have to act (Bulkeley et al. 2010).

The urgency of limiting anthropogenic climate change, as well as the need to adapt to its effects, actions now inevitable as stated by the scientific community (EEA 2012), has led Europe to adopt the Green Paper: Adaptating to Climate Change (COM (2007) 354) to define possible actions which need a European scope, while recognizing the need to cooperate with MS and regions.

The path that the EU has undertaken to implement policies and strategic plans related to the challenges of climate change inevitably, or at least should, intersect with the one related to energy and environmental sustainability, and the government of territorial changes, because of the strong indissoluble link there is between climate, energy,[1] and territorial development issues.

On the one hand, in fact, the production and consumption of energy are particularly sensitive to temperature and to extreme phenomena (for example, the change of climatic conditions can influence both the production of wind, solar and biomass, and the demand for energy in terms of increase or decrease of the winter heating and summer cooling needs) and on the other hand, saving and energy efficiency can significantly help to reduce the production of greenhouse gas emissions. In turn, most of the effects on consumption, production and distribution of energy, as well as the release in atmosphere of greenhouse gases, depend on the physical and functional organization of the territory, the distribution of assets and infrastructure, as well as flows of commuting (Grazii and van den Bergh 2008).

Because of this unity of purpose, and interdependence of the issues addressed, often strategic plans are implemented with actions that concern, even exclusively, themes of one other. So since there is this, even implicit, multi thematic approach analyzing the EU Energy policy must maintain the broadest possible perspective.

To support MS in the development of innovative approaches to research, development and implementation of actions and programs to respond promptly to urgent climate and energy challenges, the EU in November 2007, proposed the European Strategic Energy Technology Plan (September-Plan) that sets new goals for European research in the field of energy and low-carbon technologies. It is a strategic plan for the development and feasibility of use of "low carbon" technologies: such as the offshore wind, solar or second-generation biofuels. Investment in areas of "the knowledge triangle" (da Graça Carvalho 2012), reinforcing research, education and innovation in the energy sector, is considered as a fundamental condition for the achievement of environmental and energy objectives.

The SET-Plan focuses on the coordinated planning between the European level, Member States and the local level; developing the potential for European Research and Innovation and fully exploiting the possibilities offered by the market. The SET-Plan, in practice, aims to contribute to the transition to a sustainable and responsible economy and development by 2050. To better manage the actions in the long run, these objectives are divided into two time periods: by 2020 reduce GHG by 20 % compared to 1990, to improve energy efficiency to reduce by 20 % the energy consumption of the EU, to cover at least 20 % of EU energy consumption from renewable sources, covering at least 10 % of total consumption for vehicles with biofuels; by 2050, to contain the rise of the average temperature within 2 °C, aims to reduce both greenhouse gas emissions, that energy costs by 80–95 %.

In particular, the energy and reduction of greenhouse gas emissions by 2020 objectives, are closely interdependent goals, as the emission of CO_2, a leading greenhouse gas, is a consequence of the production and consumption of electricity from fossil fuels.

[1]Climate Change 2013: The Physical Science Basis (Climate Change 2013: foundation in physical science); Working Group I of the IPCC, summary for policy makers, October 2013.

The different legislative instruments that translate the objectives into operational objectives for 2020 have been defined within the "Climate and Energy Package" in 2008, which is an integrated proposal on environmental, climate, and energy issues. With this package, Europe has wanted to reiterate a concept proposed within the SET-Plan: abandon the sectorial approach to deal with the energy issue in favor of a unified approach that can better respond to the transversal nature of this theme.

The importance of "integrating adaptation into EU key policy areas" has prompted the EU to adopt, in 2009, the White Paper Adapting To Climate Change Towards A Eu Framework (COM (2009) 147): "However, even if the world succeeds in limiting and then reducing GHG emissions, our planet will take time to recover from the greenhouse gases Already in the atmosphere. Thus, we will be faced with the impact of climate change for at least the next 50 years. Therefore we need to take Measures to adapt."

Later, in March 2011, the Energy Efficiency Plan (European Commission 2011a) was presented, which stresses the need to put in place concrete measures for energy efficiency and savings, given that "recent estimates from the Commission that indicate that the EU will achieve only half of the 20 % objective". Although this plan contains information regarding the efficiency of the building industry and transport, specific actions that concern territorial transformations are not found.

In the same year, in order to contain global warming determined by climate change within less than 2 °C, the EU has developed a roadmap to reduce its emissions by 80 %, compared to 1990, by 2050 (European Commission 2011b). This new target is in line with the decisions taken both in the Copenhagen and Cancun agreements, which with the provisions by the Intergovernmental Panel on Climate Change (IPCC 2014). To reach the 80–95 % overall reduction of GHG by 2050, the Roadmap indicates that a smooth and cost-effective transition requires a 40 % reduction in the production of GHG (compared with 1990) by 2030 to achieve the target by 2050. By building on what has already been achieved, the EU must define appropriate strategies in this regard and all member states should develop national low carbon Roadmaps.

Despite the willingness to adopt a systemic and transversal approach because of the complexity of the issues addressed and the need to coordinate effectively the different levels of government of the territory, the Roadmap and the EEP, provide exclusively sectorial recommendations, failing in the attempt to ensure the cross sectorial approach suggested in previous reports.

The publication of the Roadmap has influenced the design of future energy scenarios and policies and strategies to be developed in the coming years, as "any business as usual or reference to existing policies scenarios leads to results that are totally incompatible with the purpose of policy" (REF-E 2012); the full implementation of current policies and a schedule to reach the target is necessary to ensure the achievement of greenhouse gas reductions compared to 1990.

The knowledge that the EU energy efficiency target is not being implemented and that energy efficiency is a valuable tool to greatly limit climate change, has led the

EU to adopt a new directive, the Energy Efficiency Directive (EED 2012/27) outlined by the EU that commits Member States to adopt binding measures. The latter are, in practice, called to encourage public authorities, at regional and local level, to adopt an energy efficiency plan containing specific objectives and actions of energy conservation and efficiency, issues on which the public must be informed and sensitized by promoting adequate information, awareness and training initiatives. The EED is also the instrument through which to try to fill the gaps of a structural nature, such as the lack of a common calculation methodology or a report model and the absence of an ambitious savings target, which characterized the 2006 ESD.

In 2012, in addition to "a common framework of measures for the promotion of energy efficiency in order to ensure the achievement of the objective of 20 % energy efficiency by 2020" (EED 2012/27), outlined by the EU a manual (Delivering on the Europe 2020 Strategy—handbook for Local and Regional Authorities), was drafted in the same year, to promote the adoption of an integrated territorial approach, based on multi-level governance within the regional and local policies, to facilitate the declination of the objectives and 20-20-20 priorities locally by the Committee of the Regions.

The inevitable necessity to adapt to the increase in global temperature and other changes contemplated by the climate and socio-economic forecasts, has lead, in 2013, to the approval of the Strategy on adaptation to climate change (COM (2013) 216) to ensure "mainstreaming" of adaptation measures in the EU sectorial policies. In practice, the EU tries to encourage all Member States to adopt strategies, while also providing substantial funding to help them build their adaption and action capacity. Next to specific actions to strengthen the capacity of MS to reduce the vulnerability of their territories, the strategy also provides for adaptation initiatives in the city, promoting the voluntary commitment to adopt local adaptation strategies and outreach activities within the Covenant of Mayors framework (2013/2014) (Fig. 2).

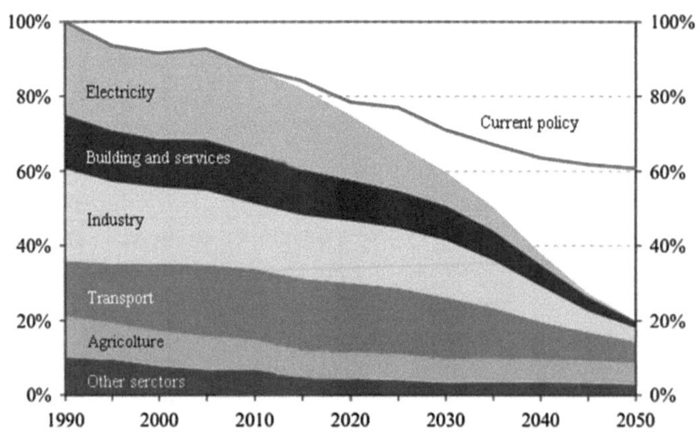

Fig. 2 Time diagram of the main documents per macro areas from 2006. *Source* European Commission (2011b)

In March 2013, the European Commission proposed the Climate And Energy Policy Framework For 2030 (COM (2013) 169), developed in agreement with what had been outlined by the Climate And Energy Policy Framework For 2020 in 2011, the Roadmap for moving to a competitive low carbon economy in 2050 (Energy Roadmap 2050) (COM (2011) 112), and the White Paper Roadmap to a Single European Transport Area—Towards a competitive and resource efficient transport system (COM (2011)114). In particular, the Framework for 2030 has set a 40 % reduction of greenhouse gas emissions compared to 1990 levels, and 27 % energy savings due to the efficiency target for Europe, making the global targets compulsory without defining the extent of the efforts that each MS will face.

In practice, for individual MS mandatory goals are not identified either for the deployment of renewable energy (RES), nor for energy efficiency. Although such a decision could have as a direct consequence the proliferation of approaches, tools and different goals defined arbitrarily by each nation, it is interesting to note the emphasis placed by the same EU on the need to ensure an integrated policy frame work to ensure a coordinated approach among the various MS. To this end the 2030 Framework proposed a governance framework based on national plans for competitive, secure and sustainable energy as well as key indicators to track the progress over time.

From the description above it is therefore evident that the EU leaves ample room for discretion both about the goals and instruments to be adopted at the national level, as already highlighted by previous studies on the subject (Rietig 2012) and by the EU itself in a 2015 document for the construction of a “resilient [Europe], accompanied by a far-sighted policy on climate change” (COM (2015) 80).

The difficulty of implementing amongst the MS a coordinated system of actions to reduce the impacts on climate change and the energy crisis, appears to be accentuated by two additional aspects: one is the difficulty of assessing, in progress, the effectiveness of long-term actions and the effect of the combined actions from the European level to the local one, and the other, the rapid succession of directives transposed into national legislation of the MS with different times and in different ways. These issues have prompted the EU to promote a number of projects designed to test the effectiveness of the policy in terms of energy efficiency using feedback to inform the processes of governance in the process of application and formulation, as reiterated in the COM (2015) 80. For example, the Energy Efficiency Watch Project, developed to monitor the implementation process of the Energy Services Directive (Directive 2006/32), which, as described above, has determined that MS submit energy efficiency national action plans. The Energy-Efficiency-Watch project through surveys and “bottom-up” analysis has compiled, for each MS, evaluating reports of the application of energy efficiency policies to monitor progress.

In the Survey Report Progress in energy efficiency policies in the EU Member States—the experts perspective Findings from the Energy Efficiency Watch Project 2012 revealed a disparity in the level of ambition of the policies and the difficulty of convincing the authorities that energy efficiency is not a "burden" for the public finances but a source of benefits for society. This report also stresses that the success of the policies is based on a multi-level governance that integrates the various levels and uses the strengths of each of these. In some MS experts see the local and regional level as the main driver of energy efficiency, as evidenced by the number of authorities willing to make it a priority, probably because of the greater ease in communicating the benefits. Instead critical factors are the shortages of staff and agencies dedicated to the issue and the difficulty of involving the financial departments.

The need to provide information to guide decision making and overcome the inherent difficulties mentioned is also the basis of the "Bottom-Up Climate Adaptation Strategies Towards a Sustainable Europe" (BASE) supporting actions for sustainable adaptation to climate change in Europe ensuring availability and accessibility to scientific information on the subject. This project shows that despite the issues of energy efficiency and climate change are inseparable, explicit integration between plans and their implementation at the local level through instruments of government land is often lacking. For example Energy Efficiency Plans studied within the project do not refer to the need to adapt to climate change or will refer to it generically without outlining explicit actions, objectives and obstacles to overcome.

3 National Energy and Climate Change Polices: Overview and Selected Examples

In the previous section, we looked into energy and climate change issues at the European strategic level. As we stated above, the success of the European strategy crucially depends on the ability of EU Member States to play their part in implementing the necessary reforms at national level. In this section, we give an overview of climate change and energy policies in EU-15 Member States, and their implementation, identifying common strategies and outcomes. Policies' analysis of EU-15 in these fields are common in the literature (e.g. Casado-Asensio and Steurer 2014; Abolhosseini et al. 2015) and we also focus our attention on EU-15 as they represent the largest emitters of GHG emission, accounting in 2012 for 84 % of total EU greenhouse gas emissions and 80 % of EU final energy consumption. After this analysis, we present two selected examples: Denmark and Italy. The two countries represent different stages of planned actions for energy and climate change and present diverse historical trends in term of GHG emission, energy consumption and share of renewable energy source. Furthermore, they will also face different challenges in terms of climate change adaptation (Ciscar et al. 2011).

The comparison between the nations is essential to determine to what degree regional planning is part of the governance and transformation processes brought about by the necessary climate change mitigation and adaption and energy related actions.

3.1 *Overview*

Stimulated by the increasing evidence of climate impacts occurring, National governments have, to varying degrees, placed climate change on the national political agenda; drafting strategies in the fields of energy, climate adaption and mitigation, through multiple approaches for each field. For example in the energy field EU policies intervene on Energy Efficiency, Renewable sources, Energy Saving.

The most recent and important policy in the energy field is the Energy Efficiency Directive 2012/27/UE (EED) that in the art. 7 has determined the obligatory EE regimes and the types of policies each member state may implement. According to this article, in fact, every Member State had to establish its own savings target to be attained between 1 January 2014 and 31 December 2020. The energy savings target, the methods and data used in the calculation, as well as policies and measures to be taken to achieve them, were notified prior to December 2013, representing, in practice, a mid-term trend of the adoption of EED in the year of adoption and the dead line delivery of the NEEAP 2014. Article. 7 of EED has also provided for the use of different regulatory, financial, fiscal, voluntary, or information tools, to facilitate the deployment of actions aimed at improving energy efficiency to achieve the objectives identified (Table 1).

Table 1 Notified measures under Article 7 (Article 7 of the energy efficiency directive: state of play Lelde Kiela Vilumsone February 2015, Brussels)

Type of measure	Number of member states
EEOS (default instrument)	17
Energy Efficiency National Fund	7
Energy or CO_2 taxes	12
Financial schemes or fiscal incentives	19
Regulations or voluntary agreements	9
Standards and norms	10
Energy labelling schemes	3
Training and education	10
Other measures	9

Because of the obvious flexibility of intervention of the MS, the EU stressed the need to establish a strong governance framework to ensure the achievement of the objectives of the European renewable energy and energy efficiency measures that will help achieve the thresholds of GHG reduction and consistent with the general principles of the European energy policy (COM/2014/015 "a policy framework for climate and energy in the period from 2020 to 2030"): "there is a need to simplify and streamline the current separate processes for reporting on renewable energy, energy efficiency and greenhouse gas reduction for the period after 2020, and to have a consolidated governance process with Member States."

Even though the EU states the need, for all MS, to use a common approach in the definition of their National climate plans, the level of implementation varies considerably.

According to the report of the Coalition for Energy Savings (2014), which analyzed the NEEAPs and the required Article 7 notifications (Fig. 3), the majority of the national plans are incomplete, and having used the maximum allowed exemptions will reduce the actual final use target from the 1.5 % per year expected to an estimated 0.8 %.

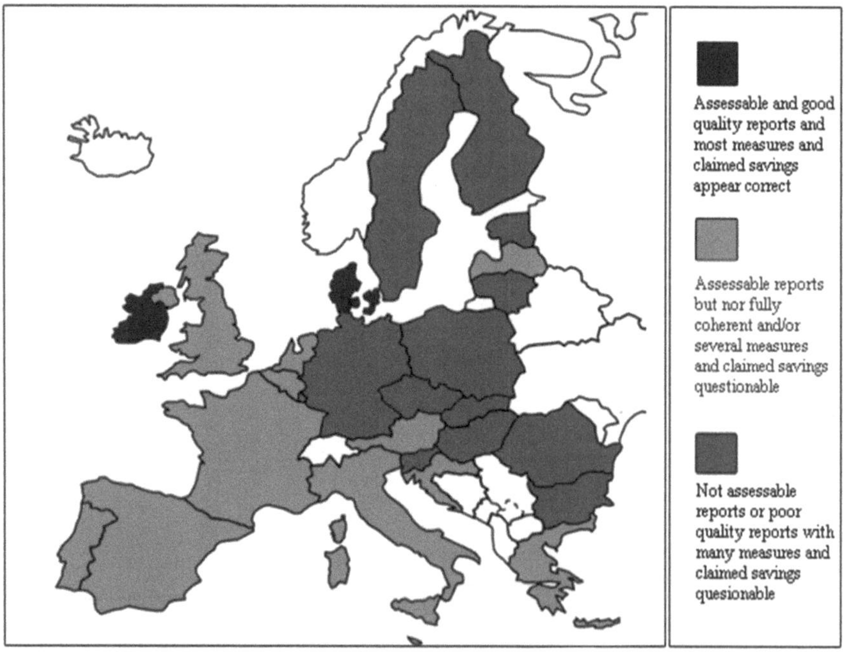

Fig. 3 Ranking of MS article 7 plans (NEEAPs). *Source* Adapted from Coalition of Energy Saving (2014) (Color figure online)

These strategic plans, although they recognize the need to reduce the rate of energy consumption attributable to buildings, identify only tax and financial measures to be implemented, without any reference to the impact that the instruments of territorial government may determine.

Although all NEEAPs reiterate the need to limit energy demand as a first step towards reducing the energy dependence of the member countries, which is exactly what the EED should achieve, the EU is unlikely to meet its 2020 target of 20 % energy savings, amounting to 368 Mtoe, with an estimated gap of 190 Mtoe.

In addition to not being able to achieve the energy saving and efficiency goals in the near future, the EU will fail to meet its ambitious targets on climate change unless significant changes are made (EEA 2015).

Until recently, the primary response to climate change has been mitigation through the reduction of greenhouse gas emissions. In this regard, 7 of 15 EU countries developed their first national strategy in the period between the 1990 and 1997. After 1997, when the Kyoto Protocol was agreed upon and the EU as a whole had to report on emissions and mitigation efforts to the UNFCCC Secretariat, other 5 EU-15 countries adopted their own national strategies. Only after the EU ratified the Kyoto Protocol in 2002, the 3 laggard countries (Portugal, Spain and Luxemburg) adopted a national mitigation strategies (Table 2). Today all the EU-15 countries have a national strategy, while those who started first have renewed mitigation strategies two or even three times (i.e. Germany and the UK).

National mitigation strategies have been developed under the responsibility of different bodies, including the Ministry for the Environment, the Ministry for Spatial Planning or ad hoc established inter-ministerial working groups. Recognizing the complexity of energy and climate policies some countries have developed integrated energy and climate national strategies (in red in Table 2), while others (i.e. Belgium, Denmark and the UK) have created governmental agencies like the Ministry for Energy and Climate which are responsible for national climate policy and international cooperation on climate change, as well as energy issues.

Despite the fact that they have been implemented by different national bodies and in different periods, national mitigation strategies presents many common points. These documents after the presentation of the necessary legal background and national GHG trends, provide an account of measures and policies organized around themes/sectors, such as: energy, building, transport, waste management, agriculture and forestry. In addition, the policy documents cover cross-sectorial issues such as financial policies, research and development, communication and information.

Adaptation practices to counter adverse impacts of the changing climate appear in policies at national levels of governance in the mid of 2000s (Table 3). Initially the most vulnerable nations and sectors started to include resilience into their planning activities, and between 2005 and 2007, 5 of 15 EU countries have developed their own national adaptation strategies. After the publications of the European Commission's Green Paper 'Adapting to climate change in Europe—options for EU action' June 2007 (CEC 2007), other 8 countries followed and today

Table 2 National mitigation strategies of EU15 (Color figure online)

	Early Stage	2004	2005	2006	2007	2008	2009	2010	2011	2012	2013	who
Austria	Austrian Strategy to Adapt to the Kyoto Goals (2002)				Adaptation of Austria's climate strategy for achieving the Kyoto Targets							Federal Ministry of Agriculture, Forestry, Environment and Water Management
Belgium	Belgian National Programme for Reducing CO2 emissions (1994); National Climate Plan 2002-2012 (2002)											Federal Public Service for Health, Food Chain Safety and Environment
Denmark	Action Plan to Reduce CO2 emissions (1996); Climate 2012 (2000)					Denmark 2020 (2008)					The Danish Climate Policy Plan - Towards a Low Carbon Society	Danish Government; Ministry of Climate and Energy
Finland	National Climate Strategy of Finland (2001)		National Strategy to Implement the Kyoto Protocol			Long-term National Climate and Energy Strategy					Review of the National Climate And Energy Strategy	Climate and Energy Policy Network
France	National Climate Programme to Combat Climate Change (2000)	Climate Plan (2004-2012)		Review of the Climate Plan		Review of the Climate Plan			Review of the Climate Plan			Ministry of Ecology, Sustainable Development and Energy
Germany	National Climate Protection Programme (2000)		Review of National Climate Protection Programme		Review of National Climate Protection Programme	Integrated Energy and Climate Change Programme						Federal Ministry for the Environment, Nature Conservation, Building and Nuclear Safety
Greece	National Climate Change Programme (1995); Agenda to Reduce Greenhouse Gas Emissions (2002)				Action Plan on Climate Change							Ministry for the Environment, Physical Planning and Public Works
Ireland	National Climate Change Strategy (2000)				Second Strategy 2007-2012				Third Strategy 2012-2020			Department of the Environment, Heritage and Local Government
Italy	National Programme for the Containment of Carbon Dioxide Emissions for the Year 2000 (1994); Review of National Climate Programme (2002)				Review of National Climate Programme							Inter-Ministerial Committee for Economic Planning
Luxembourg				CO2 Reduction Action Plan (2006)							Second National Action Plan for the Reduction of CO2 Emissions	Ministry of the Environment
Netherlands	First National Climate Policy Plan (1990); Second National Climate Policy Plan (1999)				Clean and Efficient: New Energy f or Climate Policy				Climate Letter 2050		Climate Agenda	Minister of Spatial Planning and the Environment shifted to Ministry of Infrastructure and the Environment
Portugal		National Climate Change Programme		Review of the National Climate Change Programme					National Low Carbon Roadmap 2050			Ministry of Agriculture, Sea, Environment and Spatial Planning shifted to Ministry of Environment, Spatial Planning and Energy
Spain		Spanish Strategy for the fulfilment of the objectives under the Kyoto Protocol			Spanish Climate Change and Clean Energy Strategy 2007--2012							Ministry of Agriculture, Food and Environment
Sweden	Climate Change Strategy (1993);Climate Strategy for the Energy (1997); Renewed Climate Strategy (2002) Sector	Renewed Climate Strategy				Renewed Climate Strategy	Sustainable Energy and Climate Strategy					Ministry of the Environment and Energy
United Kingdom	UK Climate Change Programme (1997); Review of Climate Change Programme (2000)	Review of Climate Change Programme		Climate Change-The UK Programme		Climate Change Act	UK Low Carbon Transition Plan					Department for Environment, Food and Rural Affairs shifted to Department of Energy and Climate Change

Luxemburg and Greece are the only two countries without a national document even if it is part of their political agenda (EEA 2015).

As is the case for mitigation National adaptation strategies have been developed by different bodies: for example in France, an ad hoc established national observatory while in the case of the Netherlands an ad hoc program has been established at the Ministry of Spatial Planning, underlying the complex relationship and the need for integration between adaptation and general planning policies. The same holds true for the similarities in the structure of the national documents.

Adaptation measures proposed by EU-15 countries range from small changes to very substantial changes that have broad implications, but very often comprise a

Table 3 National adaptation strategies of EU15

	2005	2006	2007	2008	2009	2010	2011	2012	2013	2014	who
Austria									The Austrian Strategy for Adaptation to Climate Change		Federal Ministry of Agriculture, Forestry, Environment and Water Management
Belgium						Belgian National Climate Change Adaptation Strategy					National Climate Commission
Denmark				Danish Strategy for Adapation to a Changing Climate							Ministry of Environment, shifted in 2008 to Ministry of Climate and Energy
Finland	Finland's National Strategy for Adaptation to Climate Change										Ministry of Agriculture and Forestry
France			National Strategy of Adaptation to Climate Change								National observatory dedicated to the effects of climate warning - Inter-ministerial delegate for sustainable development
Germany				German Strategy for Adaptation to Climate Change							Environmental Ministry supported by the Federal Environmental Agency
Greece											
Ireland								National Climate Change Adaptation Framework - Building Resilience to Climate Change			Department of the Enironment
Italy										Elements for an Italian National Adaptation Strategy	Ministry for Environment Land and Sea
Luxembourg							National Adaptation Strategy on Climate Change				Environment and Climate Partnership
Netherlands			National Adaptation Strategy "Make Space for Climate"								Adaptation to Climate Change in Spatial Planning Programme - Coordinated by the Ministy of Spatial Planning Housing and the Environment
Portugal						National Strategy for Adaptation to Climate Change					Ministry of Agriculture, Sea, Environment and Spatial Planning
Spain		The Spanish National Climate Change Adaptation Plan									Environmental Ministry; National Office for Climate Change
Sweden			Sweden Facing Climate Change: Threats and opportunities								Swedish Environmental Protection Agency
United Kingdom				Adapting to Climate Change in England.A Framework for Action							Department for Food, Rural Affairs and the Environment

mix of different options. The themes that have attracted the greatest attention throughout Europe in terms of risk and vulnerability assessment at national level are water, followed by agriculture, forestry, human health and biodiversity. Since the environmental aspect was central to the concerns that instigated the climate change policies, the national policies incorporate heavily these themes. Although urban planning and energy aren't the highest priorities (Fig. 4) the MS have at last recognized the key role they play to support carbon neutral communities (EEA 2015). However, as adaptation is often a multi-scale, cross-sectorial issue, cross sectorial initiatives are generally included in national strategies. In contrast with mitigation strategies, economic analysis are narrowly implemented. This is also the effect of a high level of intrinsic uncertainty that may lead to the risk for adaptation measures of being either under-scaled and insufficient or over-scaled and unnecessary, taking resources from other priority areas.

Finally a further common point in the EU-15 national strategies is the development of a research strategy and/or the establishment of national coordination initiatives aimed to expand the understanding of climate impacts, to collect and structure climatic and other types of data, to coordinate national and local efforts as well as to achieve effective stake-holders coordination.

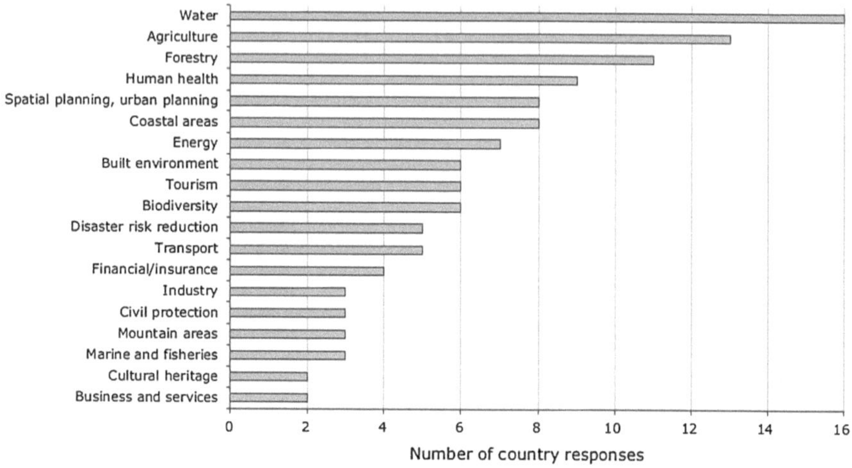

Fig. 4 Priority areas for adaptation implementation. National adaptation policy processes in European countries—2014 EEA report

Since the aim of this paper is the analysis of the role of spatial planning in climate change and energy inspired strategies and policies, it was decided to perform a comparative analysis between two MS, the main documents elaborated on the subject of climate change and energy since the 2000s. Such a comparison is, in fact, aimed at identifying common themes and lessons, as well as differences, strengths and weaknesses that emerge from the analysis of these documents and the implications in the field of multi scale spatial planning. Specifically, information on which the comparative analysis was conducted:

- the year of publication of the document;
- the scope (efficiency/energy conservation, mitigation, climate adaptation), highlighting any integrate among the issues addressed;
- the time line;
- areas of intervention and related types of proposed actions;
- the presence of guidelines for regional planning.

These elements were compared in relation to Italy and Denmark, in order to study the state of affairs of our country and the country that has achieved the best score of the Climate Change Performance Index 2014, an indicator "designed to enhance transparency in international climate politics and aimed at highlighting those countries with best-practice climate policies". This index is particularly interesting because, in addition to considering environmental and energy objective indicators, it assigns a 20 % weight to the evaluation of national and international policy. Denmark is a benchmark not only for Italy but for all the other member states, because in the ranking it is first not only amongst European but also amongst the international countries.

3.2 Energy and Climate Change Policies in Italy

Under the European Climate and Energy Package, Italy has to reduce its GHG emissions from non ETS sectors by 13 % compared to 2005 levels and increase the share of renewable energy sources in final energy consumption to 17 %. In this regard, progress towards the non-ETS emission target seems evident from Fig. 5. Compared to 2005 GHG emission are on a decreasing trajectory, while the share of RES has moved from 5.8 to 15.6 % in the period 2005–2013. Under the Energy Efficiency Directive, Italy has set its target to decrease the level of FEC to about 126 Mtoe. Probably to a significant extent due to the economic downturn, Italy FEC was lower than the 2020 since 2009 and although it has slightly increased since that, Italy is on track to reach its energy efficiency target.

The Kyoto protocol together with the more recent EU Climate Energy Package are the drivers behind the Italian commitment to climate change mitigation. Driven by the international efforts that coalesced in the ratification of the United Nations Framework Convention on Climate Change in 1994, Italy approved its first National Plan for the Containment of CO_2 Emissions, aiming at the stabilization of CO_2 emissions at the 1990 levels by the year 2000. Four years later, in 1998,

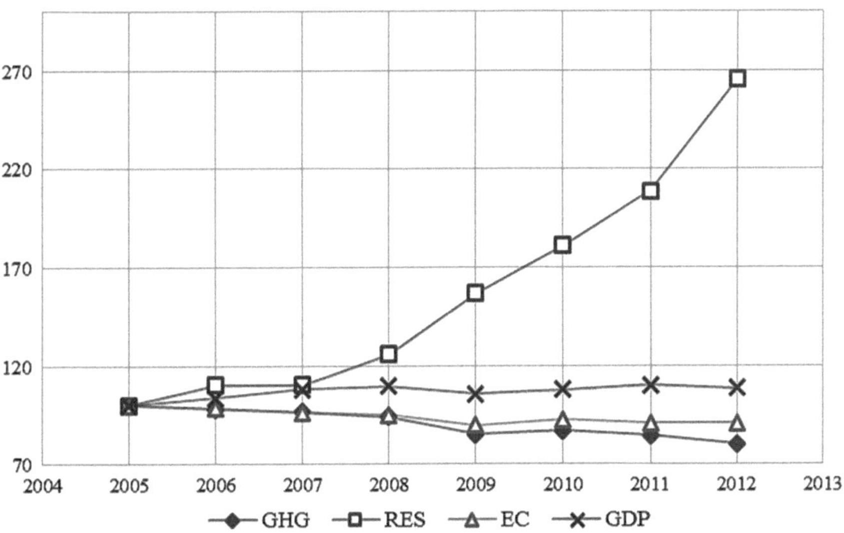

Fig. 5 Evolution of greenhouse gas emission (*GHG*), share of renewable energy source (*RES*), final energy consumption (*FEC*) and gross domestic product (*GDP*) in Italy since 2005. *Source* our elaboration on Eurostat, 2014 (Color figure online)

the Country signed the Kyoto Protocol, committing to reduce its greenhouse gas emissions by 6.5 % below the base-year levels (1990) over the first commitment period, 2008–2012. When the Kyoto protocol was ratified in 2002 an overall National Climate Change Strategy[2] was approved, update later in 2007[3] and 2013.[4]

To pursue the EU carbon neutral long-term vision (beyond 2020) the Italian Ministry for the Environment, Land and Sea (IMELS) and the Ministry of Economic Development in 2013 have approved a decree outlining a new "National Energy Strategy" that proposes plans and strategies for specific energy consumption segments (e.g. transport, agriculture, etc.…), overlooking the local level and territorial planning, since most of the actions are economic or fiscal in nature.

Specifically energy efficiency is considered important to attain the national GHG reduction targets, as indicated by the Effort Sharing Decision (2009/406/EC), for the Non ETS sectors, those that not included in the EU Emissions Trading System such as transport, buildings, agriculture and waste that are connected to the urban scale.

The 2014 NEEAP is based on the white certificates, introduced in 2004 for electricity and natural-gas distributors to achieve yearly quantitative primary-energy saving targets, minimum standard requirements for buildings, vehicles and appliances, tax deductions and direct financing of technological improvements. These measures do not include any Consumer information programs and training (as suggested by Articles 12, 17 of the EED) necessary to promote behavioral change and complete any comprehensive strategy that is aimed at the final use energy reduction. All the measures in this filed are referred to a plan called Integrated Plan for the Uptake of Energy efficiency that will be drafted, sometime in the future, to disseminate at national, regional and local level transparent information on Energy Efficiency.

Education and training, extremely important to achieve a zero carbon society, aren't considered by the Italian strategy; in fact consulting the International Energy Agency (IEA) database[5] of Energy Efficiency Policies and Measures, summarized in Table 4 with the number of EE policies per sector (e.g. buildings, industry…) and per policy type (regulation, information…), only 0.1 % of the measures are referred to research, information and education.

Regarding the growth of the renewable energy sources the Ministry of Economic Development, as prescribed by the Directive 2009/28/EC, in 2010 has drafted the Italy's National Renewable Energy Action Plan to achieve the EU target of 17 % of energy consumption from renewable sources later implemented by the Decree[6] n. 28/2011.

---

[2]CIPE deliberation 123/2002.

[3]CIPE deliberation 135/2007.

[4]CIPE deliberation 17/2013.

[5]http://www.iea.org/policiesandmeasures/energyefficiency/?country=Italy.

[6]Legislative Decree March 3rd 2011, n. 28. Correct implementation of the directive (http://context.reverso.net/traduzione/inglese-italiano/correct+implementation+of+the+directive) 2009/28/CE on the promotion of the use of energy from renewable sources and amending and subsequently repealing Directives 2001/77/EC and 2003/30/EC.

Table 4 Number of EE policies in Italy, per area (e.g. buildings, industry…) and per policy type

	Type of Measure										
Sector	(blank)	Economic Instruments	Policy Support	Research, Development and Deployment (RD&D)	Regulatory Instruments, Economic Instruments	Information and Education	Voluntary Approaches, Information and Education	Policy Support, Economic Instruments	Regulatory Instruments	Regulatory Instruments, Regulatory Instruments	Grand Total
Buildings		4			1	1			13	1	20
Buildings, Industry		1									1
Energy Utilities		4	1						3		8
Energy Utilities, Multi-Sectoral Policy							1				1
Energy Utilities, Residential Appliances		1									1
Industry		3							2		5
Lighting									5		5
Lighting, Residential Appliances									2		2
Multi-Sectoral Policy	1	4	10	1	1	1		1		1	20
Residential Appliances		2									2
Transport		8	2	1					3		14
Transport, Industry				1							1
(blank)				1					1		2
Grand Total	1	27	13	4	2	2	1	1	29	2	82

Table 5 Number of measures per sector and policy type of the Sixth National Communication under the UN Framework Convention on Climate Change—CIPE 2013

	Sector							
Type of instrument	Agriculture	civil (residential and tertiary) sector	Cogeneration	industry sector	Renewable energy sources	transport sector	Waste	Grand Total
Economic		5	4	6	5	2		22
Fiscal		4						4
Information				1				1
Other		1						1
Planning						3		3
Regulatory	2	2		1	3	3	1	12
Grand Total	2	12	4	8	8	8	1	43

To achieve this target the action plan intervenes in the transport, energy production and building sectors minimum standard requirements, information, incentives for the use of biogas and biomasses, and tax deductions.

Mainstreaming, and sectorial approaches are evident in the policies and measures included in the Sixth National Communication under the UN Framework Convention on Climate Change. A summary of the number of policies per sector (e.g. buildings, industry…) and per policy type (e.g. regulation, information…) is present in Table 4, while a summary of the type of investment is presented in Table 5.

3.3 *Energy and Climate Change Policies in Denmark*

Denmark's target for 2020 under the European Climate and Energy Package is to reduce GHG emissions in sectors that are not covered by the EU Emission Trading System by 20 % compared to 2005 levels. The legally binding EU target for gross energy consumption from renewable energy sources is 30 % for Denmark, to be

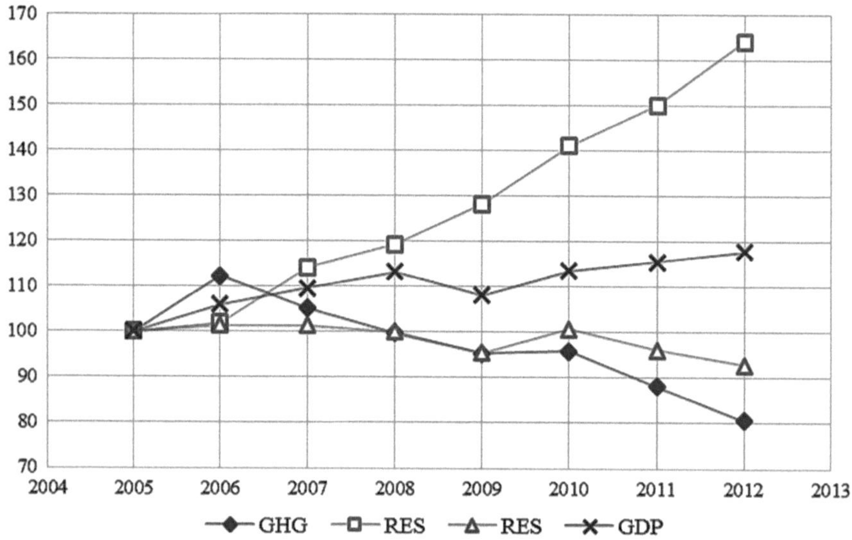

Fig. 6 Evolution of greenhouse gas emission (*GHG*), share of renewable energy source (*RES*), final energy consumption (*FEC*) and gross domestic product (*GDP*) in Denmark since 2005. *Source* our elaboration on Eurostat, 2014 (Color figure online)

reached by 2020. In this regard, progress towards the non-ETS emission target seems evident from Fig. 6. Under the Energy Efficiency Directive, Denmark has set its target to decrease the level of FEC to about 14.8 Mtoe. Although GDP has increased by 20 % in the referenced period, FEC has decreased by 8 % and in 2013 it was 14.2 Mtoe (Fig. 6). Denmark has successfully shown that economic growth and emission and energy cuts are not contradictory.

Environmental issues have been part of the national agenda since the Nature Conservation Act an environmental law established in 1973 and climate change issues since 1990.

In the sixth communication (2013) within the United Nations Framework Convention on Climate Change and the Kyoto Protocol, it is stated that climate policies should be as integrated as possible with the joint European solutions in order to ensure effective action and maintain competitiveness compared to the partners. Effectiveness is also supervised by the Climate Policy Monitor that, since 2005, monitors the impacts of climate change, prioritizes measures in implementation and finally communicates progress.

To achieve the energy efficiency goals many measures[7] have been implemented, the majority of which (Table 6) belong to the types of Support and Regulatory Policy and instruments concerning interventions within the framework defined

[7]http://www.iea.org/policiesandmeasures/energyefficiency/?country=Denmark.

Table 6 Number of EE policies in Denmark, per area (e.g. buildings, industry…) and per policy type

	Type of Measure										
	Economic Instruments	Economic Instruments	Economic Instruments Policy Support	Information and Education	Policy Support	Regulatory Instruments	Research, Development and Deployment (RD&D)	Voluntary Approaches	Voluntary Approaches Policy Support	(blank)	Grand Total
Buildings						9		1		1	11
Buildings, Energy Utilities		1									1
Buildings, Residential Appliances					1	1					2
Commercial/Industrial Equipment, Multi-Sectoral Policy, Industry			1								1
Energy Utilities					2	1					3
Industry								1			1
Multi-Sectoral Policy	3				13				1		17
Residential Appliances				1							1
Residential Appliances,Residential Appliances						1		1			2
Transport	2			1	1		2				6
(blank)	1				3					13	17
Grand Total	6	1	1	2	20	12	2	3	1	14	62

Multi-Sectorial Policy, confirming the systemic and holistic approach adopted to address these articulated and complex issues.This comprehensive approach is also evident from the structure of the Danish strategy for adaptation to a changing climate drawn up in 2008 by the Danish Energy Agency (Energistyrelsen), which emphasizes the importance of joint and voluntary action for all stakeholders encouraged by the information, research and measures initiatives to facilitate integration of the issue of climate change in spatial planning.

In particular the chosen strategy hinges on a bilateral approach to adaptation that they define as autonomous and planned. The first, considered the most effective "whereby authorities, businesses and private citizens react to the consequences of climate change on their own initiative in a timely manner" (Danish Government 2008); and the planned secondary response measure to be initiated only when a political incentive may be necessary.

Even if no specific actions are proposed for territorial planning, the efforts of the main cities are being evaluated and the need to draw up national municipal planning requirements is continuously assessed. One urban planning suggestion included, concerns high risk areas where remedies could incur significant extra expenses, and National measures under planning legislation could limit building and construction.

The analysis of the other documents (NEEAP, The Danish Climate Policy Plan, Energy Strategy 2050) shows that, on the one side, the necessity of cooperation and integration at the different scales is affirmed, but on the other, the local and regional level aren't explicitly assigned a role in the plans and strategies.

3.4 National Energy and Climate Change Policies Comparison

Both nations have made commitments to combat climate change by signing, within the same timeframe, apart from the UN Framework Convention on Climate Change, international agreements and protocols on climate change (Table 7). From these protocols and the resulting European laws documents and national standards are derived. In Italy the reduction of GHG become of national importance in 1994 from the Law n. 64, which established the "National Program to reduce emissions of carbon dioxide by 2000–1990 levels," which was followed by the "National Action Plan to combat climate change" (CIPE Resolution no. 123/2002).

The Danish equivalent document is the "Proposal for a Climate Strategy for Denmark" in 2003. These strategic documents proposed, as instruments for the resolution of the problems to reduce consumption, energy efficiency, and the introduction of new renewable energy sources, the flexibility mechanisms already proposed by the Kyoto Protocol and make no reference to territorial planning, considering the building segment as only one of the sectors in which to reduce consumption.

From our comparison of these, and the other relevant documents, several uncertainties, action gaps and policy questions characterizing national climate change and energy documents, from the point of view of spatial planning, clearly emerge. Even if it is considered as important driver for adaptation policies, both by Europe and the scientific research, none of the climate change strategies provides any specific reference. For instance, planning instruments aren't considered by the NAS as institutional promoters of multi actor, iterative, solutions to long-term challenges, and the framework to verify the effectiveness of these kinds of tools that require indicators and benchmark for a systematic review process, aren't defined.

Table 7 Status under International Climate Change Law

	Italy		Denmark	
	Signature	**Ratification**	**Signature**	**Ratification**
UN Framework Convention on CC	15/04/1994	15/04/1994	09/06/1992	21/12/1993
Kyoto Protocol	29/04/1998	31/05/2002	28/04/1998	31/05/2002
Vienna Convention	22/03/1985	29/09/1988	22/03/1985	29/09/1988
Montreal Protocol	16/09/1987	16/12/1988	16/09/1987	16/12/1988
Convention On Long-Range Transboundary Air Pollution	14/11/1979	18/06/1982	14/11/1979	18/06/1982
Energy Charter	17/12/1994	22/08/1997	17/12/1994	22/08/1997
Energy Efficiency Protocol	17/12/1994	22/08/1997	17/12/1994	22/08/1997
Espoo Convention	26/02/1992	14/03/1997	26/02/1992	14/03/1997
Quantified emission limitation or reduction commitment:	8,00%		8,00%	

Climate change is tackled in both countries by addressing the problem essentially in terms of the protection of natural resources and the reduction of greenhouse gas emissions, mainly related to energy production, postponing to the future the integration of adaptation and the evaluation of its effectiveness in the processes of regional planning.

Denmark, in fact, focuses on autonomous adaptation, according to the 2008 Strategy that "focuses mainly on the general activities to be undertaken to ensure the process" as Italy focuses on the promotion of initiatives for the exchange of experience and good practices to facilitate local authorities (regional, provincial and urban) in identifying the most appropriate measures to adapt. Both countries also argue the importance of identifying interventions and actions aimed at microclimate comfort, calling into play the scientific research in Italy and the individual building owners in Denmark. In other words, both national strategies still lack practical tools or processes to enhance capacity and adaptation actions, to ensure integration policy or measure effectiveness.

Also with regard to energy the approach seems to be, in both countries, largely aimed at improving infrastructure and efficiency, as well as the purely economic aspects. It seems that the relationship between spatial planning and energy is underdeveloped, or rather not at all taken into account, even if the first one "sets frameworks for energy consumption, production and distribution" (Stoeglehner et al. 2011), and contributes to cut energy demand of the built environments.

Both Italy's and the Danish NES, drafted on a voluntary basis, focus on the transformation of the energy markets and the strengthening of the necessary infrastructure at a supra regional scale. They differ, however, in the goals and measures proposed.

The Danish plan is more explicitly addressed to environmental issues and the achievement of international records in energy, renewable energy and the reduction of GHG, while the Italian plan is aimed primarily at lowering the cost of energy addressing the issue almost exclusively in economic terms. Although both plans indicate improved energy efficiency and the increase in the energy deriving from renewable as targets to be reached, the Italian one has three objectives concerning the expansion and modernization of production and distribution of fossil fuels.

For the pursuit of their goals both governments have proposed similar types of measures such as fiscal, financial and regulatory frameworks, in addition to which Denmark envisages actions related to education, and to pursue the three goals for the use of renewable energy sources Italy provides considerable action for the national strategic infrastructure planning.

The centrality of energy issues is also evident in the measures that the two states has adopted for the effort to mitigate the effects of climate change. Comparing, in fact, the Climate change mitigation policies and Measures in the database of the European Environment Agency (EEA) shows that measures linked to energy supply and use are 43 % for Denmark and 72 % for Italy.

Studying all the measures as subsystems shows that for Denmark, on 62 measures, the most numerous are: energy supply (15), transport (9), waste (8); while for

Italy, on 44 measures, the most numerous are: energy supply (9); energy use—residential (9); energy use—construction (10); transport (8).

In recent adaptation plans, however, the energy issue leaves room for other significant issues in the states challenge for the adaptation to the effects of climate change. In fact, both in the Danish strategy for adaptation to a changing climate and in the Italian document Elements for a National Strategy for Adaptation to Climate Change, the action areas identified are relate mostly to natural resources, rather than to the production and consumption of energy.

That awareness that adaptation to climate change is a long-term process, has led the Danish Government to develop its national strategy even before the obligations imposed by Europe, in 2008, with a time line to 2100, and to develop an Action Plan to manage cloudburst and rain water in 2012. Italy, however, despite the long process of consultations and preparation of preparatory documents which led to the adoption in 2014 of the Elements for the national strategy, appears to still lack a well-defined and formalized NAS. This guideline document contains priorities and actions which have mostly a time line to 2020, and only a few remain valid after the deadline.

Target areas of the two plans differ not in the indication of the need to preserve natural resources, but rather in considering elements such as urban settlements and tourism. This difference may be due to the fact that the European debate on adaptation has been enriched with these issues in recent years, after the drafting of the Danish plan and before the most recent Italian document. However, it is worth noting that, despite the latter's emphasis on the need to implement decisive actions to stimulate and co-ordinate at the local level, such a statement is then not reflected in the identification of actions to adapt cities Italian to climate change.

The different drafting date is also the cause of the difference of the types of actions proposed that, in the Danish case, are a heterogeneous set of interventions aimed at mainstreaming environmental concerns into sectorial approaches, and in the Italian plan, instead, refer to a more structured set of economic, regulatory, planning and management instruments as well as research, knowledge and communication.

Aware of the need to actively involve all relevant stakeholders concerned with the effects of climate change, in order to address an issue of this size and complexity, Denmark has established an electronic platform aimed at sharing information and knowledge about the newest research and development within climate change adaptation in Denmark and abroad. Even if in the Italian guidelines the importance to field communication activities, awareness and information needed to reduce the damage caused by climate change is stated, Italy is still without a dedicated portal to encourage stakeholder awareness about the vulnerability of society and natural resources to climate change.

4 Local Energy and Climate Change Polices: Overview and Selected Examples

4.1 Overview

The awareness that cities are key players in coping with climate change and in reducing energy consumption has been remarked both by scholars (Betsill and Bulkeley 2007; McCarthy et al. 2010; Bulkeley et al. 2011) and by research institutes and public institutions (IPCC 2014; Energy City 2014).

Population growth will mainly occur in urban areas "defining and shaping the 21st century" (Graham and Marvin 2001). The concentration of people and capital, be it infrastructure, housing or the production apparatus, within urban areas make the latter vulnerable to climate change impacts but, at the same time, strategic to the resilience of the cities. In the end, urban shape and design influences micro-climate conditions, affecting wind direction, precipitation patterns and can therefore exacerbate phenomena such as the urban heat Island. The weather hazards and the shortage of the available energy resources have been compelling local authorities and planners to pay attention to spatial planning as a driver of climate adaptation and energy saving.

Urban sustainability networks such as C40 or ICLEI connect an always wider number of cities. For instance, 29 % of C40 membership are in Europe, and over 4000 cities joined the Covenant of Mayors. In March 2014 within this initiative the Mayors Adapt was launched to create a more resilient Europe and to contribute to the achievement of the global objectives of the EU Adaptation Strategy Package (2013). This strategy underlines how the cities «are often ill-equipped for adaptation» (EC 2013). In fact according to a study on 200 medium sized European cities, 72 % do not have an adaptation plan and 35 % haven't yet developed a mitigation or an adaptation strategy (Reckien et al. 2014).

At the local level the integration between climate change and spatial planning seems insufficient, and the integration between energy and spatial planning is far from being straightforward (Nilsson and Martensson 2003; Palm 2006). In Italy of the 597 signers of the Covenant of Mayors, almost half are late in the presentation of the SEAP and 100 are late by at least six months; of these 24 have been temporarily suspended by the EC (Legamebiente 2011).

In Denmark, instead, more and more cities have been tackling energy and climate change issues. In fact, according to some surveys conducted since 2008, the majority of Danish municipalities have developed strategies and/or plans related to these fields (COWI 2010).

In the following section to examine the difference in the local approach to climate change issues between the cities in Italy and Denmark we've compared the capitals of each nation. Rome and Copenhagen were chosen, even if they are not necessarily the most advanced amongst each nation's cities, because they are the most populous, vulnerable, and strategic for the development of climate change strategies

4.1.1 Energy and Climate Change Policies in Rome

Rome is a Mediterranean metropolis that covers an area of 1287 km^2, bigger than the sum of six of the biggest European capitals, with a population of almost 3 million and 26,500,000 tourists per year.

The Italian capital has adopted in 2002 the Environmental Action Plan (Piano d'Azione Ambientale) financed by the 2000 Local Agenda 21 "as guideline document and strategic reference for sustainable development, which advise the program plans of the Administration" most of the issues addressed by this plan are the focus of the subsequent plans related to climate and energy; among these, those related to the mode of development of the urban built environments, albeit declared of priority, in reality they are not framed in a coordinated strategy, but rather appear to be a disorganized set of measures ranging from subsidized public buildings, urban parks to renovation of portions of the city.

In 2009 the "Environmental Action Plan for achieving the objectives of the Kyoto Protocol in the city of Rome" was formalized that identified the major areas of activity (transport, housing, waste, land use) and the actions necessary to reach the 6.5 % reduction in CO_2 emissions (Italy's goal) by 2012 attributable to the municipality of Rome compared to 1990 emission levels, in accordance with the provisions of "Aalborg Commitments" of 1994, which consider local actions to contain climate change.

To achieve this objective, the plan mainly refers to the reception of national regulations[8] concerning the reduction of energy consumption and the energy performance of buildings, for example, such as the Legislative Decree n. 192/2005.

A careful analysis of the measures to be put in place reveals two major shortcomings: firstly, the absence of a unified action framework that can systemize the different solutions that, instead, appear to be completely disconnected from each other, to achieve synergistically the identified targets; and then, the absence of references to the tools and processes of government of the territory in which such actions should fit.

These deficiencies, i.e. the definition of measures and interventions that do not seem to contribute synergistically to a specific target for the reduction of CO_2, also characterize the adaptation plan "A Third Industrial Revolution Plan Framework for the Transformation of Rome into the First Post-Coal Biosphere City", in 2012.

This plan identifies, as well as the intervention area of energy efficiency, the four action "pillars" to "decarbonize" the entire city system: renewables, buildings as power plants, hydrogen storage, smart infrastructure and transport. For each of these "pillars", scenarios of have been identified for the resolution of the various problems associated with each, looking at them as an opportunity for sustainable transformation, which assess the costs and benefits using assumptions or results of

[8]"The effect of these measures is expected to materialize in a turnaround in CO_2 emissions due to the heat consumption of housing".

international examples; these estimates are mostly related to socio-economic benefits and only a few cases quantify CO_2 savings.

In order to provide continuity to the measures to make Rome a low carbon city, in 2013 the "Plan Of Action for Sustainable Energy The City Of Rome" (Sustainable Energy Action Plan SEAP) was developed (Municipality of Rome 2013); it identifies the same themes of the action plan framework of 2012, and results in 12 possible pilot projects, feasible in the short term, in order to give an early start to the strategies contained in the plan.

Although in the SEAP, regarding the decisions to be implemented, it is stated that "among the measures envisaged in the SEAP a modern and updated system of planning [is imagined], through the development of a (conceptual and temporal) synergy between plans and programs and the inclusion of environmental and energy criteria in the process of preparation of planning and sectorial instruments", this intent does not match the corresponding implementation actions. In particular, within the chapter on regional planning, although it is stressed that all plans that concern the energy issue will have to adapt to the Provincial Energy Plan of Rome, approved in February 2008, and that the building code has already been integrated with the "Standards for energy efficiency, the use of renewable sources of energy and saving of water resources", specific actions aimed at a regional planning efficient in terms of energy are not defined.

Reference is made to a single action that regards energy: the possibility to create a small energy plants serving agglomerations of population between 24,000 and 40,000 people, without suggesting possible locations within the municipality.

If this plan is aimed primarily at mitigation, as declared by the municipality, no document regarding adaption has yet been developed, despite the creation in 2011 of the Environmental Climate Change Observatory which should be a strategic tool for the implementation of urban policies in the context of climate change and energy conservation, and to promote compliance with the commitments made by Rome with Europe in terms of energy saving and reduction of greenhouse gases.

In other words, despite the efforts of the municipality to invest in projects aimed at achieving environmental objectives such as a 20 % reduction of GHG, compared to the 2003 baseline, increase the percentage of energy production from renewables to at least 15 % and ensure economic development without significant increase in energy consumption, we do not see the use or adaptation of instruments of transformation of the territory. The transposition of the environmental guidelines is delegated to the future executive planning.

4.1.2 Energy and Climate Change Policies in Copenhagen

Copenhagen is the capital and most populated city of Denmark, with an urban population of over 1.2 million. The city, well reputed for its initiatives aimed to combine economic growth and sustainability, is currently working towards achieving carbon neutrality while also preparing for the extreme weather expected in the next decades.

In 2009, the City Council adopted the Climate Plan for Copenhagen (City of Copenhagen 2009), setting down goals for achieving a 20 % reduction in CO_2 emissions by 2015. In addition, a vision for a carbon neutral Copenhagen in 2025 was formulated. In 2011, it was decided to draw up a new climate plan setting down guidelines for achieving carbon neutrality in Copenhagen by 2025. To meet this ambitious goal, the plan supports a wide range of initiative that are categorized into four main themes: energy consumption, energy production, green mobility and City Administration initiatives.

The initiatives that fall into the first category are mainly target to reduce energy consumption of residential and commercial buildings and they will jointly contribute to a 7 % reduction in CO_2 emission. The allocation of reduction from energy consumption initiatives are highly coherent with the City of Copenhagen Master Plan, the Local Agenda 21 Plan and the Building Regulation, especially for what concerns the new building areas where strict requirements to low energy use have been set. For the existing building stock, where the retrofitting of the 33 % of housing stock and 46 % of commercial properties is needed to meet the CO_2 reduction target, the Plan recognized the need to develop future guidelines regulation and founding mechanism to secure a low carbon transition in the building sector.

The initiatives in the energy production sector will account for the 74 % share of total carbon reduction. The implementation of these initiatives will guarantee that by 2025 the production of electricity, heating and cooling in Copenhagen will primarily be based on wind, biomass, geothermal energy and waste. This target has some clear spatial implications, especially regardless the choice of the location where new power plants will be installed that are explicitly considered in the plan.

The initiatives within the green mobility theme will contribute to the 11 % share of total carbon reduction. To meet this target a shift in the modal share in favor of public transit and cycle, together with the spread of alternative fuel vehicles is needed. This vision is coherent with the Municipal Master Plan, the City of Cyclist initiative and the Action Plan for Green Mobility (City of Copenhagen 2011a, b, c; 2012; 2014) which was adopted by the city Council in 2012. Furthermore the plan recognizes the need of new and coordinated initiatives in the transport planning arena, especially for what concern the implementation of an intelligent traffic control system and the development of a congestion strategy to be implemented in collaboration with the national Governments.

The initiatives that fall within city administration theme are mainly target to reduce energy consumption of municipal-owned buildings and vehicles and of street lighting. Despite having a contained impact on total CO_2 reduction, this initiatives have a strong symbolic aim and are supposed to move the market in a climate friendly direction.

In 2011, the city approved the Climate Adaptation Plan (City of Copenhagen 2011c) aimed to prepare Copenhagen for the future by developing the Danish capital as a climate proof and green city. The climate change adaptation plan has been developed to ensure adaptation measures are undertaken in the most cost-effective and efficient way. The plan is based on the analysis of a long-term scenario which has led to the identification of the most relevant threats resulting from climate change:

- More frequent and heavier downpours. The cloudbursts over the last few years have smacked the city budget. The heavy storm in 2011 alone cost the city over one billion of euros. Precipitation in Copenhagen is expected to increase by 30–40 % by 2100.
- The rise of the sea level. With most of the city only having an average altitude of 9 m above the sea level and with a significant number of people and amount of property lying close to the water level, Copenhagen is potentially vulnerable to the effects of natural variability in sea level and, on decadal timescales, anthropogenic sea level rise. Water levels around the city are likely to rise by up to 1 m over the next hundred years.

Regardless the issue of the increasing precipitations, two main and complementary strategies have been identified. The first deals with the improvement of the drainage systems, so that they will be capable of coping with major downpours. To this end, a range of tools will be used for better rainwater management including rain and sewage reservoirs, permeable paving, filters and infiltration trenches and other sustainable urban drainage tools. The second strategy deals with the improvement and the connection of the urban green areas. The number of green areas—including 'pocket' parks, and green roofs and walls—will be increased to slow rainfall run-off. Green roofs not only will capture 60 % of rainfall, but will also improve air quality, vegetation and wildlife habitat, while reducing urban heat-island effects.

Regardless the issue of the rise of the sea level, there is an option to establish a barrier at Nordhavnen and Kalveboderne and to raise the rest of the coastline out towards Øresund. The barriers will be established so that they will protect the city against storm-surge events but without disrupting harbor operation at the same time. The plan take into consideration the provisions contained in the Municipal Master Plan regardless future urban expansion and proposes that new constructions and new buildings in areas that are at risk of flooding from the sea and rising groundwater levels will be equipped and designed considering site-specific solutions.

An interesting aspect of the plan is that adaptation is not only considered as a negative measure but also as an occasion to increase the quality of life for the city's inhabitants and create synergies with other planning initiatives. For instance, the "green" perspective embodied in the adaptation plan, while increasing the urban resilience, is expected to attract new private investments and, at the same time, expand and improve the quality of public spaces.

In general, the plans analyzed in this section show a high degree of policy coherences with the main municipal plans including the new master plan, the mobility plan and the Agenda 21 Plan (City of Copenhagen 2011a, b, 2012). They also consider the possible synergy and trade-off with national-level planning and initiatives and present roadmap, along with cost-benefit analysis and flexible mechanisms to cope with the uncertainty derived from changing climate.

4.1.3 Local Energy and Climate Change Policies Comparison

Comparing the documents produced by the city of Rome and Copenhagen, regarding the strategies and actions put in place to address global climate change and the reduction of energy consumption, in order to assess their degree of commitment, the type of actions chosen and how their implementation is put in effect through instruments of territorial planning (Table 8).

The willingness to intervene at the local level to combat climate change and adapt to its inevitable effects is evident for both municipalities. Both have signed the commitments of Local Agenda 21 and the more recent Covenant of Mayors and have adopted many plans and strategies that since the 2000s have addressed the issues of mitigation of the causes and adaption to the effects of climate change. Clearly the matter of energy is an issue that cannot be separated from mitigation as it is an indispensable tool to achieve the desired changes (issues of security of supply and the cost of energy are considered secondary in these tools).

For Copenhagen the main document for mitigation is the 2012 revision of the 2009 plan "CPH 2025 climate plan—a green, smart and carbon neutral city" consigned as SEAP to fulfill the obligations of the Covenant. Even if this document does not focus on planning at the urban scale (urban development) still it recognizes its key role in making the vision of a carbon neutral Copenhagen achievable because of the rapid expansion and transformation of the urban fabric.

In accordance with the principle of mainstreaming it is stated that "Every time decisions about new roads, buildings or transportation possibilities in a given area are taken, the focus must be on climate and environmental requirements." Such mainstreaming is indeed evident in the examples of the plans for the development areas of Nordhavn, Carlsberg and Amager Fælled that included environmental issues in their project requirements, setting the goal of the creation of the first carbon neutral neighborhoods.

The effort that the council has been making to consider any approach to urban transformation as an opportunity for sustainable urban development and the reduction of CO_2, using the sectorial plans to help achieve these objectives, for example by reading the sectorial plan "The City of Copenhagen's Bicycle Strategy 2011–2025".

Also with regard to the most recent documents that address the issue of adaptation, such as the "Copenhagen climate adaptation plan" of 2011, the continuity and coherence of the strategies of the Administration is evident; in fact the proposals of each plan include in the analysis framework, projections and actions of previous action plans, resubmitting them or extending them for the attainment of the objective, and existing urban planning (Table 8). In fact "The climate adaptation of Copenhagen requires broad cooperation. In some cases the municipality itself has to carry out climate adaptation actions, but in many areas it can induce others to adapt through its administration of laws and plans." and in the Legislation And Planning section for each challenge such as "management of more storm water" an overview is present of the applicable rules (municipal planning, local planning, building

Table 8 Urban planning relevance of the documents examined for Rome and Copenhagen

City	Name of the document	Publication date	Temporal horizon	Intervention sectors	References to territorial governance instruments
ROME	Piano di Azione per l'Energia Sostenibile della Città di Roma (Seap)	2013	2020	Local production of energy (mainly installing photovoltaics), Integrated renewable energy, Transport and sustainable mobility, Building refurbisment and energy efficiency, Public illumination.	does mention some executive plans
ROME	Piano d'Azione per il raggiungimento degli obiettivi del protocollo di Kyoto	2009	2012	Mobility, Residential buildings, Renewable energy sources,	considers the measures to adopt in the municipal planning act; proposes neighbourhood refurbishment actions
ROME	Una terza rivoluzione industriale piano quadro per la trasformazione di roma nella prima citta' biosfera post-carbone	2010	2030	(i) the "distributed generation" and distribution of renewable energy resources; (ii) the use of the buildings as power plants; (iii) the development of hydrogen and other storage technologies; (iv) the creation of a new infrastructure for smart energy and new transport system.	no
Copenhagen	Copenhagen climate adaptation plan	2011	2110	More and heavier downpours in the future, Higher sea levels, Higher temperatures and urban heat islands in Copenhagen, Climate change and groundwater, Indirect consequencs of climate change	sets the approach to adopt for the planning act and the building act
Copenhagen	CPH 2025 climate plan - a green, smart and carbon neutral city	2012	2025	Energy Consumption, Energy Production, Green Mobility, City Administration Initiatives, New Initiatives CO2 Reductions	no
Copenhagen	A greener and better everyday life local agenda 21 plan for Copenhagen 2012 – 2015	2013	2025	Resources (e.g. Increasing recycling through experiments in the homes), At Home (e.g. Saving energy in the home), Transport (e.g. Alternatives to private cars), Urban Spaces (e.g. Increasing the green areas in the city), Interdisciplinary theme (e.g. Climate conscious working culture in businesses)	no
Copenhagen	The City of Copenhagen Cloudburst Management Plan	2012	2033	Upgrading City Resilience t o Extreme Rainfall Events	The Cloudburst Management Plan is not per se legally binding. To have an effect, the contents of the Cloudburst Management Plan needs incorporating into the Administration's general planning process: primarily the Municipal Master Plan, sectorial plans (Waste water Plan) and local master plans. Furthermore, urban renewal plans and local neighbourhood facelift schemes should incorporate the Cloudburst Management Plan into their drafting

permits and other regulation) and how they can be used to prevent and deal with the various challenges posed by climate change.

In the case of Rome, mainstreaming and planning interactions are part of the mission statement in documents such as the SEAP: "For the preparation of the Plan specific issues were considered, considered as "quality indicators" for the drafting of the Plan: taking in account environmental strategies and policies, in particular those international, national and local on climate change" but it is not possible to deduce to what extent and how these have influenced the formulation of the plans.

Both cities recognize the value of educating end-users and the need to "inform and motivate citizens […], professionals and other local stakeholders on how they can use energy better" (SEAP ROMA). They have also set up an office to monitor climate change related data (Rome in 2011 Environmental Observatory on Climate Change); a necessary condition for the ongoing evaluation of the effectiveness of the measures and their adaptation to achieve the targets set. However the degree of development of the transformation process into carbon neutral cities is substantially different in the two examined cases. Indeed Copenhagen has already produced specific plans to adapt to the effects of Climate Change, such as the "The City of Copenhagen Cloudburst Management Plan" of 2012, while Rome has still only the mitigation strategy (Sustainable Energy Action Plan Of the City of Rome, 2013) that does not address the risks generated by the inevitable changes and neither does it propose the necessary measures to adapt, as happened already in previous mitigation plan of 2010 "A third industrial revolution framework plan for the transformation of Rome in the first biosphere post-coal city", also called master plan.

5 Discussion and Conclusion

The challenges imposed by the changing climate and the energy-driven development are very complex and need to be addressed from the global to the local scale. As a result, at all level of governance, policy makers are promoting integrated and adaptive strategies aimed at reducing or mitigating negative effects of climate change while reducing energy consumption and CO_2 emission.

This contribution has summarized the main efforts that have been made in Europe to secure a transition towards a low-carbon and climate proof society. To this end, an in-depth analysis of selected climate and energy policy documents elaborated at three different levels of governance has been presented.

Previous studies have analyzed the role that spatial planning may have in developing mitigation measures to reduce energy consumption and GHG emission and adaptation measures to reduce the effect of climate change. In contrast with these studies, we have focused our attention on the role that policy makers have assigned to spatial planning in the implementation of climate change and energy policies.

In order to summarize the European strategic vision on energy and climate change we have analyzed and discussed 15 policy documents elaborated by the European Commission and the European Council of Ministers during the 2006–2015 period. At the national level, we have given first an overview of the different levels of implementation of climate and energy policies in the EU 15 countries, later we have focused on two selected countries: Italy and Denmark. At the local level, we have provided an overview of the state of implementation of climate and energy policies of the European cities, focusing our attention on two European capital cities, Rome and Copenhagen.

With respect to the three levels of governance analyzed in this contribution, some conclusions can be drawn about the role that policy makers assign to spatial planning:

- Policy documents at the EU strategic level in the field of mitigation are primarily focus on the role that smart energy infrastructure, energy efficiency, renewable-energy projects, research and the deployment of new energy technologies may have in the reduction of energy consumption and CO_2 emission. While it is widely acknowledged in these documents that significant interventions are needed in the transport and the building sectors, there is no reference to the impact of spatial organization and physical planning on transport and building consumption and associated GHG emissions in the EU. Simply put, in the EU policy documents a "techno-centric approach" (Papa and Lauwers 2015) seems clear. At the EU level issues related to the role of spatial planning, in its pre-emptive capacity, to promote adaptation measures and reduce the risks associated with global climate change, are evident. In this regard, the EU Adaptation Strategy Package clearly identifies spatial planning as a critical mechanism through which climate change adaptation can be facilitated. First the EU document recognized that the climate impacts are accelerated or accentuated in urbanized areas and that current urban trends like sprawl and the continuous soil sealing can have additional negative externalities. Second, it emphasizes on the new challenges of operating infrastructure under changing climate conditions. Third, it recognizes the importance of spatial organization as determinant of urban and regional resilience and the role that spatial planning may have in mitigating the impact of the changing climate;

- At the national level it is necessary again to distinguish between mitigation and adaptation policies. While the first are more inclined towards techno-centric solutions, the latter concentrate on the physical organization of the built environment brought by spatial planning. Even if the NEAPs we have analyzed in this contribution recognize the need to reduce the energy consumption attributable to buildings, they identify only tax and financial measures as tools to be implemented, without any reference to the impact that the instruments of territorial government may determine. The same holds true for the transport sector where the primary actions proposed are the introduction of new technologies and clean vehicles. With respect to national Climate Strategies that we have analyzed, technological fiscal, financial and regulatory frameworks options, represent the most commonly implemented measures to reduce energy consumption and CO_2 emission. Again, it seems that the relationship between spatial planning and energy is underdeveloped, or rather not taken into account at all, even if the first sets the regulatory frameworks for energy consumption, production and distribution and contributes to cut energy demand of the built environments. In our view the absence of a clear reference to spatial planning seems somewhat surprising, given its potential contribution to the reduction of GHG emissions and energy consumption in the transport and in the building sectors. Technologies should be only one aspects of a more complex system where spatial organization, citizen behavior and technologies interact with each other, determining the actual and future pattern of energy consumption. In contrast, when we consider the national Adaptation Strategies, many measures tend to fall into the urban planning and development category. Indeed, the majority on national adaptation strategies identify spatial planning as a priority area for the implementation of adaptation measures;
- Since the growing recognition of the contribution of municipal areas to global environmental change (Santamouris 2001), important initiatives and programs for the reallocation of planning actions from the global to the local scale took place in Europe. As a consequence mitigation and adaptation plans at the local level are gaining significant attention in Europe. The mitigation policy documents analyzed in this contribution, include technological options, such as

improved energy efficiency and renewable energy generation, and energy saving measures. In the two documents analyzed there is a clear reference to the need of integrating spatial and energy planning. However, mitigation measures typically focus on individual sectors or urban functions (e.g. increasing bike lanes, efficient heating systems, building insulation), rather than systemic changes (e.g. zoning regulations or urban planning). We argue that the complexity of the interactions of the urban components require a more systemic approach to coordinate the public efforts and achieve maximum results. In contrast, adaptation measures tend to fall into the urban planning and development category. Urban density, the design of the built environment and the spatial distribution of green areas are considered as determinants of urban resilience and the role that spatial planning may have in organizing these elements is considered as a powerful tool to mitigate the impact of the changing climate.

Concluding, the examination of the documents and their comparison, shows that in none of the three levels of governance spatial planning is truly implemented. Its strategic importance in tackling climatic and energy challenges is affirmed but its tools and methods are overlooked. Even at the local level, where implementation should be at its most practical, measures are usually uncoordinated singular interventions that aren't part of regional, or even neighborhood, level planning. Furthermore they very rarely contemplate the specific contents of a town planners approach since they are lacking in geographic localization of the interventions, and any norm that may govern the transformation of the territory.

6 Limitations and Further Developments

Despite the new insights provided by our analysis, some limitations as well as some further research developments might be identified. First, at the European level, our analysis has been limited to the most recent document elaborated by the European Commission and the European Council of Ministers. In this regard, further developments might include the analysis of European policy documents on a longer time period, considering the evolution of the European strategy for energy and climate change. At the national level our analysis has been limited to two countries, while the number of criteria to compare the policy documents is limited if compared with other studies (EEA 2015; Casado-Asensio and Steurer 2014). This limitation however depends on the broad aim of the contribution that has analyzed policy documents across three different level of governance. Further improvements in this direction might consist in widening the sample and developing a more comprehensive evaluation framework. Finally, the above mentioned consideration might be extended to the local level of governance. In this case it would be also interesting to investigate the role of non-state actors in decision-making processes on energy and climate change at the local level (Juhola and Westerhoff 2011; Ostrom 2012).

References

Abolhosseini S, Heshmati A, Altmann J (2015) The development of renewable energy sources and its significance for the environment. Springer, Singapore

Betsill M, Bulkeley H (2007) Guest editorial: looking back and thinking ahead: a decade of cities and climate change research. Local Environ 12(5):447–456

Bulkeley H, Schroeder H, Janda K, Zhao J, Armstrong A, Chu SY, Ghosh S (2011) The role of institutions, governance, and urban planning for mitigation and adaptation. In: Hoornweg D, Freire M, Lee MJ, Bhada-Tata P, Yuen B (eds) Cities and climate change. Washington, DC: The World Bank, pp. 125–160

Biesbroek GR, Swart RJ, Van der Knaap WG (2009) The mitigation–adaptation dichotomy and the role of spatial planning. Habitat Int 33(3):230–237

Brunnengräber A (2013) Multi-level climate governance: strategic selectivities in international politics. In: Knieling J, Leal Filho W (eds) Climate change governance. Springer, Berlin, pp 67–83

Bulkeley H, Broto VC, Hodson M, Marvin S (2010) Cities and low carbon transitions. Routledge, London

Casado-Asensio J, Steurer R (2014) Integrated strategies on sustainable development, climate change mitigation and adaptation in Western Europe: communication rather than coordination. J Publ Policy 34(03):437–473

CIPE Resolution n. 123/2002. Available at: http://db.formez.it/StoricoFontiNor.nsf/bb05f9ea5aa422dbc1256a930025c290/399EF8BDE71BAFFAC1256B3B00400631/$file/Delibera%20CIPE.pdf

Ciscar JC, Iglesias A, Feyen L, Szabó L, Van Regemorter D, Amelung B, Soria A (2011) Physical and economic consequences of climate change in Europe. Proc Nat Acad Sci 108(7):2678–2683

City of Copenhagen (2009) Climate plan for Copenhagen. Available at: http://en.klimatilpasning.dk/. Last accessed July 2015

City of Copenhagen (2011a) City of Copenhagen Municipal Plan 2011. Available at: http://kp11.kk.dk/. Last accessed July 2015

City of Copenhagen (2011b) A greener and better everyday life—local agenda 21 plan for Copenhagen 2012–2015. Available at: www.sharingcopenhagen.dk. Last accessed July 2015

City of Copenhagen (2011c) Copenhagen climate adaptation plan. Available at: http://en.klimatilpasning.dk/. Last accessed July 2015

City of Copenhagen (2012) Action plan for green mobility. Available at: www.sharingcopenhagen.dk. Last accessed July 2015

City of Copenhagen (2014) Copenhagen city of cyclists. Available at: http://www.cycling-embassy.dk/. Last accessed July 2015

Coalition for Energy Saving (2014) Implementing the EU Energy Efficiency Directive. Latest analysis of member state plans for end-use energy saving targets (Article 7). Available at: http://energycoalition.eu. Last accessed July 2015

COM (2007) 354 Green Paper of 29 June 2007 on adapting to climate change in Europe—options for EU action [COM(2007) 354 final—Not published in the Official Journal]

COM (2011) 112. Office for Official Publications of the European Communities, Luxembourg. Available at: http://eur-lex.europa.eu/legal-content/EN/TXT/?uri=celex:52011DC0112

COWI (2010) Annual Report, 2010. Available at: http://www.publications.cowi.com/cowi/Annual_report_2010/. Last accessed July 2015

da Graça Carvalho M (2012) EU energy and climate change strategy. Energy 40(1):19–22

Danish Government (2008) Danish strategy for adaptation to a changing climate. Available at: http://www.klimatilpasning.dk/media/5322/klimatilpasningsstrategi_uk_web.pdf

EEA—European Environment Agency (2012) Climate change: impacts and vulnerability in Europe 2012. Available at: http://www.eea.europa.eu/publications/climate-impacts-and-vulnerability-2012

EEA—European Environmental Agency (2015) National adaptation policy processes in European countries—2014. Available at: http://www.eea.europa.eu/publications/national-adaptation-policy-processes. Last accessed July 2015
Energy City (2014). Economic impact of the energy transition at the local level. Methodologies and case studies. Available at: http://www.energycities.eu/IMG/pdf/impact_energy-transition_energy-cities_2014_en.pdf
European Commission (2001) Communication from the Commission to the Council. The European Parliament. The Economic and Social Committee and the Committee of the Regions on the Sixth Environment Action Programme of the European Community, "Environment 2010: Our future, Our choice" COM (2001) 31. Office for Official Publications of the European Communities, Luxembourg. Available at: http://eurlex.europa.eu/legalcontent/IT/ALL/?uri=CELEX:52001DC0031
European Commission (2007a) Green Paper from the Commission to the Council. The European Parliament. The European Economic and Social Committee and the Committee of the Regions. Adapting to climate change in Europe—options for EU action COM (2007) 354. Office for Official Publications of the European Communities, Luxembourg. Available at: http://eur-lex.europa.eu/legal-content/IT/NOT/?uri=CELEX:52007DC0354
European Commission (2007b) Communication from the Commission to the Council, the European Parliament, the European Economic and Social Committee and the Committee of the Regions—A European strategic energy technology plan (SET-plan)—'Towards a low carbon future' COM(2007) 723. Office for Official Publications of the European Communities, Luxembourg. Available at: http://eur-lex.europa.eu/LexUriServ/LexUriServ.do?uri=COM:2007:0723:fin:en:pdf
European Commission (2007c) White Paper. Adapting to climate change: Towards a European framework for action COM (2009) 147. Office for Official Publications of the European Communities, Luxembourg. Available at: http://eur-lex.europa.eu/LexUriServ/LexUriServ.do?uri=COM:2009:0147:FIN:EN:PDF
European Commission (2011a) Communication from the Commission to the European Parliament, the Council, the Economic and Social Committee and the Committee of the Regions—Energy efficiency plan 2011 COM (2011) 109. Office for Official Publications of the European Communities, Luxembourg. Available at: http://eur-lex.europa.eu/LexUriServ/LexUriServ.do?uri=COM:2011:0109:FIN:EN:PDF
European Commission (2011b) Communication from the Commission to the Council, the European Parliament, the European Economic and Social Committee and the Committee of the Regions. A Roadmap for moving to a competitive low carbon economy in 2050 COM (2011) 112. Office for Official Publications of the European Communities, Luxembourg. Available at: http://eur-lex.europa.eu/LexUriServ/LexUriServ.do?uri=COM:2009:0147:FIN:EN:PDF
European Commission (2011c) Communication from the Commission to the European Parliament, the Council, the European Economic and Social Committee and the Committee of the Regions. A Roadmap for moving to a competitive low carbon economy in 2050
European Commission (2013a) Communication from the Commission to the European Parliament, the Council, the European Economic and Social Committee and the Committee of The regions. An EU Strategy on adaptation to climate change COM (2013) 216. Office for Official Publications of the European Communities, Luxembourg. Available at: http://ec.europa.eu/clima/policies/adaptation/what/docs/com_2013_216_en.pdf
European Commission (2013b) Green Paper. A 2030 framework for climate and energy policies COM (2013) 169. Office for Official Publications of the European Communities, Luxembourg. Available at: https://www.kowi.de/Portaldata/2/Resources/horizon2020/2013-green-paper-climate-energy-2030.pdf
European Commission (2015) Communication from the Commission to the European Parliament, the Council, the European Economic and Social Committee, the Committee of the Regions and the European Investment Bank. A Framework Strategy for a Resilient Energy Union with a Forward-Looking Climate Change Policy COM (2015) 80. Office for Official Publications of

the European Communities, Luxembourg. Available at: http://eur-lex.europa.eu/legal-content/EN/TXT/?uri=COM:2015:80:FIN

European Communities (1987) Single European Act. Official Journal of the European Communities, L 169, 29 giugno 1987. Available at: http://eur-lex.europa.eu/legal-content/EN/TXT/PDF/?uri

European Communities (1992) Treaty of Maastricht on European Union. Official Journal of the European Communities, C 191, 29 July 1992. Available at: http://europa.eu/eu-law/decision-making/treaties/pdf/treaty_on_european_union/treaty_on_european_union_en.pdf

European Communities (1997) European treaty of Amsterdam amending the treaty on European Union. The treaties establishing the European Communities and related acts, 10 Nov 1997. Official Journal of the European Communities, C 191, 29 July 1992. Available at: http://eur-lex.europa.eu/legal-content/EN/TXT/?uri=OJ:C:1992:191:TOC

Goldthau A (2014) Rethinking the governance of energy infrastructure: scale, decentralization and polycentrism. Energy Res Soc Sci 1:134–140

Graham S, Marvin S (2001) Splintering urbanism: networked infrastructures, technological mobilities and the urban condition. Routledge, London

Grazi F, van den Bergh JC (2008) Spatial organization, transport, and climate change: comparing instruments of spatial planning and policy. Ecol Econ 67(4):630–639

Gustavsson E, Elander I, Lundmark M (2009) Multilevel governance, networking cities, and the geography of climate-change mitigation: two Swedish examples. Environ Plann 27(1):59–74

Hurlimann AC, March AP (2012) The role of spatial planning in adapting to climate change. WIREs Climate Change 3(5):477–488

IPCC (2014) Climate Change 2014: Mitigation of Climate Change. Contribution of Working Group III to the Fifth Assessment Report of the Intergovernmental Panel on Climate Change. Cambridge University Press, Cambridge, United Kingdom and New York, NY, USA

Juhola S, Westerhoff L (2011) Challenges of adaptation to climate change across multiple scales: a case study of network governance in two European countries. Environ Sci Policy 14(3):239–247

Kurukulasuriya P, Rosenthal S (2003) Climate change and agriculture: a review of impacts and adaptations. World Bank climate change series. The World Bank, Washington, DC

Labat S, White R (2007) Carbon finance: the financial implications of climate change. Wiley, Hoboken, New Jersey

Laukkonen J, Blanco PK, Lenhart J, Keiner M, Cavric B, Kinuthia-Njenga C (2009) Combining climate change adaptation and mitigation measures at the local level. Habitat Int 33(3):287–292

Legamebiente (2011) Clima in Comune 2011. Patto dei Sindaci: i piani e le buone pratiche per il 2020. Available at: www.legmabiente.it. Last accessed July 2015

Lindner M, Maroschek M, NethererS, Kremer A, Barbati A, Garcia-Gonzalo J, Marchetti M (2010) Climate change impacts, adaptive capacity, and vulnerability of European forest ecosystems. For Ecol Manage 259(4):698–709

Lundqvist LJ, Biel A (2013) From Kyoto to the town hall: making international and national climate policy work at the local level. Routledge, London

McCarthy MP, Best MJ, Betts RA (2010) Climate change in cities due to global warming and urban effects. Geophys Res Lett 37:L09705

Measham TG, Preston BL, Smith TF, Brooke C, Gorddard R, Withycombe G, Morrison C (2011) Adapting to climate change through local municipal planning: barriers and challenges. Mitig Adapt Strat Glob Change 16(8):889–909

Municipality of Rome (2013) Sustainable energy action plan. Available at: mycovenant.eumayors.eu/docs/seap/741_1320425372.pdf

Nilsson JS, Martensson A (2003) Municipal energy-planning and development of local energy-systems. Appl Energy 76(1):179–187

Oberthür S, Roche Kelly C (2008) EU leadership in international climate policy: achievements and challenges. Int Spectator 43(3):35–50

Ostrom E (2012) Nested externalities and polycentric institutions: must we wait for global solutions to climate change before taking actions at other scales? Econ Theor 49(2):353–369

Palm J (2006) Development of sustainable energy systems in Swedish municipalities: a matter of path dependency and power relations. Local Environ 11(4):445–457

Papa E, Lauwers D (2015) Smart mobility: opportunity or threat to innovate places and cities? In: 20th international conference on urban planning and regional development in the information society (REAL CORP 2015), pp 543–550

Papa R, Gargiulo C, Zucaro F, Angiello A, Carpentieri G (2014) Urban energy consumptions: its determinants and future research. In: Marchetti N, Brebbia CA, Pulselli R, Bastianoni S (eds) The sustainable city IX. Urban regeneration and sustainability. WIT Press, Southampton, UK, pp 25–34

Pasimeni MR, Petrosillo I, Aretano R, Semeraro T, De Marco A, Zaccarelli N, Zurlini G (2014) Scales, strategies and actions for effective energy planning: a review. Energy Policy 65:165–174

Reckien D, Flacke J, Dawson RJ, Heidrich O, Olazabal M, Foley A, Pietrapertosa F (2014) Climate change response in Europe: what's the reality? Analysis of adaptation and mitigation plans from 200 urban areas in 11 countries. Climatic Change 122(1):331–340

REF-E (2012) Obiettivo 2050: per una roadmap energetica al 2050. Available at: http://www.wwf.it/?1160. Last accessed July 2015

Rietig K (2012) Climate policy integration beyond principled priority: a framework for analysis. Centre of Climate Change Economics and Energy Policy. Working Paper No. 99 Available at: http://citeseerx.ist.psu.edu/viewdoc/download?doi=10.1.1.368.8302&rep=rep1&type=pdf. Last accessed July 2015

Santamouris (2001) Energy and climate in the urban built environment. New York, NY: Routledge

Scott D, Hall CM, Gössling S (2012) Tourism and climate change: impacts, adaptation and mitigation. Routledge, London

Wilson E, Piper J (2010) Spatial planning and climate change. Routledge, London

Zanon B, Verones S (2013) Climate change, urban energy and planning practices: Italian experiences of innovation in land management tools. Land Use Policy 32:343–355

Regional Local Development Strategies Benefiting from Open Data and Open Tools and an Outlook on the Renewable Energy Sources Contribution

Giuseppe Las Casas, Beniamino Murgante and Francesco Scorza

Abstract The New Cohesion Policy opens to an integrated place-based approach for the improvement of territorial and social cohesion. The issue of territorial impact assessment of regional development policies highlight that data availability, open access to datasets in "near real-time", participation, knowledge sharing, and assumed importance within the planning process. In this paper we present an application of spatial analysis techniques for evaluation of territorial effects of EU funds starting from open data by 'open-coesione'. The final scope is to have the opportunity to develop an assessment in the framework of a public debate. According to this issue we developed an empirical approach for context analysis using open-data assumed as a relevant source of information for the measurement of diffused concerns. The application regards an internal area of the Basilicata Region: the Agri Valley, a complex contest in which environmental and agricultural vocation conflicts with the recent development of oil extraction industries (the actual main territorial specialization). The research produced an integrated approach in order to highlight the energy development domain for Agri Valley. In a first stage the balance between fossil/traditional and renewable energies development should consider the traditional manufacts and practice in order to push on community identity and social commitment. This implies—for the planner—the requirement of managing social conflicts and promoting effective procurement in territorial management. The paper describes as a strategy of conflicts resolution an integrated project called an "Energy Museum". The "Conclusions" at the end of this chapter discusses further applications and perspectives for improving regional development planning, considering the exploitation of open data sources and spatial analysis in one hand and on the Renewable Energy Sources on the other.

G. Las Casas · B. Murgante · F. Scorza (✉)
Laboratory of Regional and Urban Territorial Systems Engineering (LISUT),
School of Engineering, University of Basilicata, Potenza, Italy
e-mail: francescoscorza@gmail.com

G. Las Casas
e-mail: giuseppe.lascasas@unibas.it

B. Murgante
e-mail: beniamino.murgante@unibas.it

R. Papa and R. Fistola (eds.), *Smart Energy in the Smart City*,
Green Energy and Technology, DOI 10.1007/978-3-319-31157-9_14

1 Introduction

The New Cohesion Policy opens to an integrated place-based approach for the improvement of territorial and social cohesion. Smart growth, sustainable growth and inclusive growth for EU 2020 represent overall goals to be achieved under the comprehensive approach defined by Barca (2009) as a 'place based approach'.

As the authors already discussed in previous works (Las Casas and Scorza 2009) concerning the issue of territorial impact assessment of regional development policies, the relevant instance comes from knowledge management in regional programming practice. It means data availability, open access to datasets in "nearly real-time", participation, knowledge sharing, key actors effective involvement in planning processes.

The "concentration" issue coming from EU 2020 Cohesion Policy still reflects ambiguity in interpretation (Capello 2014) and not structured implementation in Regional Programs. From a "thematic concentration" to a "spatial concentration", several attempts are going to be developed in an uncertainty framework.

If a "thematic concentration" reflects more a traditional approach considering a panel of main objectives and goals, it could represent an effective procedure if a proper context analysis identified ex-ante specific needs and priorities coming from local specializations and local communities needs (in other words "place based"). A "spatial-concentration" should produce a map of cohesion programming based on clear and informed decisions expressing the awareness of 'where' to invest in order to maximize effects of cohesion policies. There is so ex-ante solution in order to ensure the achievement of regional development results but a balance between a thematic generalization of objectives and a concrete spatial awareness of development precondition should be investigated.

Relevant information should come from lessons learned in previous programming experiences. The contribution of 'open data' to the impact assessment of EU Operative Programs appears to be mature in concept, but still weak in accuracy of available data bases. For the present research we used data from the 'Open Cohesion' project directed by the Italian Ministry for Territorial Cohesion. The Italian Ministry engaged with this unstoppable process of collecting and sharing data to improve citizens commitment on public policies. It developed a web service distributing data on investments policies developed by National and Regional Operative Programs 2007/2013 matching together data from regional and national administrations. The results are analyzed in the fourth paragraph of the paper with the application of spatial analysis techniques for the evaluation of territorial impacts and effects.

We described a process of territorial impact assessment of Regional Operative Programs investments oriented to the analysis of territorial specialization in attracting funds. This procedural approach is completely based on Open Data analysis through Open Tools (software and web services) in order to demonstrate that the integration of such resources overcomes the dependence from proprietary data formats and proprietary software towards interoperability and open information.

The application context is an internal area of the Basilicata Region: Agri river Valley (so called Val d'Agri)—we provided main framework information in the third paragraph of the paper. This area appears to be a complex territorial context, where environmental and agricultural traditional vocations conflict with a recent development of oil extraction industries. Those aspects define a peculiar condition in which energy development reflects a territorial specialization. The research produced also an integrated approach in order to highlight the energy development domain for Agri Valley under a sustainable view. In a first attempt the balance between fossil/traditional energy sources and renewable energy development should also consider the traditional manufacts and practice in order to push on community identity and commitment. This implies—for the planner—the requirement of managing social conflicts and to promote effective procurement in territorial management in order to establish a PPPP partnership (Public Private People Partnership).

The paper describes a strategy of conflicts resolution, among others, a strategy based on a project called "Energy Museum", analyzed in the context of Val D'Agri Structural Plan which connect economic development issues with environmental sustainability and social inclusion and based on the information mainly built on open data.

Conclusions regard further applications and perspectives for improving and supporting regional development planning considering the exploitation of open data sources and spatial analysis in one hand, and on the Renewable Energy Sources on the other.

2 Spatial Data, Spatial Data Infrastructures, Open Data, Open-Government, Open-Cohesion

Until the end of the 1980s, spatial data production was mainly concentrated in national organizations. In subsequent years the increase of computer performance, according to Moore's law, the decrease of GIS software prices and the increase of GIS software functions (Longley et al. 2001) generated a huge production of spatial data, because many organizations or single local authorities departments started to develop their own data. Consequently, a drastic reduction in production costs of spatial information has not corresponded to a great reduction of system implementations. The cause of this phenomenon is mainly due to a poor practice in sharing data, which generates a multiplication of resources to produce the same data.

Trying to solve this problem in 1994, Clinton's administration, with Executive Order 12906 (1994) "Coordinating Geographic Data Acquisition and Access: The National Spatial Data Infrastructure" (NSDI), established "the national data spatial infrastructure" with the aim of defining technologies, policies, standards and human resources necessary to acquire, process, store, distribute and improve the use of geospatial data. Based on this Executive Order, all Public Institutions and Federal Agencies documented all new geographic data using standards developed by the

"Federal Geographic Data Committee" and they made them accessible to the "National Geospatial Data Clearinghouse", to the network of producers, managers and all potential users of spatial data.

In addition to a redundant data production, a proliferation of GIS applications with very limited potential, often developed within a single organization, occurred in the same period creating technological barriers in data exchange. In order to solve interoperability problems an international consortium, Open Geospatial Consortium, completely dedicated to these issues, has been created. Open Geospatial Consortium (OGC) standards provide access through the internet to geo-referenced data integrating local and remote data. Spatial data can be downloaded by a simple Internet browser, or by a GIS software supporting OGC standards: Web Mapping Service (WMS), Web Feature Service (WFS), Web Coverage Service (WCS).

The European approach to spatial data infrastructures is based on INSPIRE (INfrastructure for SPatial InfoRmation in Europe) Directive (EC 2007). This Directive can be summarized in six principles:

- Data should be collected once and maintained at the level where this can be done most effectively;
- it should be possible to combine seamlessly spatial information from different sources across Europe and share it between many users and applications;
- it should be possible for information collected at one level to be shared between all different levels, detailed for detailed investigations, general for strategic purposes;
- geographic information needed for good governance at all levels should be abundant under conditions that do not refrain its extensive use;
- it should be easy to discover which geographic information is available, fits the needs for a particular use and under which conditions it can be acquired and used;
- geographic data should become easy to understand and interpret because it is properly documented and can be visualized within the appropriate context selected in a user-friendly way.

Rifkin (2000) highlights that, in the Internet age, "reticular economy" approaches have been developed with a transition from the markets intended as a physical place, centres of goods and properties trade, to virtual markets, mainly based on information transfers. Consequently there is a trend towards an economy that prefers information collection to accumulation of physical capital. This trend indicates that information is a public good, which the State must guarantee for the benefit of the entire community, as if it were a financial resource (Rey 1999).

Today data represent a big economic potential not exploited at all, because if data are available to everyone, they could create new companies and business growth for existing enterprises. A radical change in geographical information approach is necessary, based on OGC standards adoption and on implementation of INSPIRE directive. Data have to be managed and updated by the organization that has the legal responsibility and have to be shared using OGC standards.

Fig. 1 Parallelism between the different approaches that characterize the various eras of geographic information and music (Color figure in online)

Using a parallelism with music it is important to pass from the gramophone, still in vogue, or Napster syndrome, based on the accumulation of terabytes and terabytes of data, to a Spotify approach. If all geographic data produced with public resources are available online, offering the possibility to integrate remote and local data on your computer there is no longer the need to download data (Fig. 1).

INSPIRE Directive (EC 2007) and OGC standards are e-Government measures, very useful in providing faster and more efficient services through the rational use of technologies.

This approach considers the information flowing only in one direction from administration to citizens, ignoring communities potential contribution. It is important to move from e-Government to e-Governance, where the information flow from citizens to the administration is also contemplated. In recent years the distance between citizens and government has increasingly enlarged. Consequently community failing to be an active part in administrative activity becomes the antagonist of public administration, without understanding that both should converge towards the same goal. The right approach is to tend towards Open-Government policies based on a more participative method of government. It starts from the assumption that ideas of citizens have always to be collected, not only before elections. Consequently, public involvement, getting ideas and suggestions, is a daily activity aiming to have a wider inspiration in managing and to collect feedback in already started actions. Obama's administration has given a great impetus to this approach, implementing such a policy and enlarging the possibility to capture public imagination by means of social networks, blogs and all possible solutions to directly interact with citizens. This new approach is often called Gov. 2.0. Open government without a 2.0 approach is still based on a direct action. "Providers" is a sort of Right to Information where the administration tries to inform people, but having interaction just with the main stakeholders. Gov. 2.0 is a more open approach, which "enables" citizens to have an important role in defining policies. Social media and all 2.0 platforms are a key element in generating a direct contact with citizens. Extension of 2.0 philosophy completely changed the relationship between citizens and administration. People directly provide in realizing services that public administration is not interested to carry out and private sector does not consider convenient to realize. The central elements of Open-Government are therefore participation, openness and sharing data. In US experience Open-Government (Obama 2009) was based on three main principles: Transparency, Participation and Collaboration. A central element in Open-Government becomes Open-Data. Each type of activity is based on data; it is therefore essential to promote these activities where great part of data is "Open", especially public ones.

To ensure that data are really "open" they should be open and not in proprietary format, without any particular constraints of licenses, reusable and integrable, easily searchable on the web through databases, catalogues and search engines, directly accessible through Internet protocols, quickly, immediately and at any time accessible by network, communicable and directly inter-changeable on network between all users. Data must also be equipped with metadata and they should allow export, on-line and off-line use, integration, manipulation and further share, they should also be readable by any operating system. Generally speaking, Open Data means a file up-load in portable document format (pdf) on a website allowing the download. This approach is located at the lowest level of Berners–Lee's (2009) Open Data classification according to stars scoring. Two stars are assigned to data in property format, for which the owner was allowed total utilization with some interoperability problems if one is not using the specific software. Three stars are assigned to the product in non-proprietary format ("open format"). Data are in four-stars format if they follow international standards for interoperability, while five-stars data are to transfer also the meaning ("Open Linked Data").

Concerning territorial purpose, five-stars open linked data include the adoption of OCG standards (Las Casas et al. 2014).

Open-Government and Open-Data represent two faces of the same coin.

The project 'Open Cohesion' provides an open data service concerning cohesion policy effects related to planning processes. The initiative was strongly supported by the Minister for Territorial Cohesion, which stimulated a more active citizens participation in decision-making, producing also a process of social vigilance on the use of collective resources.

The publication of data in an accessible format and reusable on their corporate websites shows the willingness of the government to move in a systematic way towards a structure of transparency that encourages active participation of citizens and re-use of data. The service pursues the objective to improve Citizen Engagement on investments policies, and offers a data set with specific information concerning projects funded by the current programming period 2007–2013, matching implementation data from regional and national administrations entitled Ops management.

Open Cohesion appears to be a result of an Open Government approach. Data provided can be included in the three-stars open level of Tim Berners-Lee's classification with the opportunity to achieve spatial dimension through external elaboration. In this direction, previous researches demonstrated the added value of providing open spatial information concerning development programs. We refer to the experience developed by PIT Marmo Platano Melandro (Basilicata—IT) during the EU programming period 2000–2006 with a web GIS service for the spatialization of development policies (Murgante et al. 2011; Tilio et al. 2009).

While in that experience the main effort was in territorial data production, we have to affirm that today it is possible to develop accurate spatial analysis concerning the distribution of EU funded investments with public open data. In the next section we describe the adopted methodology.

3 The Implementation Context: A Synthetic Description of Strategic Sectors

The application context of this research is represented by an inner territory of Basilicata Region. Val d'Agri is a mountainous area whose settlement system is placed along "Agri's valley floor" (S.S. 598), a road infrastructure that connects the Valley with the provinces of Potenza and Matera and Vallo di Diano in Campania Region.

This context consists of 23 municipalities, grouped into sub-areas of specialization. It is a territory characterized by very high environmental values: approximately 50 % of territory is included in the National Park of Appennino Lucano and there are about 12.300 ha of SIC areas and 5.400 ha of ZPS.

The analysis of land use shows that most of the territory is characterized by woods, pasture lands and agricultural areas.

The territory presents a lot of physical and intangible endogenous resources linked to the forms of historical villages, to agricultural techniques and products, to cultural traditions such as the Lady Maria cult or arboreal rites. Moreover we can find unique landscapes and villages, historical and monumental riches that are not yet enhanced by a structural policy of local development.

The rural and pastoral vocation is proved by local gastronomy based on certified products (DOC, IGP) connecting agricultural traditions to social ones, geo-climatic features with traditional practices and knowledge. This is a structural dimension that should be taken into account as a primary driver of local economic development based on innovations and on the qualification of land products.

Since the beginning of fossil energy resources exploitation, Val d'Agri has obtained an "energetic vocation" that is shown in regional planning documents (especially in PIEAR Basilicata: Energetic and Environmental Framework Regional Plan of Basilicata Region). This development dimension has no endogenous foundation but represents a strategic domain that can attract investments, infrastructures, job, financial resources deriving from royalties and compensation forms.

Today we can say that the expectations of local communities are only partially satisfied in terms of public resources invested in requalification, in support of local production activities (including agro-zootechny), in territorial promotion. Unemployment represents a weakest structural dynamic in terms of local development since it delays a cross growth in economic sectors different from the industrial one and the migration of people in productive age still persists.

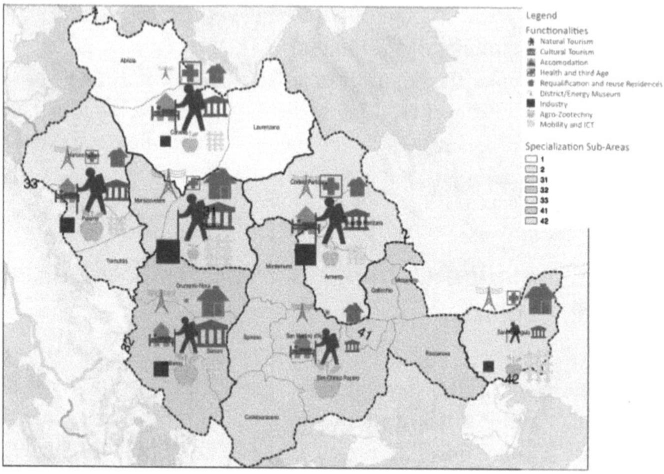

Fig. 2 Strategic functionalities and specialization sub-areas (Color figure in online)

The "Green Economy" can be considered as a prospect for the territorial system: Renewable Energy Sources (RES) have been adopted according to the Regional Programme Framework contained in the Regional Environmental Energy Plan (PIEAR) that considers the Valley as a Regional "Energy District".

Currently this policy forecast is not configured in terms of systemic interventions that are able to combine production processes and technological development with installations and implementations.

As regards oil, even if the discovery of oilfields dates back to the 80s, mining activities have been started during the 90s. The Viggiano 'Oil Center' was built in 1996 and in successive years an oil pipeline was devised in order to transport crude oil to the refinery of Taranto (ENI 2012). The growth of mining activities and the hydrocarbon processing have brought several concerns about environmental safeguards, that is a precondition for all other forms of development.

In order to describe the local specialization of the Agri Valley the whole territorial context was divided into seven territorial sub-areas and development functionalities were weighted in each sub-area.

The following figure shows four sub-areas into which the implementation context has been divided (Fig. 2).

4 Spatial Distribution of Investments: An "Open Procedure"

The innovative element of the proposed assessment is the punctual territorialisation of projects and interventions funded by EU Programs at a local level. It is beyond the traditional representation of information for aggregated items and pre-defined geographical areas (administrative boundaries, areas PIT, PIOT, program areas, etc.) according to the aim of characterizing "places of implementation."

The approach is based on an open dataset distributed according to interoperable formats, managed by open-sources software and application, with a strong relationship with web-based services.

The issue of territorial impact assessment of development policies is a domain where different approaches produce different results that often represent solutions for a specific purpose, serving a specific process of socio-economic and territorial planning without a framework methodology validated under a scientific or technical point of view.

The developed proposal aims to provide answers to the demand for territorial specialization analysis oriented to the construction of policy choices to be developed within the EU's 2014–2020 operational planning tools. The proposed approach, based on information concerning the implementation of 2007–2013 instruments, is developing an interpretation model that allows a progressive monitoring of on-going processes with a territorial monitoring system that allows a detailed scale of punctual information.

In fact, a procedure identifying punctual interventions (public and private projects in different thematic areas of intervention) through a dataset of geo-referenced points useful in developing spatial statistics, overlaying, compatibility assessments, etc. has already been developed.

Considering the structure of information sources, the implemented procedure adopts only free available tools. The flow chart presented in the following figure exemplifies the process in terms of operational components.

The first stage of the process includes the operation of extraction, analysis and preparation of data for territorialisation. The datasets provided by the project Open Cohesion contain fields relating to the 'localization' of each record; this means addresses of each individual intervention funded by different programs/funds. These are attributes that can have different levels of specialization in relation to the nature of the intervention, the type of expense, with the characteristics of the proposer or the beneficiary of resources.

Generally, the objective of intervention territorialisation is to get a precise location of each initiative. Therefore, we considered field "address" in addition to "name of the City" and "Postal Code" to generate a string "LOCATION" on which to perform an operation of geo-coding, supported by Google free tools.

Data tables, appropriately pre-processed, were uploaded within 'Google Fusion Table'. It is a web experimental application distributed within the Google application 'Google Drive', to view and share large data tables.

The application tool allows us to:

- Show online large tables of data;
- Filter and synthesize information;
- Develop online graphs, maps, graphs or layouts;
- Management of multi-user and collaborative production data;

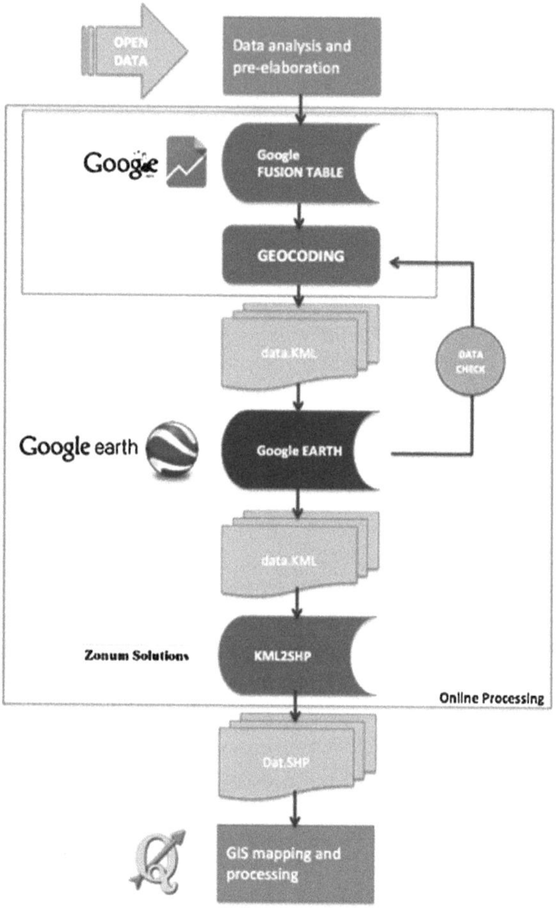

Fig. 3 Territorialisation procedure flow chart (Las Casas et al. 2014) (Color figure in online)

- Merge and cross several databases;
- Export geographic data format (.kml) and other interoperable formats (.csv).

Therefore Google Fusion Table, working online, allows us to perform a geo-coding on field "location" of the table, returning a map of points exported in KML format (Fig. 3).

These files, as verified by the operator recursively in order to reduce the uncertainty in localization, after processing by Google Earth, are exported in local . kml format and subsequently converted in .shp through an online tool distributed by Zonum Solutions. In this way you get data with a proper format, widely interoperable in GIS application software and web-GIS.

The below picture shows the investments supported by the Operational Programmes 2007/2013 falling within Val d'Agri area. GIS processing was developed using the open-source software Q-GIS.

Table 1 Accuracy assessment of territorialisation procedure

Original data (n° records)	Data referred to wider administrative area than municipality	Localized by the project address	Localized by the name of municipality of the project (main urban centre)	Not localizable as errors affect original data
551	44	419	50	38
	These data were excluded from representation as they relate to operations for which it is not possible to express a point impact	The geocoding in Google Fusion Table and the subsequent verification in Google Earth allow the unique identification of the investment point	Where the address was not recognizable by Google, location of the investment has been fixed on the main urban centre of the municipality	In the operational phase cases in which, on the basis of the information contained in the attributes of projects location, it was not possible to make an unambiguous choice of location have been reported. For example, it has been found several times that in the address field of a project, one information (also very detailed) referring to a location belonging to a municipality that did not correspond with project municipality was presented

The figure shows a total of 551 projects localized in the study context. The proposed procedure allows us to achieve high accuracy in punctual localization of interventions and projects.

Table 1 describes the accuracy level and specific uncertainty causes in percentage (Fig. 4).

The information structure obtained allows one to build several maps and elaboration that are exemplified here. In the following figure we compare the investments density map with the population density map. Both elaborations are based on a kernel estimation of spatial density with proper parameters (Fig. 5).

Investments intensity mainly follows principal population agglomerations. In such framework several planning concerns emerged in terms of development opportunities for peripheral sub-areas where we measured minimum population

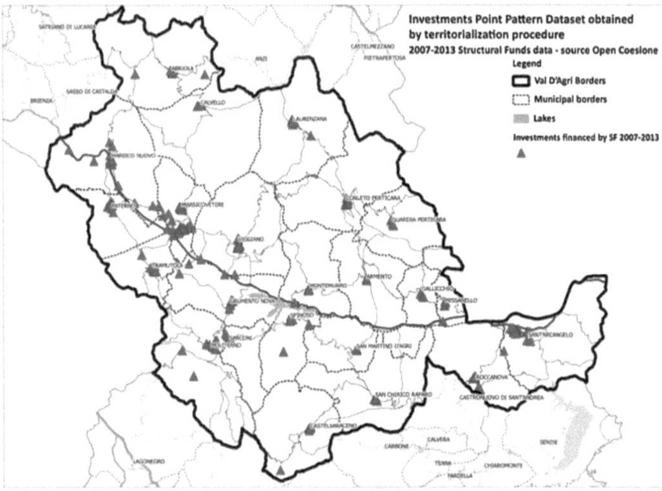

Fig. 4 Investments point pattern in Agri Valley (Color figure in online)

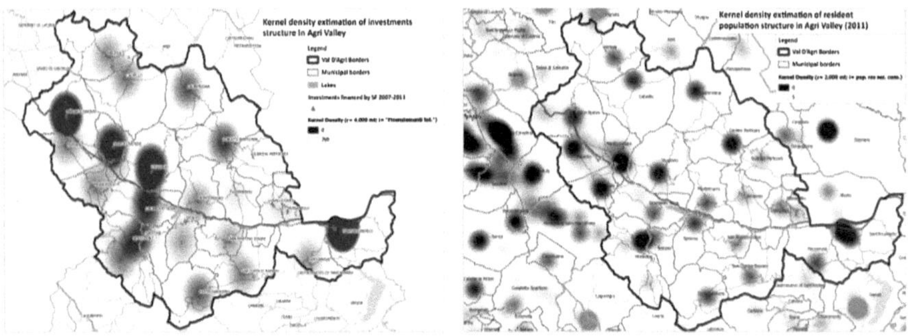

Fig. 5 Investments density estimation (2007–2013) compared with resident population density (2011) (Color figure in online)

density and marked underdevelopment. How is it possible to attract investment and to stimulate local development according to such structural weaknesses?

One possible answer lies in the definition of strategic plans articulated in ‘chains of actions’ that balance the protection of natural and cultural resources with economic development. Among the main features described in this work the theme of energy has a dual perspective: on one side the idea of environmental balance between oil industries and Renewable Energy Sources, on the other the process of redevelopment of traditional form and manufacts of agricultural energy (i.e. ancient network of water-mills) as nodes of a network of community engagement including immaterial actions (knowledge and information dissemination, community involvement and participation, etc.) and investments in local welfare.

5 "Energy Museum": A Specialization Strategy Looking at Sustainability

The research developed several strategic project according to the functionalities described in Table 2. The project "Energy Museum" intends to integrate the resources belonging to the cultural and environmental heritage, by enhancing the diffusion of knowledge on the historic relationship between man, environment and energy.

It also intends to link the undeniable environmental sacrifices (lack of natural features, pollutions etc.) by prospects of replacing fossil energy sources with clean energy (renewable ones), promoting a wider use of economic resources derived from oil extraction in research and large scale testing of RES compensation.

According to this purpose it will exhibit technologies showing their evolution and benefits in everyday life, together with fallout on economy and environment.

We think of a territorial museum consisting in a number of nodes and exhibitions structured in itineraries that represents also a form of integrated touristic supply.

Key elements of the project are also the industrial plants, in particular those related to the process of oil extraction as well as those related to energy production from renewable sources, immaterial actions and researches, experimentation, and training/information activities addressed to local communities and tourists.

It is a project focused on generating results in terms of awareness on on-going processes.

The objectives relate to actions for improving the benefit of local communities aimed at understanding and to increase their awareness on on-going processes in the energy sector. The establishment of a center of excellence direct to a rigorous study of social and environmental phenomena and the territorial projection is an operative tool to guarantee local assessment.

Table 2 Strategic development domain and specialization degree of territorial sub-areas (Las Casas and Scorza 2015)

		Sub-area						
	Functionality	1	2	3.1	3.2	3.3	4.1	4.2
1	Nature tourism	5	5	5	3	5	3	1
2	Cultural tourism	3	5	5	5	3	3	3
3	Tourist accommodation	4	4	–	3	3	3	–
4	Health and third age	5	5	3	1	3	1	3
5	Requalification and reuse residences and services	3	3	5	5	3	3	5
6	District/energy museum	1	3	5	3	3	2	3
7	Industry	1	4	5	2	2	1	1
8	Agro-zootechny	3	3	1	4	4	5	3

The bond between local traditions and future scenarios is the network of old water-mills spread all over the area. These have to be catalogued and redeveloped according to the new functions and needs related to "Energy Museum".

The project involves all sub-areas of specialization according to specific interventions connected with existing artifacts, energy plants settlements and environmental resources.

This activity is based on the active role of local administrations that, for each specific objective and theme will set up a process of synergies development and integration of ongoing projects and activities. This role will be enhanced by wider involvement of local communities including online communication/interaction strategies and tools.

This defines a cooperative approach between institutions, research centres and local communities as a precondition for an effective operative implementation of the project "Energy Museum".

6 Conclusions

Place based approach claims for innovations in EU cohesion policy management. It represents an umbrella term and opens to the selection and assessment of several tools and procedure in order to realize concrete applications at territorial level.

This paper considers territorial assessment of an on-going process and integrated development strategies as two necessary components for enhancing local innovations in policy implementation.

Where, at local level, outcomes indicators measure the implementation of cohesion operative program (Barca and McCann 2011), other efforts should be addressed to the identification of local specialization. This appears to be a synthetic way to face the opportunity to renew cohesion policy development under "place-based" ensign. This effort in representing and describing in details local implementation areas could generate not a fix picture of a context. Places and communities continuously evolve, especially as a reaction to the huge changes brought by economic crisis.

The main issues connected with instances of the New Cohesion Policies are:

- the need of a clear identification of the combined place-specific characteristics in each region;
- a clear identification of the appropriate territorial context in order to effectively implement "smart specializations".

Open data represents a useful process that has already driven research from data production to exploitation of the informative value of several data sources available to everybody.

Regional Science has the task to produce effective 'places' interpretation in order to support public decision in incoming generation of EU ROPs. We are in the case where it is relevant to use numerous data sources and indicators assuming a variable rate of approximation in the accuracy of datasets.

In the field of Regional development the ontological approach provided alternative interpretation models of the interaction among context, program and beneficiaries (Scorza et al. 2010, 2012).

Specialization analysis should be developed through an integrated set of techniques, oriented to generate descriptive geographies of EU region at a variable scale.

This work described geographical elaborations compared to strategy design concerning 'energy' considered as an opportunity for redevelopment of peripheral areas. The perspective concerns the application of such processes in the framework of managing Regional Operative Programs in order to involve beneficiaries and citizens in the process based on territorial consciousness.

References

About W3C Standards. http://www.w3.org/standards/about.html. Accessed 13 July 2015

Barca F (2009) An agenda for a reformed cohesion policy: a place-based approach to meeting European union challenges and expectations. Independent report prepared at the request of the European Commissioner for Regional Policy, Danuta Hübner, European Commission, Brussels

Barca F, McCann P (2011) Methodological note: outcome indicators and targets-towards a performance oriented EU cohesion policy and examples of such indicators are contained in the two complementary notes on outcome indicators for EU2020 entitled meeting climate change and energy objectives and improving the conditions for innovation, research and development. See: http://ec.europa.eu/regional_policy/sources/docgener/evaluation/performance_en.htm. Accessed 1 Oct 2011

Berners-Lee T. 5 ★ Open Data. http://5stardata.info/#addendum4. Accessed 13 July 2015

Berners-Lee T (2009) http://www.w3.org/

Capello R (2014) Smart specialisation strategy and the new EU cohesion policy reform: introductory remarks. Scienze Regionali

EC (2007) Inspire directive. http://eur-lex.europa.eu/LexUriServ/site/it/oj/2007/l_108/l_10820070425it00010014.pdf

ENI in Basilicata, Local report (2012) ENI

Executive Order 12906 (1994). http://www.archives.gov/federal-register/executive-orders/pdf/12906.pdf

Ginsberg WR The Obama Administration's Open Government initiative: issues for Congress https://www.fas.org/sgp/crs/secrecy/R41361.pdf

Las Casas G, Lombardo S, Murgante B, Pontrandolfi P, Scorza F (2014) Open data for territorial specialization assessment territorial specialization in attracting local development funds: an assessment. Procedure based on open data and open tools. Tema. Journal of Land Use, Mobility and Environment

Las Casas G, Scorza F (2009) Un approccio "context-based" e "valutazione integrata" per il futuro della programmazione operativa regionale in Europa. In: Bramanti A., Salone C. (eds) Lo sviluppo territoriale nell'economia della conoscenza: teorie, attori strategie. Collana Scienze Regionali, Vol 41. Milano: FrancoAngeli

Las Casas G, Scorza F (2015) Discrete spatial assessment of multi-parameter phenomena in low density region: The Val D'Agri Case. In Computational Science and Its Applications–ICCSA 2015. Springer International Publishing, pp 813–824

Longley PA, Goodchild MF, McGuire DJ, Rhind DW (2001) Geographic information systems and science. Wiley

Murgante B, Tilio L, Lanza V, Scorza F. (2011) Using participative GIS and e-tools for involving citizens of Marmo Platano – Melandro area in European programming activities special issue on "E-Participation in Southern Europe and the Balkans". Journal of Balkans and Near Eastern Studies 13(1):97–115. Taylor & Francis, London. ISSN:1944-8953, doi: 10.1080/19448953.2011.550809

Obama B (2009) Memorandum for the Heads of Executive Departments and Agencies on Transparency and Open Government. http://www.whitehouse.gov/the_press_office/TransparencyandOpenGovernment/

Open Government Initiative. http://www.whitehouse.gov/open

Rey GM (1999) "L'informazione è un patrimonio come il denaro. Non sprechiamolo", in Telèma Burocrazia elettronica Società più civile, n.18 http://www.fub.it/telema/TELEMA19/Rey19.html. Accessed 15 Feb 2011

Rifkin J (2000) L'era dell'accesso Mondatori editore, Milano

Scorza F, Las Casas G, Murgante B (2010) Overcoming interoperability weaknesses in e-Government processes: organizing and sharing knowledge in regional development programs using ontologies. In: Lytras MD, Ordonez De Pablos P, Ziderman A, Roulstone A, Maurer H, Imber JB (eds) Organizational, business, and technological aspects of the knowledge society, communications in computer and information science, vol 112, XXVII, pp 243–253, Springer-Verlag, Berlin, ISSN: 1865-0929. doi: 10.1007/978-3-642-16324-1_26

Scorza F, Las Casas G, Murgante B (2012) That's ReDO: ontologies and regional development planning. Lecture Notes Compu Sci 7334/2012, 640–652. doi: 10.1007/978-3-642-31075-1_48

Springer SocPsych Style: Based on the reference list style that was established by the American Psychological Association (APA)

Tilio L, Larosa S, Sansone A, Scorza F, Lanza V, Las Casas G, Pontrandolfi P, Murgante B (2009) E-gov tools for evaluation in programming processes. In: Rabino G, Caglioni M (eds) Planning, complexity and New ICT. pp 167–177. Alinea Editrice, Firenze. ISBN: 9788860554154

Transparency International Georgia, Ten Open Data Guidelines. http://www.transparency.ge/en/ten-open-data-guidelines

Reshaping the Urban Environment Through Mobility Projects and Practices: Lessons from the Case of Palermo

Ignazio Vinci and Salvatore Di Dio

Abstract Transport and mobility are crucial factors in the process of adaptation of contemporary urban areas to the challenge of sustainable development. In this perspective, however, cities need to perform a more effective integration between a wide range of different policies and planning practices. Together with a more effective integration between transport policy and land-use planning, for instance, cities could take growing advantage by organisational factors, as well as the spreading of new technologies and the related processes of social innovation. Through the description of several project and planning initiatives taking place in Palermo, the fifth Italian city, this paper attempts to explore the potential sinergies between the "hard factors" of mobility—in this case given by the huge programme of infrastructure redevelopment under realisation in the city—with some other dimensions, such as urban regeneration processes, governance and community-led processes of innovation.

1 Introduction

In the last two decades, the relation between mobility and urban development has undergone significant changes from both theoretical and practical points of view. While infrastructure projects for the movement of people and goods is historically functional to the spatial, demographic and economic growth of cities, transport policy today is perceived as an instrument to achieve a broader range of social objectives.

The reasons for such a changed perspective are various: two processes, in particular, have reshaped the question in cultural and political terms.

I. Vinci (✉)
DARCH, University of Palermo, Palermo, Italy
e-mail: ignazio.vinci@unipa.it

S. Di Dio
PUSH Design Lab, Palermo, Italy
e-mail: s.didio@wepush.org

R. Papa and R. Fistola (eds.), *Smart Energy in the Smart City*,
Green Energy and Technology, DOI 10.1007/978-3-319-31157-9_15

A first process regards the growing attention given to sustainability in the diverse forms it can be conceived at the urban scale, from the protection of the environment to the fight against climate change. After decades during which car-dependent urban development has been considered an unavoidable consequence of economic growth in the Western countries, the international debate agrees that planning compact settlements (UN-Habitat 2011; World Bank 2011) as well as more efficient public transport systems (Schäfer et al. 2009) are essential requirements for achieving environmental sustainability goals. As a consequence of this process, both at an academic (Banister 2008) and political level (European Commission 2007) different approaches and paradigms to planning have emerged in the attempt to merge the question of mobility with sustainable development.

The second process regards innovation in the transport policy in response to the spatial rearticulation of cities which, in turn, is influenced by technological factors and people's lifestyle change. An increased mobility, it is well known, constitutes a distinctive element of contemporary society since its transformation in a post-industrial society (Castells 1996; Urry 2007). The rationalist paradigm founded on the separation of spaces destined to different functions (live, work, leisure) has faded into urban systems within which a growing mixity of activities coexist in the same places.

While the new technologies enable enterprises to displace their activities with increased flexibility than ever before, the regeneration of urban spaces formerly dedicated to other functions (productive, logistic) requires transport policies not simply to ensure accessibility, but also contributing to the creation of new livable places. Contemporary mobile communication technologies (social networks, gps applications), also, by increasing the flexibility of people's moves over time and space, tend to provide a new relational geography of cities (Grieco and Urry 2012), with the effect of binding inextricably any policy for mobility with the issue of social innovation and the broader concept of the smart city (Albino et al. 2015).

Over the last two decades, the idea of fostering a stronger integration between mobility and urban development, on the background of a broader attention to environmental sustainability, has represented the source of inspiration for new cultural movements and the creation of alternative approaches to urban planning.

A first recognisable strand in the international literature can be referred to the concept of *smart growth* (Bullard 2007; Inam 2011). It was first used in North America to designate a pragmatist approach as an alternative to the most radical views (pro-growth vs no-growth) on urban sustainability in the early nineties. The *smart growth* approach has the objective of creating more sustainable communities through the spreading of some planning principles—including mixed land-use, creation of communities with a strong sense of place, promoting of a variety of transportation choices—in the implementation of spatial plans from the regional to the local scale. Other movements in North America such as the *New Urbanism* (Frug 1999) are theoretically linked to the smart growth concepts, providing instruments for the creation of more livable communities grounded on the quality and extent of open spaces and 'soft mobility'. The concept of *compact city*, which is being developed also in Europe and Asia (OECD 2012), tries to promote the advantage of highly concentrated urban settlements in order to reduce land consumption and to capture the positive socio-economic externalities given by activities integration. A common ingredient of the compact city concept with the previous approaches is the role given to accessibility in all its forms as a prerequisite for achieving together sustainability and quality of life in the urban environment.

A prominent place in all these research strands must be given to the concept of *Transit oriented development* (TOD) (Cervero 1998; Curtis et al. 2009; Suzuki et al. 2013). The TOD is meant as a theory and a set of techniques that, more explicitly than others, emphasize mobility networks as the vital framework around which to address urban development.

Together with some applications driven by the aim of creating new highly attractive settlements along the main transport corridors, in the recent years the TOD experiments are increasingly facing the question of the existing city. In Europe, particularly, when dealing with the reshaping of existing districts touched by new transport nodes, the TOD objectives cannot be easily separated by the overall objective of generating more complex processes of urban regeneration. For the planning projects it implies controlling a broader range of constraints and variables to development, but also can count on a broader range of tangible and intangible resources for the place-making process.

With this conceptual framework on the background, this paper discusses the challenges given by the reshaping of the mobility system in the metropolitan area of Palermo, the fifth Italian city and Sicily regional capital. In particular, it attempts to explore the potential connection between the "hard factors" of the transport policy—given by the reshaping of the transit system in the city—with some "soft factors" provided by organisational efforts in the public sector, forms of public-private partnership and ITC-based projects that attempt to link mobility with social innovation. In the final section of the paper integration among these different factors will be critically discussed from the perspectives of planning, social innovation and governance.

2 The Context for Transport and Mobility

With a population of 678.492 inhabitants (Dec., 2013), Palermo is the fifth Italian city by demographic size. It is the functional and economic centre of an urban regions that has around 1 million of residents within the first ring and around 1.27 millions of inhabitants if considering a larger zone coinciding with the province (Cittalia 2014). In the last three decades, the loss of residents in the core city (around 6 % between 1981 and 2011) has resulted in a process of suburbanisation, leading to a significant growth of population in the suburbs of the first ring, particularly along the coastal areas.

This process, however, has happened with less intensity respect to other Italian urban regions and with no significant redistribution of relevant functions between the core and the outer urban centres. As a result, the population trend in the municipality is almost steady—+0.5 % in the decade 2004–2013 (Cittalia 2014)—while the core city has maintained all his level of attractiveness towards the metropolitan area.

As a consequence of the concentration of functions in the core city and its role as regional capital, the urban area is a destination of thousands of commuters every day. Due to the lack of an efficient public transport system, the congestion rate in the main road corridors in the peak hours are among the highest in the country. According to the TomTom Traffic Index (2014)—computed upon the time spent in car for a same length trip—Palermo is the 4th more congested European city. The car density is the fourth in Italy in relation to the land area (around 3500 cars per square kilometers), while it is second only to Naples in terms of ratio between cars and population (ISTAT 2012) (Table 1).

The city's bus network—which extends for 341 km and has a quite good distribution of bus stations (14.7 stations per square kilometers, above the average of the largest cities)—is still the most used transit system within the urban area. However, the low commercial speed due to congestion and the narrow number of seats available due to the limited number of buses effectively in use on an average day (around 51 % of the entire fleet), does not make it attractive to a broader share of population (Di Dio et al. 2015). The only significant alternative to the bus system has been in the last two decades an urban rail service operated by Trenitalia (the national rail system operator) that runs across the urban area with 12 stations, but with a frequency (30 min) that makes it not comparable to a metropolitan service. As a consequence, the modal split in transport is still sharply unbalanced towards the private motorised vehicles, with the city that is first ranked among the major cities both for the use of cars (37 %) and motorbikes and scooters (12 %) (Di Dio et al. 2015).

The vicious circle between an high car-dependent transport model and the lack of an efficient public transport implies, among other things, a series of negative effects on the urban environment. According to *Ecosistema Urbano* (Legambiente 2014), a report published annually on the environmental quality of the major Italian cities, Palermo has one of the worst national performance in terms of rate of nitrogen dioxide in the air. The presence of this gas in the air is mainly related to the diesel engines and the age of vehicles is generally proportional to the level of emissions. In accordance to a survey on the age of vehicles in the major cities (ACI 2011), the city of Palermo leads the national ranking in terms of buses with more then 20 years (22.5 %) and it is second ranked in terms of circulating trucks with the same age (32.1 %). The city's

Table 1 Vehicles speed, density and age in the first five Italian cities (ISTAT 2012)

	Rome	Milan	Naples	Turin	Palermo
Car average speed (km/h)	23	22	21	26	20
Public Transport average speed (km/h)	12	12	10	13	12
Cars per 1000 inhabitants	565	529	597	565	590
Cars older than 8 years (%)	49	47	48	72	60

bus network, also, is served by one of the oldest bus fleets in Italy, of which 50 % belongs to the Euro 0 class (before 1992) and Euro 1 class (1993–1997 period).

After several decades of inertia, the municipality is defining measures in order to rebalance the modal split of mobility in the urban area and to reducing the impact of transport on the environment.

In 2013 the city council adopted the first Urban Traffic Plan (PUT), an integrated programme of interventions that is compulsory, among other things, in order to introduce restriction to the private mobility. The strategy of the plan responds to the general objective to lower the level of congestion and increase walkability by reducing the access to the central areas to the private cars. The flagship projects are therefore a series of zones with traffic restriction (ZTL) which are mainly targeted to the historic centre and its surroundings expected to enter in service in april 2016. Other measures regard the increase of public transport efficiency through bus lanes, the reorganisation of parking and a broader use of ICTs to manage the mobility system. An additional operative component of the PUT is being developed for the extension of the bike paths network along the main city's mobility corridors.

The Sustainable Energy Action Plan (SEAP), also launched in 2013, stresses the uses of shared vehicles and the conversion of part of the bus fleet to natural gas as conditions to reduce the impact of transport in the urban environment. Other policy tools, such as the city's programme within the National Programme for the Metropolitan cities (currently under implementation in the framework of the 2014–2020 EU's planning cycle), provides additional investments for sustainable modes of transport, including new bike lanes, GPS-based remote control for bus transit and natural gas fueling stations.

3 The Reshaping of the Infastructure System

For many decades the city's public transit system has been characterised by a chronic lack of investments with the result of being based uniquely on the road network. The rail system itself—conceived in the XIX century to serve primarily the regional connections—has been only marginally reshaped to serve urban and metropolitan mobility. As a result, the private cars and an inefficient bus system have assumed a dominant role on urban mobility with a growing impact on city's environment both in terms of congestion and pollution.

A sharp change of direction takes place in 2002 with the approval of the "Integrated Plan for Mass Public Transport", which provides a wide programme of redevelopment of the transit system mainly based on the rail networks. The plan can be considered an integration of new projects and an upgrading of existing networks promoted by several public authorities, including RFI (Rete Ferroviaria Italiana, the national owner of the rail system), the former provincial authority and the municipality. It is based on the interconnection of four major infrastructural projects (Fig. 1):

- the upgrading of the Railway ByPass (Passante ferroviario);
- the completion of the Railway Ring (Anello ferroviario);

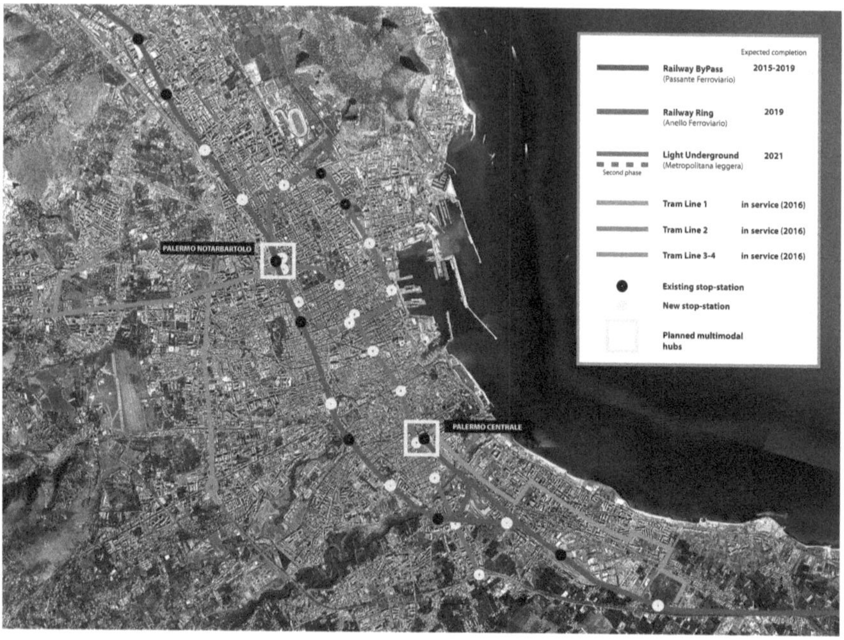

Fig. 1 Transport networks under construction in the Palermo urban area (Color figure online)

- the realization of the Tram system and;
- the planning of an Automated Light Metro (Metropolitana leggera automatica).

The Railway ByPass is expected to be the project with greater impact both on the urban and metropolitan scale. It consists on the redevelopment of around 30 km of existing rail lines between the South-Eastern suburbs and the international airport (Punta Raisi) located 25 km North-West the city centre. The logic of the intervention is to transform the existing regional line into a double-track and high frequency railway able to connect the city's major functional nodes with the urban region. For this purpose, the project provides the undergrounding of 7 km of railway and the creation of 10 new stops, including 5 underground stations in the most dense and populated areas of the city.

Some of this new stations are about to be opened in proximity of places and urban functions that already catalyse amounts of commuters and city-users, such as the Courthouse, the historic district, the palaces of regional government, the university campus and the hospital district Civico-Policlinico. In other cases, however, the new stops seem not giving a direct response to the most attractive urban functions, being located in less dense neighborhoods as well as in urban areas with poor conditions in terms of urban quality. This fact, mainly given by the priority of reducing the construction costs by the rail operator, poses the question of the interface between the city and the infrastructure and is being increasingly perceived as a planning issue in the near future. For two sections of its route the completion of the Railway ByPass (started in 2008) is scheduled for 2016, while the whole infrastructure is expected to enter in service in 2019.

The Railway Ring, as well as the Railway ByPass, is an adaptation of an existing and partly underground rail line that linked the Central station to the Northern side of the port. After being dedicated for decades exclusively to the freight traffic, in 1990 the line has been reconverted to metropolitan service with the opening of four new stations. The project under realisation (started in 2015) provides the extension of the line up to 6.5 km with a planned route as a circle line with a link to the Railway ByPass and the regional service in the North-Eastern station of Palermo Notarbartolo.

The most interesting aspect of the project is that it provides the opening of three new underground stations in some of the most dense and attractive places of the city until now served only by the bus network. The Piazza Politeama station, particularly, will bring the rail service within the commercial and touristic core of the city, with an expected impact both on the local flows and the connection with the airport through the link of the line with the Railway ByPass. The line in the new configuration and the new stations are expected to enter in service within the end of 2018.

The Tram system is a project launched at the beginning of the nineties which has faced several problem of financing before its implementation started in 2007. Consisting of four lines, for a total extension of around 15.5 km, it was conceived with the aim of connecting two remote and popular suburbs with the central area of the city and the two main rail stations. The line 1, with 16 stops, connects the Central Station with the industrial suburb of Brancaccio, with a termination in correspondence of the ByPass station of Roccella and a major shopping districts (the Forum) opened in 2010. The lines 2 and 3 (with 12 and 26 stops, respectively) run between the popular neighborhoods of Borgo Nuovo and CEP and the North-Western transit hub of Palermo Notarbartolo. A section of the initial line 3 has been later devoted to a fourth line serving several neighborhoods along the highway (Circonvallazione) surrounding the urban area to the West. A project under discussion provides for the extension of the tram network through three lines with the aim of ensuring a better linkage of the existing lines and the penetration in the city centre.

The project of the fourth infrastructure, an Automated Light Metro (with driverless trains), is currently under development. The extended version of the project provides a line running from the Southern gate of the city (Svincolo Oreto) to the district of Partanna-Mondello, a residential suburb located in the North of the urban area. The entire planned line extends for around 17 km, with 23 stations along the main development axis of the city from South to the North. A first section of the project—running from the Southern terminus to the rail station Notarbartolo—has been funded in 2014, with completion expected in 2021. It extends for around 7 km, with 10 stations covering the central area including the Central Station and the historic district. Beside that, the line should create a series of links with the other transport projects under implementation (Railway ByPass, Railway Ring and Tram system).

The four projects are expected to reshape the city's transport system at different territorial scales. On the one hand, it is hoped an impact on transit at metropolitan level, by reducing the amount of people using the road transport to reach the core city from the urban region. At this scale, also, it appears relevant the impact that they could have on connecting the main transport nodes of the metropolitan area, such as the airport, the port and the two rail stations within the city. On the other hand, they hopefully will contribute to change the mobility behaviours of citizens within the urban area, since now served only by an inefficient bus service and congested by private cars.

There are also other perspectives, however, than simply observing the impact of this networks from the point of view of transport policy. The extent of these transformations, for example, is massively reshaping urban space in several parts of the city. The Tram system is changing the face of kilometers of urban roads, most of which characterised by decay and low economic activity. The route of the line 1, particularly, passes across one of the most deprived neighborhoods of the city (Brancaccio), with the new infrastructure inserted on old roads and abandoned urban spaces. Some of the sites interested by the new stations of the Railway ByPass are largely underused and the space around them appears degraded despite their closeness to the core city.

An open question for the future of city's development, however, remain how these spatial and infrastructural transformations could affect other dimensions of urban change, such as economic regeneration and neighborhoods revitalisation.

4 Innovating Mobility Through Soft Factors: Technology, Social Innovation and Governance

The intersection between social innovation and new communication technologies are deeply influencing also the transportation domain, so as to contribute to the creation of new models of urban sustainable development (Patier and Browne 2010; Gal-Tzur et al. 2014). One of the main reason of a wide range of experiments made in this field is the capability to reach the motivations that push people to choose one urban transportation system instead of another (Moore 2011; Nasrudin et al. 2013). Despite people's transport choices are mainly dependent on the presence of good and diversified infrastructures (Urry 2012), habits are also a very important key in the mobility patterns. In fact, contributing to the change of individual behaviors can be a strategic factor not only in terms of efficiency, but also from the perspective of sustainability (Ben-Elia et al. 2008).

The urban transportation domain, therefore, it's not only about top-down policies (Singapore Land Transport Authority 2008), but also increasingly dependent from the bottom-up effects driven by citizens' lifestyles and different ways to live and perceive the city itself (Gatersleben et al. 2013). If we take as a reference the 'Urban Metabolism' metaphor (Pincetl et al. 2012), improving the "body" performance of a city could imply healing its 'nervous system' and, in turn, the need to interfere with people that actually 'use' the city too often improperly (Wamsler and Brink 2014).

While all over the world designers continue to deal with urban traffic issues in the old and consolidated cities through conventional ways, new approaches and tools based on a social innovation perspective are hybridizing the urbanism field. American private enterprises such as Nuride (http://nuride.com), Zimride (http://www.zimride.com/) and LYFT (https://www.lyft.com/), and European enterprises like moovel (https://www.moovel.com/en/US/), Mobility (http://www.mo-bility.com/mo/home.html) and Covivo (http://www.covivoturage-dynamiq.eu/) are quickly moving the debate on changing cities by switching their attention from the necessary modifications of the urban structure (the hardware) to the changes that can be induced by working on citizen behavior and the urban communities' habits (the software).

All of these initiatives try to stimulate the urban communities to change their (bad) habits, fostering more responsible behaviors through the use of ICT networks and smartphone apps. Essentially, the idea is to trigger a social and cultural change with the "dialog" tool provided by social media technology. A common aim, therefore, is creating new services of public interest but, possibly, also a new market considering that the majority of the above-mentioned projects were developed by start-ups.

In the city of Palermo the most relevant initiatives in this direction are three projects funded by the Italian Ministry of University and Research in the framework of the competition "Smart Cities and Communities and Social Innovation". All these initiatives aimed to stimulate local communities to change their habits fostering more responsible mobility behaviours with the help of smartphone applications. The three groups that won the fellowship are young multidisciplinary teams with a "social entrepreneurs" approach to the theme. They are working on different softwares—*Muovity*, *CityFree* and *TrafficO*$_2$—that provide services of carpooling logistic (mostly focused on short distance commuters) and inter modality logistic to foster sustainable mobility values.

Muovity aims to give a contribution to the empowerment of the social and technological networks to improve sustainable and energy efficient transports in the sicilian territory. The mobile application offers a platform where is possible to easily plan carpooling trips and having real time information about the city mobility. The team—who is currently testing the main features of the application—is broadly the same who run "Mobilita Palermo", a very popular blog on mobility and city's urban transformation started in 2008.

Cityfree has developed an application which main feature will be an easy way to plan dates to reach the university facilities through the local public transport system. The application do not give the user another social network platform but instead wants to be a mobility information system that supports the existing social networks. The main goal of the tool is reaching the critical mass able to change the behaviors of college students. Currently the group activities are focused on the creation of the mobile application and on disseminating actions of carpooling concepts to the University of Palermo community.

TrafficO$_2$ is an info mobility decision supporting system that tries to foster a better modal split through gaming policies and tangible incentives for each sustainable choice. The idea is to match the interests of two complementary actors on the city traffic scene: communities workers (communities that already need a mobility manager) and local business communities (places on the community workers daily-paths). The goal is to decrease traffic and pollution by creating an equal agreement for both communities: prizes in exchange for a respectful behaviour towards the environment. So, all of the local businesses that belongs to the platform (as sponsors) became the stations of a new kind of transport system that foresee only moving by foot, by bicycle, by local public transport and by carpooling. Each trip from station to station gives O_2 points to the user, which is the virtual money users can collect to get prizes from the sponsors. A first test of the mobile application (an alpha version) has started during December 2013 with about 80 students selected through a workshop from three different university departments (computer science, design and marketing) and has shown about the 55 % of reduction of CO_2 emissions (Di Dio et al. 2015) (Fig. 2, Table 2).

All of the funded projects are based on a strong involvement of the university community, a decision taken individually by the teams motivated by the fact that young and curious students are probably the best targets in order to experiment behaviors' changing project driven by social media technologies. Furthermore, it has to be mentioned the supporting role in the projects given by ARCA, the University of Palermo incubator who is playing a major role in the development of startups and more sustainable business models.

Beside the domain of social innovation, the city is also experiencing a phase of spreading of new modes of transport through the construction of public-private

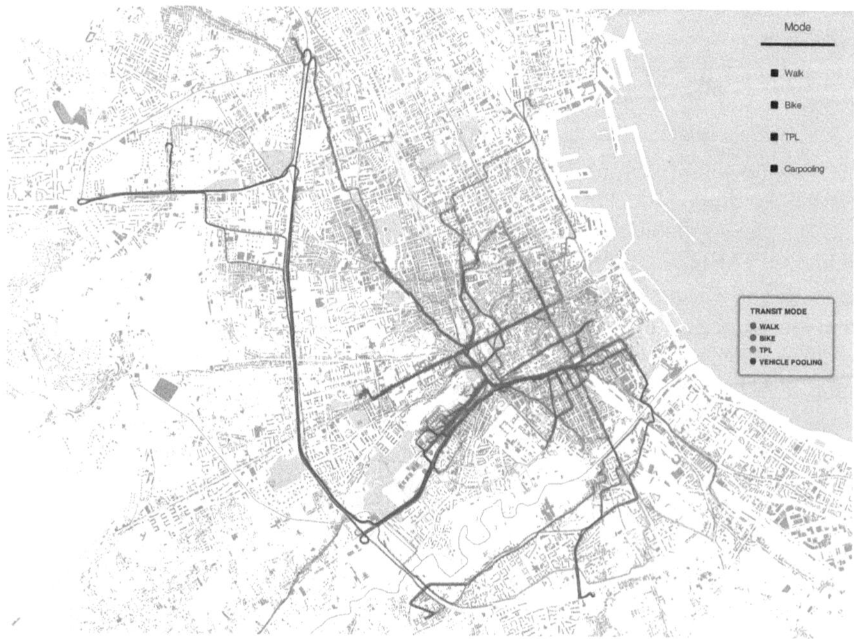

Fig. 2 Home to University flows by transport mode (elaborated by PUSH) (Color figure online)

Table 2 Comparative analysis of the main features of the mobility social innovation projects

	Cityfree	Muovity	TrafficO_2
Events logistic management	x	x	–
Intermodality system logistic	–	–	x
Infomobility	x	x	–
Extra-urban scale	x	x	–
Payment service	x	x	–
Benefits for users	–	x	x
Sponsors	x	x	x
Social network	–	x	x
Feedbacks	x	x	–
Gaming	–	x	x
Testing	x	x	x

collaboration. A car-sharing service, launched for the first time in 2010, has been redeveloped in 2015 with a fleet increased up to 140 cars, while a bike-sharing service has been made available in 2015 with a fleet of 420 bikes.

While there are concerns regarding the impact of the car-sharing service due to the limited number of cars respect to the potential demand, the bike-sharing service is raising growing expectations in the public. In April 2015, in fact, the municipality has approved a plan for the extension of the bike lanes network up to around 145 km with the aim of serving the main functional nodes and tourstic attractions within the urban area. These policies have also stimulated new bottom-up initiatives and connections with the broader domain of social innovation: "B.I.C. Bici in città" and iMove, for example, are local examples of smart community projects carried out internationally that encourage and organise people to make critical mass in terms of sustainable transports.

The linkage between all these bottom-up, public-private and governmental activities is the aim of fostering people to abandon unsustainable mobility habits, by taking greater advantage from data communication (through high tech smartphones, social networks) and principles of the shared economy.

It is a process that place the city within a field of innovation that, from the energy consumption (Chen Lillemo 2014; Shimokawa and Tezuka 2014) to the waste management (Lee et al. 2014) fields, is starting to give encouraging results. In this era of "zero" resources change bad behaviors could be a very cost effective set of solutions to improve the city liveability without the involvement of infrastructural actions (Ceder et al. 2013).

On the other hand, the bigger limits of those approaches are the constraints given by the chosen technologies itself. These kind of services need, as example regarding the mobile apps mentioned before, powerful smartphones always connected to the web and this means an high consume of the battery and of the personal band connectivity. Considering that Palermo has not yet a powerful public service for the

Internet connectivity, a challenge for the near future will be finding an effective governance system in order to integrate investments on the hard factors and soft resources in the urban community.

5 Process of Change and Policy Challenges: Conclusive Remarks

Looking at the diverse, and only partially interconnected, planning and organisational processes taking place in the city, we certainly can get the impression of a dynamic and fast-moving context. After decades of lack of investments, with a mobility model dominated by road transport, mobility is apparently becoming a transversal policy issue able to catalyze the interest of a broader range of local actors. In general terms, we can identify at least three separate processes from whose interaction the mobility domain of the city should be shaped in the near future.

- First of all, a process of creation of a new urban infrastructure for mobility, beyond the road networks that have characterised the city's first, and long lasting, stage of development.
- Secondly, the spreading of bottom-up and community-led experiments, mainly based on self-organisation initiatives, aimed at stimulating changes in the citizens behaviours as well as creating links between private and public ways of transport.
- A third process could be described as a public-private domain, given by a series of initiatives, driven both by public interest and the market, requiring effective governance mechanisms in order to take advantage from the cooperation between top-down and bottom-up initiatives.

To which extent these factors will rapidly contribute to define a more sustainable and effective transport system in the city is a largely open question depending on several political and organisational factors. The point we would like to discuss in the conclusion, however, is how to shift the attention from the narrow domain of transport to the wider perspective of mobility as a strategy for creating a smarter and qualitative place for citizens and city-users. In fact, the city resulting from this processes will be an interaction between a series of "hard factors" in support of mobility—mainly infrastructures with their material impact on the built environment—and the outcome of a variety of policies and organisational practices (including plans, projects and new services), promoted by public and private actors, that can be defined as the "soft" components of city's mobility.

This perspective implies that the question of transport and mobility cannot be separated by the broader question of a city-making process as a result of a more integrated and community-led vision of urban development. It means recognising the "systematic nature" of mobility (Bertolini 2012) within the question of urban development, as well as the cognitive, planning and organisational challenges it implies in the policy-making process at the local level (UN-Habitat 2013).

In this specific context it seems there are at least three, interdependent, policy challenge toward this perspective:

- a challenge of effectiveness of the transport system;
- an urban regeneration challenge;
- a governance challenge.

The first challenge arises from the basic need of making really effective the new network of infrastructures under completion. The question could be faced at different levels: on the one hand, it is needed to increase connectivity among the different networks, considering that they have been conceived and will be managed by different operators. Arranging forms of coordination among the operators might be a first basic response to the question (an integrated ticketing system is under discussion), but also avoiding lack of efficiency in the service upon the single network could be relevant in order to tackle users expectations. An investment on the ICT resources, in the short term, could significantly help to increase the added value given by the intersections of the different modes of transport, as well as in reducing the negative impact of the missing links existing in the networks. The social networks and self-organised practices, as well, could play a relevant role in stimulating formal government to provide services more efficient and shaped on city-users' expectations.

The challenge of urban regeneration is linked to the objective of making the nodes and corridors of the new transport system as a framework for the future city's sustainable development. It appears as an ambitious challenge if considered that the question of interfacing the city and the new infrastructures has been often neglected, both from spatial and functional perspectives.

It is widely recognised, instead, that the increased accessibility provided by new transport nodes can mean a wide range of urban regeneration opportunities, including functional reorganisation, social revitalisation, as well as processes of place-making and creation of new public space. From this point of view, a challenge for future policy-making would be perceiving the nodes and corridors generated by the new transport networks not only in terms of mobility hubs but also as the epicentres of urban regeneration processes for the involved neighborhoods and the city as a whole.

The two previous challenges are intimately bound by the need of a more effective governance system at different scales able to treat an evolutionary mobility-led strategy of urban development. From a more sectoral perspective, it would mean simply creating the room for stronger cooperation among the different transport operators and between them and local government. But looking at the objective of a creating a smarter development model for the city in the next future, the identification of an holistic and integrated strategy is required in order to take full advantage from all the (material and immaterial) resources for urban mobility, as well as the different expectations of citizens and stakeholders.

References

ACI—Automobile Club d'Italia (2011) Analisi dell'anzianità del parco veicoli in alcuni comuni capoluogo di provincia. http://www.aci.it/fileadmin/documenti/studi_e_ricerche/dati_statistiche/Anzianita/Capitolo_4.pdf. Accessed 2 Apr 2015

Albino V, Berardi U, Dangelico RM (2015) Smart cities: definitions, dimensions, performance, and initiatives. J Urban Technol 1–19. doi: 10.1080/10630732.2014.942092

Banister D (2008) The sustainable mobility paradigm. Transp Policy 15(2):73–80

Ben-Elia E, Erev I, Shiftan Y (2008) The combined effect of information and experience on drivers' route-choice behavior. Transportation 35(2):165–177

Bertolini L (2012) Integrating mobility and urban development agendas: a Manifesto. disP 188:16–26

Bullard RD (ed) (2007) Growing smarter: achieving livable communities, environmental justice, and regional equity. The MIT Press, Cambridge

Castells M (1996) The rise of the Network Society. Blackwell, Oxford

Ceder A, Chowdhury S, Taghipouran N, Olsen J (2013) Modelling public-transport users' behaviour at connection point. Transp Policy 27:112–122

Cervero R (1998) The transit metropolis: a global inquiry. Island Press, Washington

Chen Lillemo S (2014) Measuring the effect of procrastination and environmental awareness on households' energy-saving behaviours: an empirical approach. Energy Policy 66:249–256

Cittalia (2014) Taccuino Metropolitano. Fondazione Anci Ricerche, Rome. http://www.cittalia.it/images/taccuino_metropolitano_Cittalia.pdf. Accessed 2 Apr 2015

Curtis C, Renne JL, Bertolini L (eds) (2009) Transit oriented development: making it happen. Ashgate, Farnham

Di Dio S, Lo Casto B, Micari F, Rizzo G, Vinci I (2015) Mobility, data and behavior: the TrafficO2 case study. In: Vesco A, Ferrero F (eds) Handbook of research on social, economic, and environmental sustainability in the development of smart cities. IGI Global, Hershey.

European Commission (2007) Green paper: towards a new culture for urban mobility. European Commission, Brussels

Frug G (1999) City making: building communities without building walls. Princeton University Press, Princeton

Gal-Tzur A, Grant-Muller SM, Minkov S, Nocera S (2014) The impact of social media usage on transport policy: issues, challenges and recommendations. Procedia—Soc Behav Sci 111:937–946

Gatersleben B, Murtagh N, White E (2013) Hoody, goody or buddy? How travel mode affects social perceptions in urban neighborhoods. Transp Res Part F: Traffic Psychol Behav 21:219–230

Grieco M, Urry J (2012) Mobilities: new perspectives on transport and society. Ashgate, Farnham

Inam A (2011) Smart growth. A critical review of the state of the art. In: Banerjee T, Loukaitou-Sideris A (eds) Companion to Urban design. Routledge, London, New York

ISTAT (2012) Indicatori ambientali urbani. Istituto Nazionale di Statistica, Rome

Lee D, Offenhuber D, Biderman A, Ratti C (2014) Learning from tracking waste: how transparent trash networks affect sustainable attitudes and behavior. 2014 IEEE World Forum on Internet of Things Proceedings, pp 130–134

Legambiente (2014) Ecosistema Urbano. Annual Report, Rome

Moore S (2011) Understanding and managing anti-social behaviour on public transport through value change: the considerate travel campaign. Transp Policy 18(1):53–59

Nasrudin N, Nor ARM (2013) Travelling to school: transportation selection by parents and awareness towards sustainable transportation. Procedia Environ Sci 17:392–400

OECD (2012) Compact city policies: a comparative assessment. OECD Publishing, Paris

Patier D, Browne M (2010) A methodology for the evaluation of urban logistics innovation. Procedia Soc Behav Sci 2(3):6229–6241

Pincetl S, Bunjeb P, Holmesc T (2012) An expanded urban metabolism method: toward a systems approach for assessing urban energy processes and causes. Landscape Urban Plann 107:193–202

Schäfer A, Heywood JB, Jacoby HD, Waitz IA (2009) Transportation in a climate-constrained world. The MIT Press, Cambridge
Shimokawa M, Tezuka T (2014) Development of the "Home Energy Conservation Support Program" and its effects on family behavior. Appl Energy 114:654–662
Singapore Land Transport Authority (2008) Land transport masterplan 2008: a people centered land transport system. https://www.lta.gov.sg/content/dam/ltaweb/corp/PublicationsResearch/files/ReportNewsletter/Connect/Connect_April_FA.pdf. Accessed 13 Mar 2015
Suzuki H, Cervero R, Iuchi K (eds) (2013) Transforming cities with transit: transit and land-use integration for sustainable urban development. The World Bank, Washington
TomTom Traffic Index (2014) http://www.tomtom.com/en_gb/trafficindex/#/. Acessed 17 Apr 2015
UN-Habitat (2011) Global report on human settlements 2011: cities and climate change. Earthscan, London
UN-Habitat (2013) Global report on human settlements 2013: planning and disegn for sustainable urban mobility. Earthscan, London
Urry J (2007) Mobilities. Polity Press, Cambridge
Urry J (2012) Changing transport and changing climates. J Transp Geogr 24:533–535
Wamsler C, Brink E (2014) Interfacing citizens' and institutions' practice and responsibilities for climate change adaptation. Urban Clim 7:64–91
World Bank (2011) Cities and climate change: an urgent agenda. The World Bank, Washington

Geodesign: Lost in Regulations (and in Practice)

Michele Campagna and Elisabetta Anna Di Cesare

Abstract Recent outcomes of a trans-disciplinary debate spreading in America, Europe and Asia among scholars and practitioners in spatial planning, urban design, landscape architecture, and Geographic Information Science, propose the concept of Geodesign as a framework for planning and design aiming at addressing some of the most urgent issues of sustainable growth. Geodesign entails the application of methods and techniques for planning sustainable development in an integrated process, from project conceptualization to analysis, simulation and evaluation, from scenario design to impact assessment, in a process including stakeholder participation and collaboration in decision-making strongly relying on the use of digital information technologies. As such, Geodesign may help to put into practice the methodological innovation brought by Strategic Environmental Assessment in planning process that often, so far, failed to be properly implemented. In the light of the above premises, this paper aims at offering critical insights as a contribution to bridge the gap between Geodesign concepts and both spatial planning regulations and practices, with reference to the Italian case study. The analysis is intended to contribute to offer a better understanding of both normative and methodology issues for a fruitful application of Geodesign principles. Making a more explicit linkage between policy principles and planning, may possibly contribute to foster innovation in education, governance and practices to move towards more sustainable processes of savvy growth, which is one of the main goals in smart city governance.

M. Campagna (✉) · E.A. Di Cesare (✉)
DICAAR, University of Cagliari, Cagliari, Italy
e-mail: campagna@unica.it

E.A. Di Cesare
e-mail: elisabetta.dicesare@unica.it

R. Papa and R. Fistola (eds.), *Smart Energy in the Smart City*,
Green Energy and Technology, DOI 10.1007/978-3-319-31157-9_16

1 Introduction

Since the middle-Eighties in Italy, territorial governance faced an evolution towards more environmentally savvy approaches to urban and regional planning. Since the same period, sustainability, transparency, communication and participation are all components of further cultural innovation to the planning paradigms underpinned by a broader scope in the sustainability of development discourse. More recently, innovations in environmental sustainability policies at the European level, affected the innovation in spatial governance as their main principles were transposed into national and regional legislation frameworks of the Member States. Firstly, the European Directive 2001/42/EC on Strategic Environmental Assessment (SEA), shifted the scope of urban and regional planning towards decision-making processes aimed at governing territorial development according to sustainability principles, introducing a new approach to plan-making. Secondly the European Directive 2007/02/EC, currently under implementation, started to foster an innovative change in the spatial planning media through the creation of an INfrastructure for SPatial InfoRmation in Europe (INSPIRE) which enabled the sharing and the public access to spatial data according to common standards. However, the path to achieve full innovation in urban and regional planning towards more sustainable spatial governance has been demonstrated to be neither short nor easy. Indeed, many pitfalls are still being reported with regards to SEA application in the Member States (Parker 2007; COWI 2009). Many difficulties can also be found in the SEA of spatial planning at the regional and local level. More precisely, they include—but are not limited to—unsatisfactory explanation of how the environmental sustainability objectives inform the plan options, difficulties in properly evaluating relevant impacts, unclear explanations of uncertainties and difficulties in analysis, and last but not least unclear explanations of the impact of public participation in the decision-making process (Fisher 2010). Many of these problems are likely to be related to the lack of a clear and shared vision on how to implement SEA in spatial planning in terms of principles, methods and tools.

In order to address these evident and common issues in a fruitful implementation of SEA in spatial planning, Geodesign—as a digitally aided design approach for the creation of change proposals and impact simulations in their geographic contexts (Flaxman 2010)—may help to bridge the gap between SEA policy principles and their implementation in practice with an operational methodology. This approach appears to be particularly actual with regard to the Italian spatial planning governance which recently was affected in many regions by innovation drivers thanks to the current developments in regional Spatial Data Infrastructures (SDI), as enabling technical platform supporting spatial governance.

In the light of the above premises, the aim of this study is to analyse the Italian national and regional legislative framework and investigate where in the past and current regulations it is possible to find elements of coherence with the Geodesign principles, methods and tools in order to supply an explicit and clearer framework for its application. Afterwards an overview of methods and tools currently used in planning practice at the municipal level (i.e. Local Land-Use Planning, LLUP) is given, with particular reference to selected case studies in the Sardinian context (Italy), in order to understand to what extent in practice such methods, supported by the development of a regional SDI (RSDI), are already used.

The paper is organized as follows: the first section gives a brief explanation of the concept of "smart city", focusing on the importance of smart governance—and spatial governance in particular—to overcome fragmentation and supply a holistic approach to cities' governance. The second section gives a clearer definition of Geodesign based on a literature review with the aim of supplying an operational framework for the following analyses. The third section illustrates the study methodology, which focuses on the analysis of the relationships between the Steinitz' Geodesign Framework (GDF) (2012) and selected European, national and regional regulations shaping the planning process in Lombardy, Tuscany and Sardinia. The results of the analysis aim at making these relationships more explicit in order to reduce the gap between policy principles and their technical implementation. Eventually the last section presents a systematic critical review of selected Sardinian case studies of LLUP's SEAs in order to check for any Geodesign method and tools application in practice.

Current results do not claim to be systematically exhaustive at the present stage of development, although they already offer some interesting critical hints for addressing the issues at hand. Indeed, many concepts entailed in the Geodesign Framework can be found in past and current planning legislative frameworks defining the planning systems in many countries. To mention but one example, the exercise of Environmental Impact Assessment (EIA) which is central to the GDF, is concerned by United States' regulation since 1970 with the National Environmental Policy Act (NEPA). Similarly EIA was introduced in European environmental aquis since the middle-Eighties (i.e. Directive EIA Directive 85/337/EEC). In a similar vein, relationships between many concepts and methods, which found integration in the GDF, and normative rules defining the planning systems and procedures guiding the practice can be found in national and regional planning regulations. However, these relationships may not be always evident to the practitioners, thus creating weaknesses and pitfalls in their implementation. The reasons are many including the lack of reliable expertise (this issue having implications also with regard to planning education) and contextual socio-cultural and political settings (i.e. role of actors, level of rationality of the process, availability of data and tools, and the like). Thus, it is argued that contributing to make a more explicit link between good planning and design methodology approaches, and that actual planning regulations may eventually help to achieve more effectiveness in implementation of the underlying principles of a planning system.

2 Smarter Spatial Governance for Smarter Cities

The rapid growth of urban population and, as a consequence, the uncontrolled expansion of urban areas, generates the urgency to find new ways to manage several emerging problems. These issues can be identified in two major domains: technical and material problems—such as air pollution, traffic congestion and waste management—and social and organizational problems imputable to the complex level of interdependence between the different types of stakeholders (Chourabi et al. 2012).

The concept of smart city arises from the need to improve the quality of life in the big cities in terms of environment, safety, transportation, mobility and energy, or, in other words, to make the urban environment more liveable for citizens. Sustainability, efficiency and competitiveness are the main targets of a smart city and they should be achieved through the integration of the multiple aspects involved in social, cultural and urban development.

The term "smart" does not refer to a rigid concept of city, but to one that is flexible and in evolution, able to face the new urban development challenges quickly in an adaptable way (Schaffers et al. 2011). The understanding of the multidimensional network of the city supports the planning of an adaptable framework necessary to meet all the different needs (Toppeta 2010).

According to Nam and Pardo (2011a), smart city innovation includes three main dimensions strictly related: Technology, Organization and Policy. Technology is a necessary condition for innovation and relies on integrated computing technologies to provide the tools enabling the creation of improved urban services for citizens. In particular technological components offered by Information and Communication Technologies (ICT) and Web 2.0 represent important key drivers for innovation. However technology itself does not guarantee smart city initiative success because the ultimate aim of a smart city is not a system-driven, but a service-oriented city in which urban society needs are in the spotlight. As noted by Nam and Pardo (2011a) all adopted technological tools require the managerial capability to manage them in the smarter way (i.e. organization innovation) and an enabling government environment in which to implement them successfully (i.e. policy innovation). In literature many different perspectives can be found about the importance of each of those aspects, according to Belissent (2011) smart governance represents the core of a smart city. Governance can be defined as a kind of concerted action which tries to solve complex conflicts among actors (van Winden 2008) and includes collaboration, cooperation, partnerships and citizen's participation issues (Nam and Pardo 2011b) to identify potential solutions and overcome fragmentation. One of the challenges of the new smart cities is to work like an integral part of a large system, also encouraging a more transparent, equitable and efficient governance process (Papa et al. 2013). The correct use of technological instruments helps to speed up bureaucratic processes (Toppeta ibidem) and to create a shared knowledge for democratic city governance (Batty et al. 2012) because, supplying new channels of communication, allows the cooperation among institutional and economic stakeholders, data exchange, service integration and even making community members active users in decision-making processes (Giffinger et al. 2007). Within this context, according to Papa et al. (ibidem) the role of urban planning should be identified in the ability to supply a holistic approach to cities' development, supporting the coordination of urban policies, and make cities smarter in a more liveable perspective.

3 The Emerging Geodesign Paradigm

The term Geodesign, introduced by Jack Dangermond in 2005 to indicate the design activity in the geographic space, is relatively new (Artz et al. 2010). Nevertheless, its application has old origins. In fact, "any design-related activity which depends on or in some way changes the context of our surroundings can be considered geodesign" (Miller 2012).

Geodesign employs a multidisciplinary and synergic approach to solving complex problems that involve not only territorial and environmental issues, but also social and economic concerns (Dangermond 2010).

Integration of different information in a design workflow, both at the local or at the global scale, is achieved through methods and tools borrowed from Geographic Information Science (GIScience—Goodchild 2010, 2015). The growing interest in Geodesign is demonstrated by an increasing number of conferences and even more by many new curricula on Geodesign flourishing within the school of planning in the United States (Foster 2013).

In order to apply Geodesign to regional landscape design studies Steinitz (2012) recently synthesized a complete methodology framework oriented to understand how the context should be transformed. Steinitz's Geodesign Framework (GDF) represents a particular adaptive methodology for decision-making in urban and regional planning and design informed on (digital) spatial information. The framework enables the planning (or Geodesign) technical team to develop a holistic view of the multiple issues involved in a planning process and, using geospatial modelling and impact simulations, to get real-time feedbacks on the performance of alternative development scenarios in the form of maps, charts and reports (Ervin 2011). An example of this approach has been recently implemented by Campagna and Matta (2014) on the case study of the SEA of local land-use planning (LLUP) in Sardinia, which implements among other concepts Harris's (1989) idea of interactive sketch planning in Planning Support Systems, which is also central to Geodesign.

The core of the GDF relies on six models that allow designing of future development scenarios and identifying the possible consequences of those changes, through territorial reference context description, analysis of its own dynamics and evaluation of its potentiality. The first three models describe the present situation of the territorial context: the Representation Model (RM) abstracts information into a set of digital spatial data layers, the Process Model (PM) combines them to describe how spatial phenomena evolve in time, and the Evaluation Model (EM) supports exploration of which implications these processes have in the area. Then, in the light of the results of the above analysis, the last three models describe how the territorial context could develop or become in the future: the Change Model (CM) delineates possible alternative options for transformation, the Impact (IM) Model evaluates the presence of beneficial or harmful impacts on natural and human environment deriving from those alternatives, and eventually the Decision Model (DM) will drive the stakeholders in expressing preferences on alternative and eventually in making final decisions.

In addition, to properly implement the framework in practice, a complete Geodesign study should consist of three iterations along the six models: in the first iteration (i.e. scoping) the framework steps are considered from model one to model six and the intent is to understand the scope of the analysis (i.e. scoping process); during the second iteration models are conceived in reverse order with the aim to define in details how to carry out the study (i.e. metaplanning). Lastly, in the third iteration (i.e. implementation) the design/planning study method and models are carried out from the first to the last one again. Linearity along the iterations is not strict and feedback loops or shortcuts may be necessary several times before the study is completed.

One of the elements that most make Geodesign useful in terms of spatial planning is that the analysis informs design since the early stage of planning process and goes along with it until the end in a non-linear course enriching it through many iterative possible loops. With this respect the GDF shows a consistent logic with SEA which should run since the early stages of the planning process in order to inform decisions at any stage, and it may contribute to address many current SEA pitfalls encountered in the urban and regional planning practices (e.g. among and above others how to inform design alternatives). Moreover, in the emerging Geodesign debate its practical implementation is closely connected to the use of digital data and technology in planning and design, which after a few slow years is slowly starting to characterize current planning governance and practices. Recently in Italy, thanks to development in regional geographic information systems according to a Spatial Data Infrastructure paradigm, the process is eventually started.

As a result Geodesign implementation may lead towards a new systematic approach to spatial planning and SEA application both in terms of operational methodologies and technologies, possibly bringing innovation into the practices.

4 Methodology

This study springs from the assumption that design in geographic space, or Geodesign, may become a reliable way to drive the planning process towards more sustainable spatial decision-making and development processes. Many of the ideas underlying the concept of Geodesign are not new though, and references to them can be found by looking back to the evolving urban and regional planning regulatory framework, and this is done here with regard to the Italian planning system. Nevertheless, while several ideas may be already there, their practical implementation as discussed earlier often lacks fulfilment of underlying principles. To this end, understanding the relationships between methods and norms may help to ensure a more responsible and proactive application of the regulations. Starting from this assumption, the research methodology involves a critical review of selected Italian national and regional planning laws in order to detect links to Geodesign key concepts into them. After a brief review of the evolution of the national planning legislation from a Geodesign perspective, the study focuses on three specific regional case studies: namely Lombardy, Tuscany, and Sardinia. Hence, relating the six models of the Steinitz's GDF with each of them, existing (or missing) links are identified and analysed critically. The results are outlined in a summary matrix shown in Sect. 5, which is aimed at contributing to put light on which issues are of particular relevance in order to properly take into account in an integrated way Geodesign principles in the planning regulations.

In addition, to complement the analysis we explore elements of coherence between the principles underlying the GDF and methods and tools currently used in planning practice this time focusing on the Sardinian context (Italy). To this purpose, a systematic critical review of 15 case studies of SEA of LLUP is given. The results of this review are presented in Sect. 6. The final objective of this study is to eventually contribute to inform guidelines to foster the Geodesign diffusion and its proactive implementation as a mean to address current pitfalls in urban and regional planning SEA.

5 Geodesign Principles in Planning Regulations

5.1 Geodesign Principles in European Directive 2001/42/EC

The European Directive 2001/42/EC officially introduced the Strategic Environmental Assessment (SEA) in all Member States for plans and programmes at any level (including regional, urban and sectorial plans) with the aim to provide the integration of environmental considerations during the design phase.

Article 5 of SEA Directive requires the preparation of an Environmental Report (ER), the contents of which are listed on the Directive's Annex I. Currently, many relationships between the GDF models and the Environmental Report contents may be already found as synthetized in Table 1 (adapted from Campagna 2014b).

Indeed at a closer look, the GDF models represent the process of analysis, design and choice that always forms the basis of the draft plan, but that today must be enriched by the contents of the Environmental Report. In this sense, the application of Geodesign methods and tools achieves the desired synthesis between urban planning and SEA.

Table 1 Relationships between environmental report contents (ex Annex I of directive 2001/42/EC) and GDF models

Environmental report contents	GDF model
(a) An outline of the contents, main objectives of the plan or programme and relationship with other relevant plans and programmes	Representation Evaluation Decision
(b) The relevant aspects of the current state of the environment and the likely evolution thereof without implementation of the plan or programme	Representation Process
(c) The environmental characteristics of areas likely to be significantly affected	Representation
(d) Any existing environmental problems which are relevant to the plan or programme including, in particular, those relating to any areas of a particular environmental importance, such as areas designated pursuant to Directives 79/409/EEC and 92/43/EEC	Representation Change
(e) The environmental protection objectives, established at international, Community or Member State level, which are relevant to the plan or programme and the way those objectives and any environmental considerations have been taken into account during its preparation	Representation Change
(f) The likely significant effects on the environment, including on issues such as biodiversity, population, human health, fauna, flora, soil, water, air, climatic factors, material assets, cultural heritage including architectural and archaeological heritage, landscape and the interrelationship between the above factors	Evaluation Impact M
(g) The measures envisaged to prevent, reduce and as fully as possible offset any significant adverse effects on the environment of implementing the plan or programme	Change Decision
(h) An outline of the reasons for selecting the alternatives dealt with, and a description of how the assessment was undertaken including any difficulties (such as technical deficiencies or lack of know-how) encountered in compiling the required information	Decision
(i) A description of the measures envisaged concerning monitoring in accordance with Article 10	–
(j) A non-technical summary of the information provided under the above headings	Synthesis

Such a methodological framework may be implemented in many ways in practice, through the use of a variety of methods and tools—never specified by the Directive—that can be chosen depending on the contextual conditions.

5.2 *The Italian Planning System*

The skeleton of the Italian planning system dates back to the 1940s. Since then it has evolved to the present through incremental steps of innovation. Since 1942 the Italian National Planning Law n. 1150 promotes a top-down three tiers hierarchical system which basically includes territorial plans, local land-use masterplans (usually at the municipal level), and sub-municipal implementation sectorial plans. Initially the system was intended as a tool to deal with strong urbanization pressure in the first half of the XX century and urban reconstructions in the aftermath of World War II. After a few decades, in a period of considerable residential expansion, the issue of urban quality standard came to stage as a consequence of the need to provide cities with modern social services and infrastructure facilities. Therefore, land-use plan design priorities became related to the need to ensure a proper balance between housing and services for residents. Law n. 765/1967 and the Ministerial Decree n. 1444/68 addressed these issues introducing standards for land-use design, as well as physical parameters for urban development. The zoning instrument was also somehow institutionalized by the same law, encoding in a standard classification land-use categories for future development. The attention paid to such services as education, cult, health, as well as recreation, parking and green areas contributed to shift the scope of planning from quantity to quality and community well-being. While the latter can be considered one of the perspectives of Geodesign, the solution proposed at that time (i.e. the introduction of minimum quantity standards) can be nowadays considered an unfledged approach which should be further innovated.

A new breath of innovation was brought to the Italian planning system in 1985, when the Law n. 431 introduced the contemporary conception of territory as a unitary system and finite resource storage to be wisely safeguarded. Modern landscape plans were introduced, bringing to the stage mandatorily an object of analysis. The territory started to be considered as an environmental, economic and socio-cultural system, the savvy management and development of which had to be based on sound knowledge and informed decision-making. More recently, with Code of Cultural Heritage and Landscape n. 42/2004 the planning model undergoes an important qualitative evolution towards a slight different concept of landscape as brought by the European Landscape Convention. Environmental or landscape compatibility assessment became a significant element of contemporary urban and regional plans. Environmental systems carrying capacity are becoming the starting points of contemporary territorial planning, where the planner is a coordinator of a multidisciplinary pool of experts ranging from architecture to engineering, from history and archaeology, from geology to hydraulics, from sociology to economy, to which Geodesign adds a new figure of the spatial information specialists.

The evolution of the spatial planning approaches brought by the Italian legislation briefly outlined here, is intended not to be exhaustive, rather it is proposed as a general background framework for the analyses of the Geodesign key principles found in selected Italian regional case studies, as discussed in the next section.

5.3 *Geodesign Principles in Italian Regional Planning Regulations*

In this section a more detailed discussion about the relationships between each GDF model and some Italian regional planning regulations is given. Regional regulations analysed so far include:

- Lombardy Regional Territorial Government Law n. 12/2005 (L-LR12/2005);
- Tuscany Regional Territorial Government Law n. 1/2005 (T-RL1/2005);
- Sardinian Regional Spatial Planning Law n. 45/1989 (S-RL45/1989);
- Sardinian Regional Guidelines for SEA of LLUPs (S-RGuidelines 44/51-2010).

The first stage of the GDF is the **Representation Model (RM)**, whose purpose is to understand how the geographic study area should be described. The description should be as complete as possible in space and time. There is not a predefined set of data to analyse, but it is important to select those needed in relation to the case study at hand, and to organize them in geographic space (Steinitz, ibidem).

The RP is perhaps the one among the other GDF models which is never missing in a planning and design study. However looking back to the first Italian Law 1150/42 and following modification and integrations, its content has deeply evolved along time. Directive 2001/42/EC requires to include in the ER a complete description not only of the environmental, economic and social data, necessary to represent the local Environmental Framework, but also all the other existing local regulations and projects affecting territorial context (or, the Normative Framework). A correct Geodesign implementation would imply treating all this relevant information spatially: this is a requirement that is not necessarily always properly addressed in practice. Also there is not a single methodology to collect and display data, but it is recommended to choose visualisation methods and spatial scale of representation which provides the variation of different ways to look at the territorial context (Steinitz, ibidem).

In the first iteration the RM should help to define boundaries of the study area, its geographies, and list existing sources of (digital) information and services. The latter part is something that has currently started to be addressed by spatial planning law thanks to the recent development in RSDIs, as demonstrated by the examples discussed in this section. In the second iteration, however it should be decided in the light of the input need of all the GDF models, exactly what data, what scale and accuracy, what processing services are required. Eventually, in the third iteration GDF data are collected and integrated for use. This issue seems to be successfully addressed as current RSDI download and invoke services usually allow online open data access.

Since recently, the representation of spatial data is more generally concerned by the INSPIRE Directive. According to INSPIRE interoperability principles, public administration at all levels in the Member States should give public online access to 34 spatial data themes according to common specifications. Moreover, many planning regulations and spatial plans require using public authorities' data resources in plan-making, thus affecting the format of the representation model with important implications for the planning and design processes. This is the most challenging part of the process, for data in RSDI are not necessarily created for planning purposes and their reliability should be carefully evaluated. One important implication is that SDI implementing rules, which define common data models, are eventually affecting the content and format of the planning knowledge in a European wide process. This issue started to be addressed by Plan4All (http://www.plan4all.eu) an eContentplus EU project and carried on by a consortium of 24 partners from 15 European countries, aiming at harmonizing spatial planning data and related metadata according to the INSPIRE principles, and it should be further analysed in order to understand the technical implications for the planning practice.

Looking at the regional level in Italy, the L-LR12/2005 in Article 3 institutes the Territorial Information System as a new tool to coordinate local information. The Lombardy RSDI (L-RSDI) integrates thematic datasets from multiple sources, including public authorities at all levels and the scientific community. Public authorities also provide approved plans and projects in digital form updating dynamically the L-RSDI knowledge base and making it available. Article 8 specifies the territorial knowledge framework contents in municipal Territorial Government Plan (PGT): the historical development of settlements, geological, hydrogeological and seismic characteristics, mobility infrastructures, identification of risk vulnerability sites, natural habitat sites, agricultural landscape structures, and urban fabrics, among other spatial data themes. With the possible limitations explained earlier in this section, the L-RSDI allows us to acquire relevant data to be used in the GDF iterations for the representation model.

In Tuscany T-RL1/2005 specifies, thanks to Article 28, that the regional government together with the provinces and the municipalities are responsible for the management of the Regional Geographic Information System (T-RGIS) in order to integrate the geographical knowledge, to organize, update, advertise and make it available to all the stakeholders involved in the planning processes. The T-RGIS also features a geoportal which gives access to download services. In the same law article the T-RGIS is also clearly defined as a key reference point for construction of the knowledge framework useful in developing decision-support tools for spatial planning and environmental impact assessment. According to Article 29 moreover, the T-RGIS collects information from the public authorities and the scientific community, which includes the following themes: topography and geology, orthophotos, satellite images and historical maps, thematic databases on the state of the local resources as well as the local knowledge as resulting from local planning tools. The Regional Government is responsible for regulating the information management process, the technical specifications and the standards to be adopted for preparation and diffusion of geographic data, affecting the first and especially the second iteration of the RM.

In Sardinia conversely, the S-RL45/1989, does neither refer to any specific RSDI, nor does it recommend any specific data management technology unlike the former. This is not surprising, for dating back to 1989 this law could have barely foreseen what would have been the evolution of the paper map based information system available at that time. Nevertheless integrations to Article 19, specifys that a landscape and environmental compatibility study attached to Local Land-Use Plan (LLUP) must contain the municipality cognitive framework with indication of geologic, geomorphologic, hydrologic, landscape and historic contents, and settlement and infrastructure transformations, somehow affecting implementation of the RM and the overall GDF. Nevertheless, in Sardinia, recent S-RGuidelines 44/51—2010, while defining the ER contents, requires a description of the current status (i.e. do- nothing alternative) of 11 given environmental components (e.g. air, water, waste, soil, flora and fauna biodiversity, landscape, settlement system, economic system, transport, energy and noise) through specific summary data sheets which lists the information to be included, the indicators to be developed and the maps to be produced. The same indicators should be compiled at a later stage (i.e. in the impact model) for design alternatives. This way, the S-RGuidelines 44/51–2010 affects both the RM and the IM.

The second stage of the GDF is the **Process Model (PM)**, whose purpose is to understand how the study area operates without the implementation of the plan. In the first iteration it should be identified as ongoing physical, ecological, human and geographical processes and relationships among them. In the second iteration it should be defined what analytical, simulation or forecast models should be used to describe selected environmental or anthropogenic processes. Finally the last iteration consists in implementing models and visualises them to be shared. To the latter respects, it is not easy to find clear links to specific steps of the PM within the body of legislation under analysis, possibly because no guidance was found on the methodology to be adopted for the plan preparation, of which this model is an essential part according to the GDF. However, it is possible to find some references to the first iteration in terms of which processes should be concerned in the analyses.

T-LR1/2005 Article 33 specifies that the Regional Landscape Framework Plan should contain the analysis of the territorial transformation dynamics through the identification of risk and vulnerability factors, whereas in the Sardinian LR45/1989 the socio-demographic analyses and forecasts, giving information about the housing requirements, represent a mandatory step in local land use plan-making.

With regards to SEA regulations, both in Tuscany (i.e. RL10/2010) and in Sardinia (i.e. Reg. Dept. Env. Act n. 33/34—2012) it is specified that ER should contain information about the likely evolution of the environment without the plan implementation. No reference is given to any method or tools to comply with this requirement, which is actually transposed from Annex I of Directive 2001/42/EC.

The third stage of GDF is the **Evaluation Model (EM)**, which concerns such questions as whether the current study area is working well, why or why not, what are the main problems and the possible opportunities. This is a knowledge step in which, during the second iteration, it is important for decision-makers and stakeholders to choose suitable evaluation method (e.g. scientifically or judgement based), criteria related to the geographical context and to assign them weights and values. Finally in the third iteration there will be result visualization and communication showing the location of those areas which need to be changed, to be preserved or suitable for a specific use.

In L-LR12/2005 Article 8, it is required for the PGT to identify and portray in a suitable representation scale several functional suitabilities with regard to such objectives as environmental, landscape, historic, geologic and ecologic preservation, or requalification, indicating also the possible actions and their purposes. In T-LR1/2005, Article 53 states that the LLUP should indicate the functional and territorial systems and sub-systems which define the territorial identity structures, the structural values as defined in Article 4, the criteria for the use of resources, the instructions for landscape safeguard and for environmental and cultural heritage protection, as well as significant public interest areas and properties. Article 33 moreover specifies that Regional Landscape Framework Plan contains analysis of relationships among historic, natural and aesthetic characteristics and as a consequence the definition of landscape value to be considered as attitude or vocation to change. In S-LR45/1989, Article 19 specifies that the LLUP should define the areas which need special protection and safeguard rules as well as the areas which need extensive urban rehabilitation.

Thus, in all cases territorial systems or sub-systems to be evaluated are somehow pre-determined by the regional planning legislations. As in the case of the PM, also for the EM methods and tools are not mentioned. Nevertheless they might be addressed by technical guidelines complementing regulations. For example the S-RGuidelines 44/51—2010, in order to represent in a synthetic manner the results of the environmental analysis, suggest the application of the SWOT analysis (do Rosário Partidário 2012), which allows the identification of Strengths and Weaknesses of an area and as a consequence Opportunities and Threats. This analysis aims to detect possible environmental critical issues that may be affected by the plan, and to highlight the vocations of the territory.

The fourth stage of the GDF is the **Change Model** (**CM**), whose purpose is to understand how the geographic study area might be altered. This is the first stage of actual design as traditionally intended in which strategic and physical development policies are proposed and simulated. While in the first iteration the major changes foreseen for the territorial context should be identified, in the second one development scenarios and the models to implement them should be devised. Finally in the third step the typical products are data that will be used to simulate and represent future possible conditions (i.e. the zoning map in a land-use plan). The representations of chosen scenarios are then input for the next steps of the GDF and could be used during communication and participatory phases.

Article 8 of L-LR12/2005 specifies that the PGT should identify strategic development goals according to higher level territorial policies, quantitative development goals considering the importance of territory requalification, the soil consumption reduction, the rational use of environmental and the savvy use of energy resources. T-LR1/2005 in Article 53 claims that LLUP sketches territorial development strategies defining territory governance orientation and in the Article 11 on integrated assessment, it refers to possible alternative plan solutions. In the S-RGuidelines 44/51—2010 it is specified that the SEA process should support the design of one or more possible alternatives for the development of the municipal area, also including participatory consultations at this stage. Likewise, T—RL10/2010 in Article 24 describes the ER contents referring to the identification and description of the sustainable plan alternatives implementing the system of the objective. In Lombardy, a regional dispatch on the SEA of PGTs n.13071/2010 refers to reasonable alternatives to be designed and documented during ER elaboration. No Regulation however does describe how to design plan alternatives, neither which methods to use.

The fifth stage of the GDF is the **Impact Model (IM)**, whose purpose is to understand which effects the changes might cause in the whole area. In the first iteration there is a broad set of phenomena to be considered, and their choice depends on the expected changes and on the local context to be analysed. In the second iteration each phenomenon being investigated need a different impact model, focusing on a specific potential consequence that will have to be evaluated: economic and demographic impacts, environmental impacts related to such parameters as water and air pollution, energy use, biodiversity, public safety, noise and so on. Different impacts are often linked to each other in networks and they need to be combined into geographic context and summarized to identify their interactions. The outcomes of this step generally include a set of thematic maps showing qualitative, quantitative and spatial differences between the state of the place with or without the plan. Also impact matrices, check-lists and network models can be used, as well as complex overlay or spatial multi-criteria methods.

In order to promote resources preservation and sustainability development, Article 4 of L-LR12/2005 requires the SEA of all territorial and urban plans since the early stage of decision-making. The environmental impact study should supply the identification of potential impacts resulting from alternative plan choices, verify if they are beneficial or harmful and foresee mitigation measures if needed. Article 8, which specifies content of the local plans documents remarks the necessity to demonstrate the compatibility of intervention policies with public administration economic resources and with the effects generated in the affected territory. The main regulation never refers to the way of carrying out the analysis and does not specify which models are needed to assess and compare impacts deriving from potential changes.

T-LR1/2005 on Article 11 introduces the Integrated Assessment as a mean for evaluation of territorial, environmental, social, economic and human health effects before planning tool adoption since the first stage of the plan preparation. Integrated assessment should include compatibility assessment in relation to the use of essential territorial resources as defined in Article 3 (i.e. air, water, soil, natural ecosystems, cities, landscapes, cultural heritage and infrastructural systems).

S-LR45/1989 clearly states in its first article that planning should guarantee territorial resources and environmental protection, artistic and cultural heritage safeguard, and on the integration of Article 19 refers to the SEA Directive specifying that landscape and environmental compatibility studies should include the environmental assessment of local plans. More recently, S-RGuidelines 44/51—2010 specify that the assessment of the effects that the implementation of the plan on the environment should be carried out for all the design alternatives. The guidelines also suggest the use of such methods as impact matrices and overlay mapping.

The sixth stage of GDF is the **Decision Model** (**DM**), whose purpose is to understand how the study area should be changed through the identification of the best balanced possible alternative among those designed during the CM and evaluated during IM. Firstly, the main stakeholders involved in the decision making process should be identified as well as the major concerns that influence decision. Then, the level of importance of the elements on which the choice is based should be specified by decision makers. Finally, the comparison of the impacts of each alternative in the light of the complex system of objectives should be carried out in order to select the preferable choice. The DM also entails a series of activities which are relevant for the SEA of regional and urban plan for it should eventually demonstrate that the final choice is preferable in the light of the complex system of objective at stake. In S-RGuidelines 44/51—2010 it is generically specified that evaluation of the plan alternatives aims at identifying the one that, while ensuring the achievement of development objectives, determines the lower environmental impact.

Summarising the result of the analyses, the matrix in Table 2 shows in shades of grey (i.e. the darker the stronger) the relationships between the Geodesign framework models and the analysed regulations.

Table 2 shows that RM is one of the most complete models in planning regulations because Regional Laws always require a knowledge step of data collection and analysis with the aim to create the local cognitive framework. Moreover Tuscany and Lombardy laws consider the RSDI as a base platform for data finding. Conversely Sardinian Regional Law never recommends any specific data management technology. This is not surprising for this law dates back to 1989, while Sardinian RSDI dates back to 2006. The same goes for the IM: in all Regional Law

Table 2 Relationships between GDF models and regional legislations

		RL Lombardy	RL Tuscany	RL Sardinia	Sardinian Guideline
RM	I				
	II				
	III				
PM	I				
	II				
	III				
EM	I				
	II				
	III				
CM	I				
	II				
	III				
IM	I				
	II				
	III				
DM	I				
	II				
	III				

there's a reference to the environmental assessment of the effects of the plan but there are not specified methods and tools to implement impact analysis, except for within the S-RGuidelines.

There are some references to the EM while talking about finding vocation of the context, individuation of risks and vulnerability of the area and evaluation of the zero alternative. However, it is not easy to find clear links to specific steps of the PM within the body of legislation under analysis, possibly because no guidance was found on the methodology to be adopted for the plan preparation, of which this model is an essential part according to the GDF.

CM and DM are the least represented: sometimes it is mentioned that the plan should design possible alternative scenarios, but it seems like a simple transcription of the ER contents according to 2001/42/EC Directive and there are neither specified methods for their preparation nor the way to carry out the choice among them. The latter should be considered as one of the possible reasons why the design alternatives step is often missing in planning practice, as it was found in the following part of this study.

6 Geodesign Principles in Sardinian Practices

This section presents the analysis of the elements of coherence between the GDF models and methods and tools currently used in planning practice, in 15 case studies of Sardinian LLUPs and ERs of LLUPs. It is worth recalling that LLUPs in Sardinia must be in compliance with the Regional Landscape Plan (Piano Paesaggistico Regionale, PPR) and the Regional Hydrological Risk Assessment Plan (Piano di Assetto Idrogeologico, PAI) according to the RL No. 8/2004. At the same time they are also subject to SEA, according to shared standards and procedures as required by the S-RGuidelines and other relevant regional regulations. For these reasons the case studies analysed so far often present a degree of commonality.

The results of this critical review are presented in Fig. 1, reporting for each GDF model, and respective iterations, the number of case studies showing any relevant practice (i.e. application of methods and tools) related to them.

In the ERs and LLUPs analysed to date, the early stage of the process is always represented by the preparation of the Normative Framework and the Environmental Framework which, in a GDF perspective, can be considered as part of the **RM**. The Normative Framework construction is oriented to identify the main objectives of local legislations affecting the territorial context under analysis, so as to inform the plan consistently. Use is made of tables or matrices showing the consistency/inconsistency of the plan objectives with the higher-level plans.

On other hand, the Environmental Framework construction aims to depict not only information related to environmental characterization, but also to the cultural heritage and the settlement system of the municipality, as required by the RLP. Geared towards this purpose, the RSDI is accessed as a major resource to

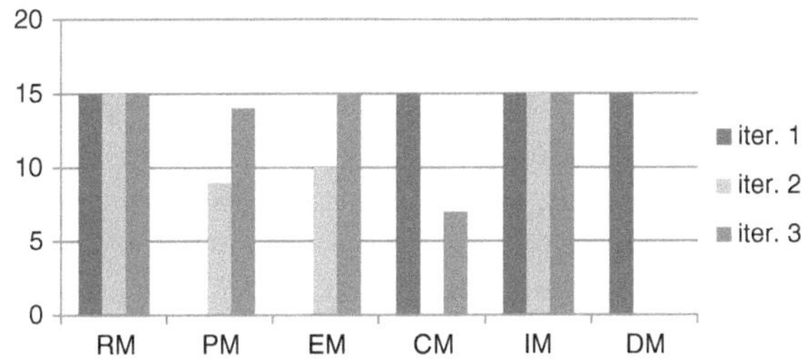

Fig. 1 Relationships between GDF models and Sardinian case studies

retrieve digital datasets relating to these issues. Spatial data from the RSDI are often integrated with further large scale data provided by the local authorities and governmental organizations or agencies: Geographic Information Systems seem to be used mostly, and limitedly, to visualise and overlay multiple levels of information in the geographic space. As a result, a set of thematic maps in a suitable representation scale are provided, portraying territorial characterisation (e.g. land use, geology, hydrology, settlements and land cover characterisation). Together with the maps a set of 11 environmental analysis sheets is admissibly given, describing every single environmental component as defined by S-RGuidelines (e.g. air quality, water, waste, soil, flora and fauna, landscape, settlement system, economic system, mobility, energy, noise) through a predefined set of indicators representing their current state, showing a sort of bureaucratic approach which routinely offers a standard representation of the territorial system irrespectively of the local features. Finally, all the collected data are reported in the ER and in the enclosed annexes (i.e. the geological and hydrological reports), also making use of histograms, graphs and tables when useful to better describe the phenomena.

The RM can be surely considered as the most complete GDF model in the examples analysed so far, both in terms of contents, portrayal methods and technological tools employed.

Since SEA is intended as a process to improve strategic decision-making on what concerns the future, use is often made of forecast models based on extrapolations of historical data and trends of the past years, trying to understand possible future scenarios before the implementation of the plan (**PM**). In a complete PM step these trends should be blended with spatial analysis, combining data to describe processes geographically (e.g. soil erosion, vegetation changes, traffic flows). In the case studies the most common forecast model found is the socio-demographic dynamic analysis, which consists on an investigation of demographic and economic trends as a mandatory step for housing requirements in plan-making according to the S-RL45/1989 and it is usually related only in the form of a chart or report. In most cases it also described the ongoing trend of a specific resource or environmental component (e.g. water, soil, vegetation). Most LLUPs include studies, reports and thematic maps of the analysis of danger in terms of hydrogeological risk, landslide risk and potential slopes' instability made by geologists, in compliance with the Regional Hydrological Risk Assessment Plan. Nevertheless PM usually is not exhaustively investigated because often the totality of the phenomena in evolution in the territory is not considered.

PM thematic maps should be useful to identify possible strengths and weaknesses of the area during the next stage of evaluation of its past and present conditions (**EM**). These considerations strongly influence the decision-making process because they determine area attractiveness, vulnerabilities, risks, and, as a consequence, those areas which need to be changed or to be preserved. The most common method found in the case studies in this step is the SWOT analysis, a structured procedure aiming at evaluating the Strengths, Weaknesses, Opportunities and Threats in the municipal area. The critical interpretation of the SWOT analysis results helps to formulate plan objectives during the next stage. In a few cases, the suitability and capability modelling are also used for the creation of maps representing the territory criticalities and suitability to change. While maps such as those representing the land capability, the level of criticality and the degree of naturality are commonly found, the suitability maps (e.g. maps representing the suitability for farming or breeding use) can be found only in a few examples of those analysed to date. No suitability analysis was found with regard to other land-uses such as residential or commercial.

The **CM** in the analysed case studies is definitely the less developed among the GDF models. In spite of proposing and simulating alternative future changes represent a relevant and strategic step of a complete SEA process, as indicated by European Directive and good practice guidelines (do Rosário Partidário 2012), nevertheless maps showing possible future land use scenarios are mostly missing. Only a few exceptions indicate two alternatives are given: they are often not sufficiently described, representing only the zero-scenario and the planned one. This lack represents the bigger methodology pitfall in SEA of local land-use planning processes, for the lack of clear explanations on how decision-making is informed by the results of the analysis, failing to fulfil the transparency principle which should be at the bottom of the process.

Since the **IM** goal is to identify, locate and assess the environmental effects of the different plan alternatives proposed, to enable the identification of the preferable solution during next step, this model is always only partially achieved. Indeed, the impact analyses are always carried out, but only for the unique plan scenario to identify possible impacts to minimize or mitigate. The most popular impact method used is the evaluation matrix, as recommended in the S-RGuidelines. Usually the evaluation matrixes show in their x-axis all plan choices and in y-axis all environmental components, as listed by the S-RGuidelines. For each intersection it is indicated if that particular action has positive or negative effects on the selected environmental component. Sometimes check-lists and state indicators are also used, showing the expected variation of some values as a result of plan choices (e.g. air pollution, energy consumption, soil consumption). Only in few cases these considerations are transposed in form of maps: for example through the overlay among the planned land use scenario and specific sensitive areas in the territory (e.g. sites with high natural value, Natura 2000 sites, areas susceptible to hydrogeological risk).

The identification, in the ER, of the major stakeholders and the people to be involved during scoping, consultation and participation processes (e.g. professionals, public authorities, private stakeholders, agencies responsible for the environmental, public) could be considered part of the first iteration of **DM**. Finally the defined plan scenario is always represented in the form of a map to visualise the future land-use scenario for the municipality, and often a paragraph is given illustrating the expected mitigation for the plan's actions showing negative effects on the environment. However it is not possible to consider the DM well developed in practice, because of the lack of a real set of assessed alternative plan scenarios among which we must make a choice.

In this review several elements recalling the Geodesign approach were identified, however not all the identified analyses are treated spatially in their portrayal results and not always do the GDF models exhaustively consider all phenomena taking place in the territory which have relevance for future change planning according to the Geodesign approach.

7 Conclusions

As a result of this study we can observe the relationships between Geodesign framework models and urban planning both from a normative perspective (i.e. normative framework analysis) and from a practical perspective (i.e. LLUPs and ERs of LLUPs analysis).

The first part of the study highlights the correlations and the gaps between underlying principles and the planning regulations, in particular with respect to design methods and tools. Indeed, most recent regional territorial government laws in Lombardy and Tuscany carefully take into account the role of the RSDI as a base platform for knowledge management in planning, affecting substantially the implementation of representation models in urban and regional planning. In Sardinia, a similar trend is not being addressed by the Regional Spatial Planning Law, for its dating back to 1989, but by the new Regional Landscape Plan which sets directions for LLUP including technical rules for plan content and portrayal.

Nevertheless, fewer connections are found between planning regulations and applied methods, which should be also defined in the second iteration of the GDF. Indeed, this iteration is strictly related to metaplanning, that is the activity of shaping the planning process in an operational workflow of activities (Campagna 2014a), which is often missing although it may have great relevance to the overall process assessment as required by SEA. Only Sardinian Guidelines are an exception because, giving some suggestions on tools and methods to use for implementation, should be partially considered as a methodological support for the decision process. However, more references are found to the first GDF interaction, where the scope of the Geodesign study is defined, and to the third, which defines the practical implementation of the study up to the output, whose format is often partially pre-defined by law, with a risk of counter-effective standardisation.

A clearer understanding of these relationships by educators, public administrators and practitioners may contribute to set up operative guidelines and tools to better interpret and to proactively inform the principles brought by current regulations.

The second part of this study highlights that some typical Geodesign methods and tools are already starting to appear in practical case studies, even though they are not always specified in the legislation framework. In particular the widespread dissemination of the Regional SDI use during planning processes and Strategic Environmental Assessments, especially during the first stage of the methodology, is one of the most relevant observed phenomenon. This trend, together with the increasingly frequent collaboration among experts in multidisciplinary teams, may facilitate the diffusion of innovative Geodesign approaches also in other stages of the planning process. Furthermore, the process leading from analysis to plan choices is not always clear: indeed a systematic shared and shareable methodology is often missing from this perspective.

The implementation of Geodesign principles may hopefully bring innovation into the practice contributing to define a more clear and systematic vision on how to enhance the issues at the base of a smart city not only in terms of technological tools, but also providing a transparent and shared methodology to organize the design process, and thus improving urban governance.

Acknowledgements Elisabetta Anna Di Cesare gratefully acknowledges Sardinia Regional Government for the financial support of her PhD scholarship (P.O.R. Sardegna F.S.E. Operational Programme of the Autonomous Region of Sardinia, European Social Fund 2007-2013—Axis IV Human Resources, Objective l.3, Line of Activity l.3.1.)

The work presented in this paper was developed by the authors within the research project "Efficacy and efficiency of landscape and environmental management in Sardinia: the role of SEA and of SDI" CUP: J81J11001420007 funded by the Autonomous Region of Sardinia under the Regional Law n° 7/2007

References

Artz M, Dangermond J, Ball M, Abukhater A (2010) Changing geography by design. Selected readings in geodesign. Esri Press, Redlands

Batty M, Axhausen KW, Giannotti F, Pozdnoukhov A, Bazzani A, Wachowicz M, Ouzounis G, Portugali Y (2012) Smart cities of the future. The European Physical Journal Special Topics 214(1):481–518

Belissent J (2011) The core of a smart city must be smart governance. Available on http://blogs.forrester.com/jennifer_belissent_phd/11-05-15-the_key_to_being_a_smart_city_is_good_governance_smart_governance. Accessed 15 Nov 2014

Campagna M (2014a) Geodesign from Theory to practice: from metaplanning to 2nd generation of planning support systems. TeMA J Land Use Mobility Environ

Campagna M (2014b) Geodesign: dai contenuti metodologici all'innovazione nelle pratiche. In: AA. VV., Atti della XVII Conferenza Nazionale SIU—Società Italiana degli Urbanisti 15–16 May 2014. L'urbanistica italiana nel mondo. Planum Publisher. pp 71–76

Campagna M, Matta A (2014) Geoinformation technologies in sustainable spatial planning: a Geodesign approach to local land-use planning. In: Second international conference on remote

sensing and geoinformation of the environment (RSCy2014). International society for optics and photonics, pp 92290T–92290T
Chourabi H, Nam T, Walker S, Gil-Garcia JR, Mellouli S, Nahon K, Pardo TA, Scholl HJ (2012) Understanding smart cities: an integrative framework. In: 2012 45th Hawaii international conference on system science (HICSS), IEEE, pp 2289–2297
COWI (2009) Study concerning the report on the application and effectiveness of the SEA Directive (2001/42/EC), DG Environment European Commission, Report no. p-67683-a, Issue no. 2
Dangermond J (2010) Geodesign and GIS–designing our futures. In: Peer reviewed proceedings of digital landscape architecture. Anhalt University of Applied Science, Germany
do Rosário Partidário M (2012) Strategic environmental assessment better practice guide. Agência Portuguesa do Ambiente e Redes Energéticas Nacionais, Lisboa
Ervin S (2011) A system for GeoDesign. In: Proceedings of digital landscape architecture. Anhalt University of Applied Science, pp 145–154
Fischer TB (2010) Reviewing the quality of strategic environmental assessment reports for English spatial plan core strategies. Environ Impact Assess Rev 30(1):62–69
Flaxman M (2010) Fundamentals of geodesign. In: Proceedings of digital landscape architecture. Anhalt University of Applied Science
Foster K (2013) Geodesign education takes flight. Arcnews Fall 2013. ESRI Press, Revue international de géomatique n. 2
Giffinger R, Fertner C, Kramar H, Kalasek R, Pichler-Milanovic N, Meijers E (2007) Smart cities-ranking of European medium-sized cities. Vienna University of Technology
Goodchild MF (2010) Towards geodesign: repurposing cartography and GIS? Cartographic Perspect 66:7–22
Goodchild MF (2015) Twenty years of progress: GIScience in 2010. J Spat Info Sci 1:3–20
Harris B (1989) Beyond geographic information systems. J Am Plann Assoc 55(1):85–90
Miller WR (2012) Introducing Geodesign: the concept
Nam T, Pardo TA (2011a) Smart city as urban innovation: focusing on management, policy, and context. In: Proceedings of the 5th international conference on theory and practice of electronic governance. ACM, pp 185–194
Nam T, Pardo TA (2011b) Conceptualizing smart city with dimensions of technology, people and institutions. In: Proceedings of the 12th annual digital government research conference. University of Maryland, College Park, 12–15 June 201
Papa R, Gargiulo C, Galderisi A (2013) Towards an urban planners' perspective on smart city. TeMA J Land Use Mobility Environ 6(01):5–17
Parker J (2007) Strategic environmental assessment and SFs operational programmes: an assessment, DG environment European commission presentation at the Aarhus workshop on public participation in strategic decision-making (PPSD). Sofia, Bulgaria, 03–04 Dec 07
Schaffers H, Komninos N, Pallot M, Trousse B, Nilsson M, Oliveira A (2011) Smart cities and the future internet: towards cooperation frameworks for open innovation. Springer, Berlin, Heidelberg, pp 431–446
Steinitz C (2012) A Framework for geodesign. Changing geography by design. Esri Press, Redlands
Toppeta D (2010) The smart city vision: how innovation and ICT can build smart, "livable", sustainable cities: the innovation knowledge foundation. Available at http://www.intaaivn.org/images/cc/Urbanism/background%20documents/Toppeta_Report_005_2010.pdf. Accessed 16 Nov 2014
Van Winden W (2008) Urban governance in the knowledge-based economy: challenges for different city types. Innovation 10(2–3):197–210

Using Citizen-Provided Information to Build Purposeful Knowledge for Planning: Principles, Requirements, and Three Examples

Alessandra Antonini, Ivan Blečić, Dario Canu, Arnaldo Cecchini, Giovanna Fancello and Giuseppe A. Trunfio

Abstract We present three tools as examples of how information and contributions collected from citizens through online services may be meaningfully employed in evaluation and decision-support tools for planning. The whole idea of using "citizens as sensors" has seen remarkable advancements with the advent of the Web 2.0 and mobile technologies. Yet many tools lack explicit, transparent and publicly accessible evaluation models useful for guiding decisions, for prioritising issues, and for assigning resources by the municipal government. We argue that such evaluation models are necessary for the principles of publicity, accountability and equity to be observed by the public authority. The three tools we present here are attempts, in three different fields, to show how this lack may be overcome. The tools (one for urban maintenance, one for evaluation of walkability and one for energy management) are intended to enhance public authority's capabilities to use "social energies", to make their actions more efficient, fair and accountable, giving citizen control and oversight of the allocation of resource.

A. Antonini · I. Blečić · D. Canu · A. Cecchini · G. Fancello · G.A. Trunfio (✉)
DADU, University of Sassary, Alghero, Italy
e-mail: trunfio@uniss.it

A. Antonini
e-mail: aantonini@uniss.it

I. Blečić
e-mail: ivan@uniss.it

D. Canu
e-mail: dacanu@uniss.it

A. Cecchini
e-mail: cecchini@uniss.it

G. Fancello
e-mail: gfancello@uniss.it

R. Papa and R. Fistola (eds.), *Smart Energy in the Smart City*,
Green Energy and Technology, DOI 10.1007/978-3-319-31157-9_17

1 Introduction

Presently available information and communication technologies (ICT) offer unprecedented possibilities for a systematic use of information, proactive contributions and evaluative judgements coming from citizens collected through online services. This possibility opens up a promising prospective for policy making in urban and spatial planning. The complexity of policy processes in spatial and urban planning requires "adequate methods and suitable processes for creating purposeful knowledge, aiming at supporting those specific type of decision processes assuring within a policy cycle" (Tsoukiàs et al. 2013). Within such a thrust to build "purposeful knowledge", in this paper we attempt to exemplify how information and contributions collected from citizens through online services may be meaningfully employed in evaluation and decision-support tools for planning.

We will present as examples three tools we have developed. The first is "Wall-E", a system for citizens' reporting, evaluation and management of issues of urban maintenance. The second is "Walkability Explorer (WE)", an evaluation, design and decision support tool centred on walkability and pedestrian accessibility. The third is "VerdingTool platform", a decision support tool for citizen, businesses and public authorities for the implementation and optimization of solutions related to energy and raw materials consumption, and to the production of wastes.

But before getting at the examples, two general remarks are due. The first is related to the complicated and interesting debate on public participation (Irvin and Stansbury 2004), even more so if we think about the so-called e-participation (Coleman 2007; Charalabidis et al. 2009; Castells 2012). A starting point we want to pick is the much cited quote by Arnstein (1969): "[Participation] is the redistribution of power that enables the have-not citizens, presently excluded from the political and economic processes, to be deliberately included in the future. It is the strategy by which the have-nots join in determining how information is shared, goals and policies are set, tax resources are allocated, programs are operated, and benefits like contracts and patronage are parceled out. In short, it is the means by which they can induce significant social reform which enables them to share in the benefits of the affluent society."

Arnstein's paper has since become much renowned also for the so-called "ladder of citizen participation", so many times debated and revisited (Connor 2007; Dorcey et al. 1994; Pretty 1995; Rocha 1997). So, to put our three examples in perspective, we will, following Cecchini (2010), also use a slightly revisited ladder from Arnstein's original, shown in Fig. 1.

The examples we present here should in this context be then placed under the ladders of information and consultation. But with a twist. In Arnstein's original intention, "information" for instance is intended in the sense of public authority informing citizens. In our examples, as we will see, the information also flows in the opposite direction, and the whole process resembles much more something we could call an integrated information-consultation cycle, where the information from

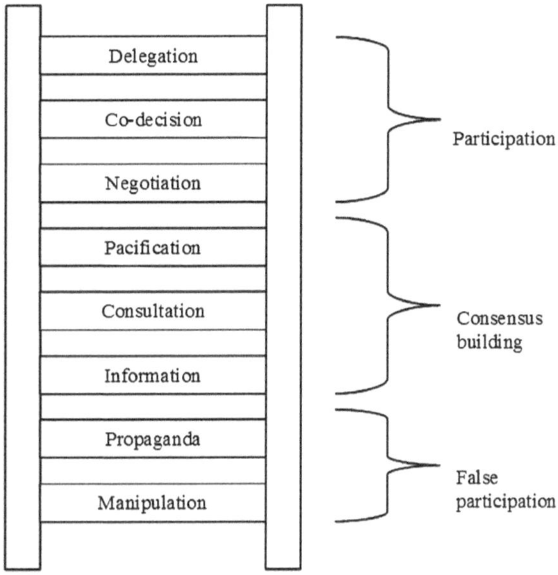

Fig. 1 Arnstein's ladder of citizen participation revisited (Cecchini 2010)

citizens is collected, processed, made meaningful using appropriate evaluation models, and finally delivered back to citizens.

A set of principles should, we hold, be observed by local authorities for such information-consultation cycles to contribute for a better, that is to say more democratic, decision-making; which brings us to the second general remark. These principles serve as the general meta-requirements for the design and integration of features in our support tools:

- openness and inclusivity: citizens should be given a clear, publicly known and non-discriminatory access to the tool;
- transparency: all the reports, proposals, alternatives, constraints and any other information relevant for decision-making should be known to citizens, should be easily accessible, clearly presented and made understandable;
- publicity principle: in the general Rawlsian conception (Rawls 1971), the publicity principle bans government from selecting a policy that it would not be able or willing to defend publicly to its own citizens; in our cases, this relates to ensure that the reasons for a decision (what to do, where to do it, with what priority) should be explicit, as much as possible non-arbitrary and grounded on some "public reason";
- accountability: the decision-makers should openly acknowledge and assume the responsibility when they exercise their discretionary power of choice and decision:
- equity: distributional considerations among people and neighbourhoods should count; given the inevitable constraints of resources and time, these should be distributed among citizens living in different neighbourhoods according to some

principle of equal treatment; this principle has one important corollary: the need to take into account the diversity of urban populations, the fact that the city should "work" for all of them, the fact that the point is seizing the grip on the "right to the city" for all.

Our support systems are an attempt to provide a software infrastructure with application logic which in principle allows the local authorities to adhere to those principles.

The whole field of using "citizens as sensors" has seen remarkable advancements with the advent of the Web 2.0 and mobile technologies. Many Web applications and mobile apps already exist to let citizens for instance report neighbourhood issues. Yet many also feature a crucial lack. While they have by and large successfully settled the technicalities of how citizens could contribute, report problems, comment, discuss, vote and track issues, the missing link is the lack of an explicit, transparent and publicly accessible evaluation model useful for guiding decisions, for prioritising issues, and for assigning resources by the municipal government. To speak in terms of general principles, the systems mentioned may well grant greater openness and inclusiveness, possibly a somewhat better transparency, but the publicity principle, accountability and equity may only be assured if the criteria and the constraints for choosing which issues to address when they are publicly known (possibly after a public debate). The three tools we present here are attempts, in three different fields, to show how this missing link may be provided.

In this sense, in what follows we present three online tools which are meant to enhance public administrations' capabilities to use "social energies", to make their actions more efficient, fair and accountable, giving citizen control and oversight of the allocation of resource. All these tools use open data and/or community data and try to consider the individual diversity, taking account of individual preferences, needs and abilities. Also, our research benefits from development and evolution of GIS software, open source tools and applications as well as publicly available data which allow more precise measurements, better reproducibility and easier transferability and comparability of results.

2 Wall-E: An Online Citizen Reporting and Decision Support Tool for Urban Maintenance

WALL-E is an online tool for urban maintenance which allows (1) direct citizen's reporting of neighbourhood issues related to urban maintenance services; (2) evaluation of priority of reported issues; (3) allocation and management of resources and workforce on solving reported issues; and (4) public tracking of their status.

Through an online form (Fig. 2c), citizens can report the location and the type of the issues, provide a description and upload photos. The types of issues currently contemplated in the online form are waste (uncollected or damaged waste containers, littering, unauthorised dumps, abandoned vehicles), infrastructures (water

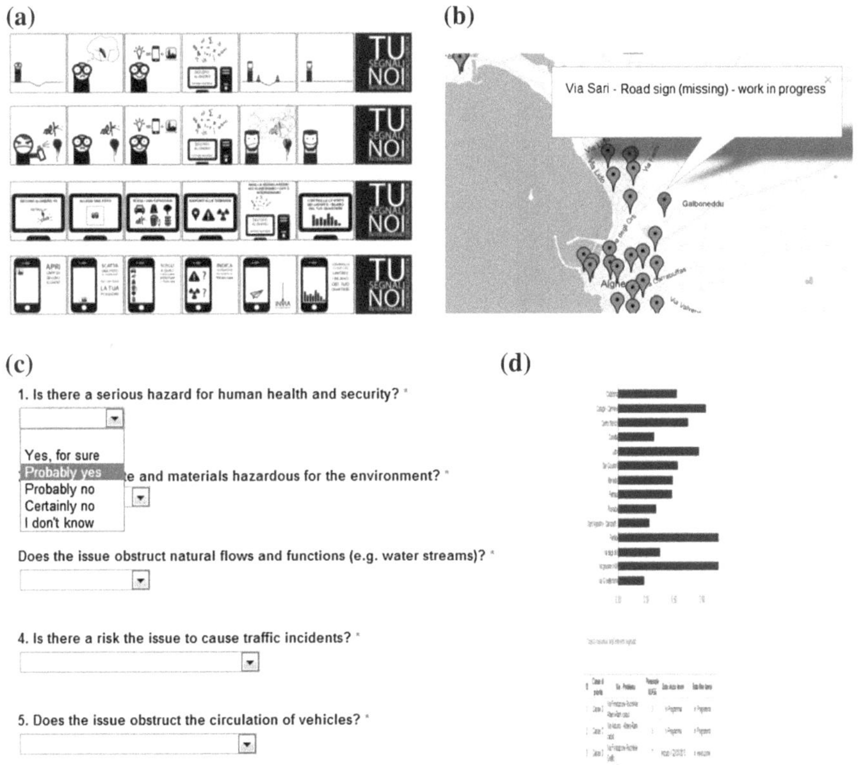

Fig. 2 Screenshots from WALL-E frontend. **a**. Visual tutorial. **b**. Map of reported issue. **c**. Detail of the online form for issue reporting. **d**. Past distribution of workforce among neighbourhoods (*above*) and issues' priority classes and statuses (*below*)

and sewage pipes damage or leaks), transportation (unauthorised parking, damaged, incorrect or missing signs and traffic lights), maintenance (fallen branches and trees, damaged flowerbeds, damaged roads and footpaths), acts of vandalism (graffiti, unauthorised billposting), environment (pollution, request for disinfestations, bad smells, stray animals).

In the online form we further ask citizens to answer several multiple-choice questions, which are essential for the subsequent evaluation and rating of the issues. There are, of course, two standard problems with this approach. The fuzziness of evaluative judgments, when different citizens have different perception of the problem and different understanding of the evaluation scale. For this, we tried to construct the questions and the possible answers (scale) in a language as much as possible natural and comprehensible.

Other problems may come from citizens behaving opportunistically: knowing that different answers will induce different responses from the maintenance services, citizens may overemphasise the gravity and urgency when reporting issues. For this reason, the information provided by citizens is not directly fed into the evaluation model, but first validated by an operator. The operator takes in new reports, revises, validates and if necessary corrects user's evaluations, or asks for further clarifications to the reporting user; if required, it may send out inspectors for direct observation on the field. All this leads to a validated record of the report, which is then made publicly available and rated by the evaluation model.

Based on the information provided by citizens and validated by operators, the evaluation model assigns a priority rating to each issue, following an evaluation procedure. Again, once attributed, the priority class of each issue is made publicly visible (Fig. 2d).

As we said, the purpose of the evaluation model is not to automatically provide a complete ordering of issues. It is rather an aiding and hinting tool. That is why we held it more appropriate to have an evaluation model for the classification (rating) of issues in a series of priority classes. Among the methods for multiple criteria evaluation of ratings (Bouyssou et al. 2006), the so-called ELECTRE TRI model (Roy and Bouyssou 1993) is a prominent approach. It allows a complete sorting of issues in priority classes and the aggregation of criteria is fairly flexible, permitting to account for their importance (weights), coalitions (majority rule and threshold) and possible veto powers. It is also reasonably easy to communicate and be intuitively understood by citizens.

The specific instance of ELECTRE TRI used in WALL-E defines four classes of issue priority (low, medium, high, urgent) and assigns reported issues based on eight evaluation criteria corresponding to eight questions answered by citizens through the online report form (1) Is there a serious hazard for human health and security? (2) Are there waste and materials hazardous for the environment? (3) Does the issue obstruct natural flows and functions (e.g. water streams)? (4) Is there a risk the issue to cause traffic incidents? (5) Does the issue obstruct the circulation of vehicles? (6) Does the issue obstruct the pedestrian routes and footpaths? (7) How many people daily visit the place on average? (8) How visible is the issue?) (For more details on the evaluation model see Blečić et al. (2014).

Once the issues have been classified for priority, the system offers to the operator the "Balance of neighbourhoods" showing the current demand and the past resources assigned to different neighbourhoods. This is to assist the management to allocate resources also taking the distributional issues among neighbourhoods into account.

Citizens can track the status of issues on the public front-end as they are being processed by the operators and evaluated by the system. Once the maintenance works have been completed, the operators flag the issue as "resolved" and the system registers the end date. This then allows the system to update the total workforce allocated to that neighborhood.

3 Walkability Explorer (WE)

Walkability Explorer is a software tool for design and planning support centred on the evaluation of urban walkability.

In the last decade the topic of walkability has gained a growing attention of the research community, especially in the fields of urban design, transport planning and public health (Capp and Maghelal 2011; Saelens and Handy 2008). Walkability should be seen as an attempt to move beyond the studies that assess the mere accessibility of places and to take into account also the *quality* of the accessibility and how the urban environment (built environment, social practices, etc.) is *conducive* for walking: factors such as physical features, urban design qualities and individual reactions may influence the way an individual feels about the environment as a place to walk (Ewing and Handy 2009; Porta and Renne 2005).

Many researches (Ewing and Cervero 2010; Iacono et al. 2010; Krizek 2003) analyse walkability exclusively through physical features at a macro scale, neighbourhood or city block. However, these physical features are not exhaustive of the experience of walking down a particular street because they tend to miss qualitative factors like sense of enclosure, imageability (Lynch 1960), and liveliness (Ewing and Handy 2009; Jacobs 1961; Porta and Renne 2005).

Without presenting here the details of the formal evaluation model used by Walkability Explorer (for that see Blečić et al. 2015a), it is useful to outline some of its peculiarities. The WE software uses a multi-criteria assessment model to assign a walkability score to all points in the city. Our construction of the walkability in the model proposes a certain change of perspective with regard to the methods suggested thus far: rather than evaluating how a place is walkable in itself, the walkability score we calculate reflects how and where to one can walk from that place, in other words, what is the walkability the place is endowed with. Therefore, the walkability score combines three components: (1) the number of available destinations (urban "opportunities") reachable by foot; (2) their distances; and (3) the quality of pedestrian routes towards those destinations. The quality of pedestrian routes is evaluated on different attributes relevant for walkability, related to the characteristics of the streets and their surrounding environment which influence objectively and subjectively people's spatial behaviours and choices (inclination for walking) (Ewing and Handy 2009) and contribute to render the route pleasant, secure and attractive.

We are particularly interested to capture urban elements relevant for the comfort and efficiency of the path, elements that generate the sense of safety and influence the pleasantness of walking and the attractiveness of the path and its surroundings. Comfort concerns physical components of the path that make easy and simple to walk through. Urban design features support people interaction and help pedestrians to feel safe. Moreover, attractions in proximity and landscape views make the path more enjoyable and attractive to walk.

In addition to this, the evaluation model explores how people can by foot reach some attractors (urban opportunities) from all the point in the city: health services; cultural and educational services; transportation facilities; administrative services; commercial activities; areas for recreation, leisure and play, tourist services; and so on. Their availability, quality and spatial distribution and the easiness to access them, influence the effective possibilities of a person to develop in and within the city.

This detailed urban analysis produces a set of raster maps representing both the geography of urban constraints and walkability scores in urban space.

One the evaluation model and the WE software tool for assessing walkability in the town of Alghero of about 40.000 inhabitants, in North-East Sardinia (Italy). The analysis of the street network and of urban opportunities was done between August and October 2014. A team of urban planners carried out the field research by direct in situ observation and with the support of dynamic images provided by Google Street View. Furthermore, in order to have a more exhaustive dataset of urban facilities we supplemented the OSM (Open Street Maps) data with georeferenced addresses of urban attractions provided by the Yellow Pages and Bing Maps services.

In Fig. 3 we report the essential input data and an intermediate elaboration. The first map includes information useful for the policy makers ad illustrates computed costs of street edges (left): the higher the cost, the lower the quality/walkability of the arc of the road network computed.

The second map (central) represents the spatial distribution of the relevant urban opportunities considered for this study: commercial attractions (food shops, bars, butchers, bakeries, fish shop, supermarkets, tobacco shops), service attractions (health services, schools excluding university, banks and public offices) and leisure and green urban areas attractions (city parks, beaches, sport services).

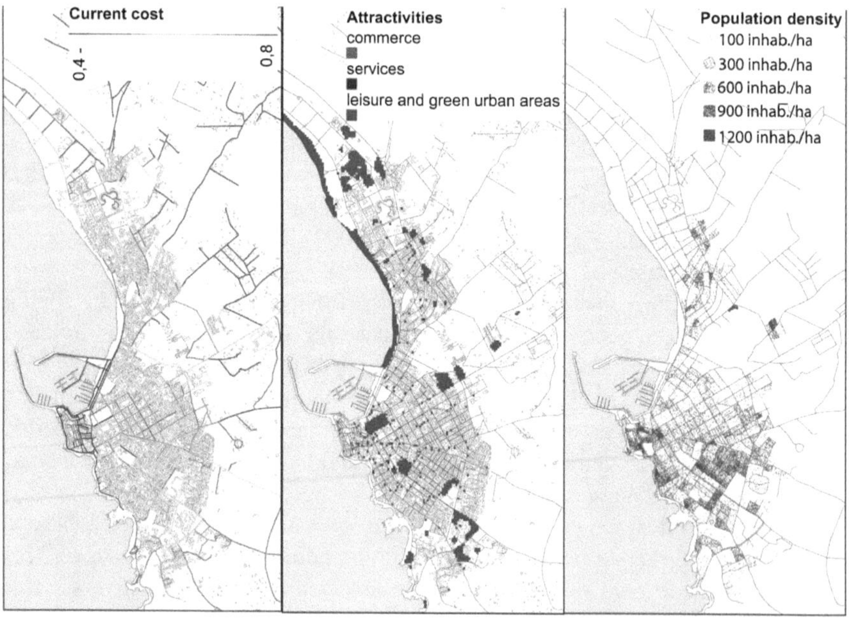

Fig. 3 Computed "costs" of the street edges (*left*), map of attractions (*central*), map of population density (*right*)

Finally, the third map (right) describes the spatial distribution of the population density in respect to the census data (ISTAT 2001), useful for evaluate the amount of population that can walk to urban opportunities from the whole city points.

Then, WE combines the cost of the paths, the spatial distribution of destination and the distance and calculates the walk score for each node of the street network. These data are interpolated to obtain a synthetic map representing the spatial distribution of the walk score, shown in Fig. 4.

As we have shown, Walkability Explorer may serve as a decision aiding tool especially to evaluate the effects of walkability policies and for comparing different planning and urban design scenarios by analysing their impacts on the respective walkability spatial distribution. Another important opportunity of the software is the possibility to take into account of the abilities and behaviours of different categories of individuals (age, gender, disabilities, …) by adjusting the parameters (as the weights of attributes) of the assessing model. There is finally one substantial promising prospect opened up by the decision support tool: the potential of developing not only evaluative, but also generative procedures. We envision the possibility, under the user-given objective function and constrains, the decision support tool itself generates hypotheses of projects. Given the in-principle limitless combinatorial alternatives and a vast search space, this of course calls for devising specific search heuristics, which is a stimulating, though intricate, challenge (Blečić et al. 2015b).

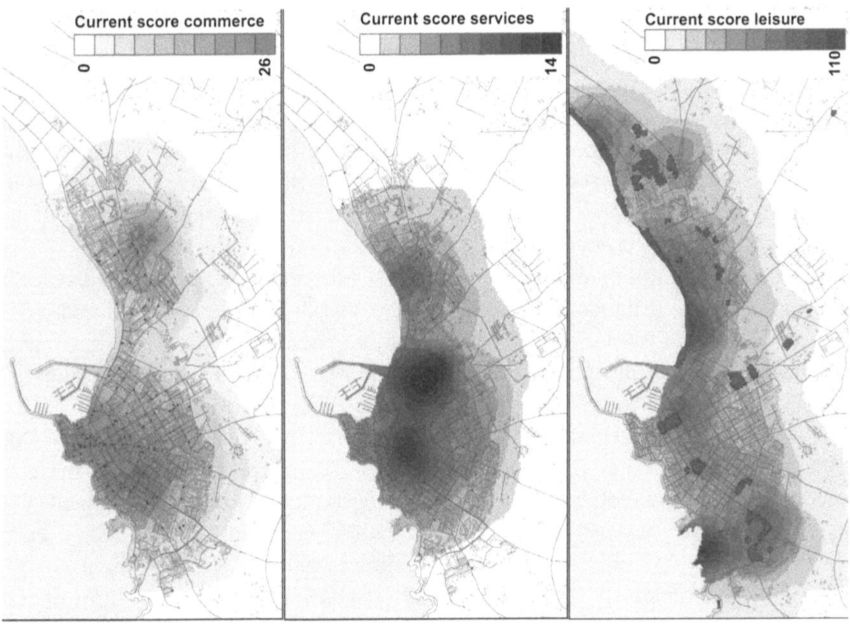

Fig. 4 Walkability scores for the central area of Alghero

4 Verdingtool Platform: The Analysis of the Implementation of Sustainability in Everyday Life

VerdingTool platform[1] is a decision support tool for citizen, companies and public authorities for the choice and implementation of optimization solutions associated with energy consumption, production of wastes and unnecessary consumption of raw materials.

Most people still show resistance to the use of green products and services (Wilson et al. 2015): they often think these solutions to be more expensive than the other ones, only focusing on initial prices. VerdingTool tries to reverse these concepts and makes the total cost of ownership explicit in order to lead the platform users to more informed choices.

VerdingTool analyses goods related to basic issues of the everyday lives of citizens or of SMEs and Public Authorities (Mont et al. 2014). Following this idea, VerdingTool works on widespread goods and services and proposes a simplified approach: the core challenge is not supporting innovative stakeholders for pilot projects, but guiding non-expert citizens towards more sustainable choices.

In doing that, the tool aims to respond directly to the need for unbiased information by many citizens (Wilson et al. 2015) and small businessmen (Woodward 1997). As for the public administration, the tool has a twofold objective of orienting small Public Authorities about specific technological choices and supporting Public Authorities in the dissemination of the concept of sustainability to citizens and companies.

VerdingTool analyses the user's profile of consumption, providing guidance on impacts and cost-cutting strategies and selects appropriate technologies for this purpose. The first function provided by VerdingTool is the assessment of the operating costs of the goods and services selected by the users. In the section function devoted to the specifics, more detailed economic and energetic parameters are supplied, in order to complete the procedure. The clarification of these parameters gives the users feedback for a more informed choice.

The detailed definition of the interventions is left to a second phase: the users can independently ask usual consultants for advice, or ask Verding for a tailored-made support via the platform. For this purpose, some "social tools", such as messaging and thematic groups, have been integrated in the web platform. These "social tools" can also foster behavioural change (Jain et al. 2013).

VerdingTool operates according to the approach of Total Cost of Ownership (TCO), which takes into account the total cost of a good during its life cycle: it considers not only direct, but also all indirect costs (hidden costs) as maintenance, management and administration, implementation, technical support, fixed costs (services, electricity, etc.), environmental impacts and the eventual final cost of disposal. TCO approach allows to ground purchasing decision not only on the

[1]Project co-funded by *Sardegna Ricerche*—Support Programme for Innovative StartUps—II call —POR Sardegna 2007–2013—Line of Action 6.2.1.

initial prices of the products (Wouters et al. 2005), but also on their costs throughout their life-cycle. According to TCO logic investment in innovation can become sustainable.

Considering the current economic environment, VerdingTool looks for saving measures in small steps each of which can have a significant value within a coherent medium-term strategy.

In the final platform set-up, the providers of "Smart Solutions" are expected to play an important role in the system. The "Smart Partners" provide technical information about their products through the platform, and could become partners for the refinement of the evaluation procedure.

The platform is structured into thematic modules (Fig. 5a). The basic function of each module is "Search for savings" (Fig. 5b). It analyses the user's spending profile and identifies existing margin of economic and environmental savings in that field. Each module provides the indication of possible savings and CO_2 emissions

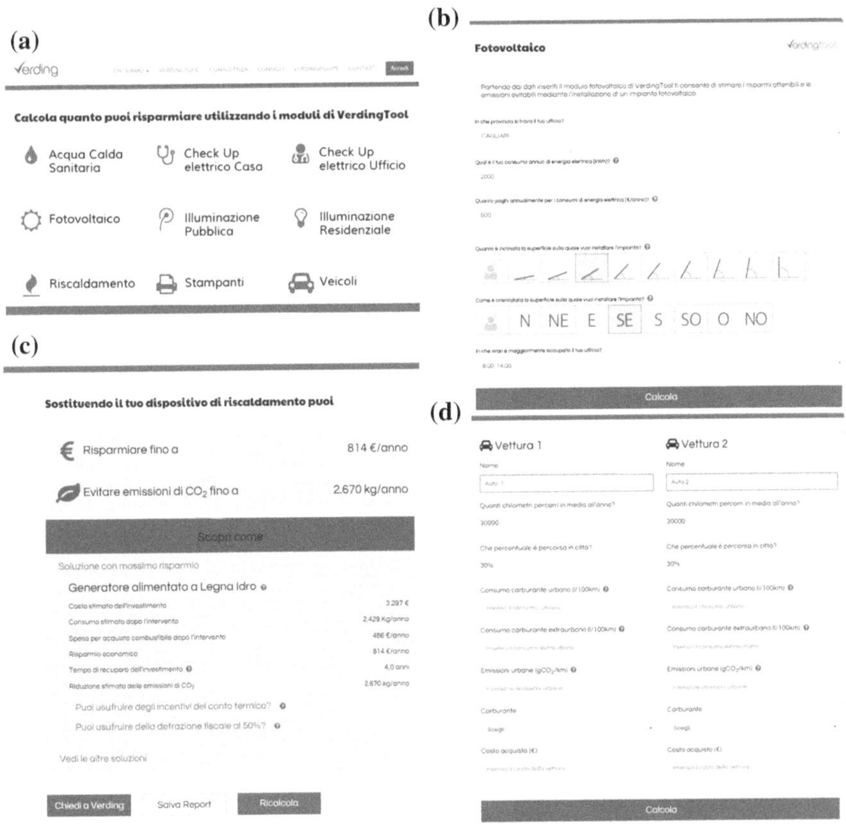

Fig. 5 Screenshots from VerdingTool front end. **a** The main modules of VerdingTool. **b** "Search for savings" function. **c** "Find out more" section. **d** "Compare solutions" function

avoided. In the next section "Find out more" (Fig. 5c) the user can find a more detailed overview of the performance of the possible alternatives. This function has been developed for the existing nine modules.

Once the existing savings margins have been calculated, in some modules another function has been experimentally implemented. It is called "Compare solutions" (Fig. 5d) and it is aimed at comparing two products defined by the user in order to identify the most convenient one. In the section "Find out more" more details of the two solutions and references for further investigation are available.

In the development step of VerdingTool further functions have to be integrated. The most important is called "Test a product": it is designed to assess the economic and environmental effectiveness of specific products available on the market. This function may provide the opportunity for a direct dialogue between users and manufacturers of "Smart solutions".

Beyond functions embedded in each module, the platform also includes the implementation of several cross-cutting services. The main function is "Ask Verding": it is an online help desk for qualified consultancy on issues related to sustainable projects. In connection with this section a glossary about sustainability issues will be created on the platform ("Verding Tips").

In further steps of the platform development two additional cross-cutting functions are implemented: "Design your VerdingPlan" and "Finance Your VerdingPlan", aimed at creating and funding a tailored sustainability plan. Each of them, assesses the potential economic savings and emissions reduction thanks to the substitution of the actual appliances with more efficient ones, comparing the performances of different technologies.

Considering the kind of the processed data, VerdingTool could evolve into a more complex platform aimed at the monitoring of CO_2 emissions in relation to Carbon Footprint assessment or at supporting the building and monitoring of Energy Management Systems (EMS).

5 Conclusion

By presenting these examples, we have stressed the importance of having at one's disposal operational tools for using citizens' contributions for the purpose of supporting public policy processes. For them to allow observing the general principles we have presented in the introduction (openness and inclusivity, transparency, publicity, accountability, equity), there are several technical and functional requirements we believe are important: simplicity, cheapness, universality, customisability and modularity.

The set of tools we proposed tries to take into count these requirements, although there is still much to do to make these tools operational for daily practices useful for the environmental government and planning, especially in respect to their usability and customization.

Also, the tools benefit from the development and evolution of open source tools and applications as well as publicly available data which allow more precise measurements, better reproducibility and easier transferability and comparability of results.

But a crucial question, also in this case, is how to develop a communication strategy for effective involvement of all citizens. Although our tools make an attempt in that direction, questions remain as to what to do about those who do not participate and how to involve Arnstein's "have-nots" in the process, and which strategies to devise to reach them.

There are three groups of people in general who do not participate. Those who do not show interest, do not feel like participating, do not have the necessary capacities nor tools (among whom we find Arnstein's "have-nots" citizens). Then, there are those who hold that the "system" doesn't deserve people's involvement and that the only right way to "fight" it is to "stay out" of it. Third, there are those who don't participate because they have no interest in making decision-making mechanisms more transparent and accountable, quite the opposite, their true interests would not be safeguarded in democratic processes.

Precisely because it doesn't require adhesions and commitments to predefine and pre-chart processes, the mechanisms we have proposed in this paper may be useful in involving some from the first and the second group. There remains the problem of how to kindle and support the participation of the first group, those who neither have the necessary tools nor the capabilities, yet may have a "deep" knowledge of the city and territory.

References

Arnstein SR (1969) A ladder of citizen participation. J Am Plann Assoc 35(4):216–224

Blečić I, Canu D, Cecchini A, Trunfio GA (2014) A web-based participatory management and evaluation support system for urban maintenance, computational science and its applications—ICCSA 2013. Lect Notes Comput Sci 7973(2013):594–605

Blečić I, Cecchini A, Congiu T, Fancello G, Trunfio GA (2015a) Evaluating walkability: a capability-wise planning and design support system. Int J Geogr Inf Sci. doi:10.1080/13658816.2015.1026824

Blečić I, Cecchini A, Trunfio GA (2015b) Towards a design support system for urban walkability. Procedia Comput Sci 51:2157–2167

Bouyssou D, Marchant T, Pirlot M, Tsoukiàs A, Vincke P (2006) Evaluation and decision models with multiple criteria: stepping stones for the analyst, vol 86. Springer Science & Business Media

Capp CJ, Maghelal PK (2011) Walkability: a review of existing pedestrian indices. J Urban Reg Inf Syst Assoc 23(2):5–19

Castells M (2012) Networks of outrage and hope: social movements in the internet age. Polity Press, Cambridge

Cecchini A (2010) Il paesaggio come bene comune (e un esempio di gioco per imparare i beni comuni), Quaderni dell'Istituto Alcide Cervi n.6 "Il paesaggio agrario italiano protostorico e antico" (2010)

Charalabidis Y, Koussouris S, Kipenis L (2009) The 2009 report on the objectives, structure and status of eParticipation initiative projects in the European Union. MOMENTUM Whitepaper
Coleman S (2007) E-Democracy: The history and future of an idea. In: Avgerou C, Mansell R, Quah D, Silverstone R (eds.) The Oxford Handbook of Information and Communication Technologies, Oxford University Press, Oxford, 362–382
Connor DM (2007) A New Ladder of citizen participation. Natl Civic Rev 77(3):249–257
Dorcey A, Doney L, Rueggeberg H (1994) Public involvement in government decision making: choosing the right model. B.C. round table on environment and economy, Victoria
Ewing R, Cervero R (2010) Travel and the built environment. A meta analysis. J Am Plann Assoc 76(3):265–294
Ewing R, Handy S (2009) Measuring the unmeasurable: urban design qualities related to walkability. J Urban Des 14(1):65–84
Iacono M, Krizek KJ, El-Geneidy A (2010) Measuring non-motorized accessibility: issues, alternatives, and execution. J Transp Geogr 18:133–140
Irvin RA, Stansbury J (2004) Citizen participation in decision making: is it worth the effort?, Public Admin Rev 64(1):55–65
Jacobs J (1961) The death and life of great American cities. Random House, New York
Jain RK, Gulbinas R, Taylor JE, Culligan PJ (2013) Can social influence drive energy savings? Detecting the impact of social influence on the energy consumption behavior of networked users exposed to normative eco-feedback. Energy Build 66:119–127
Krizek KJ (2003) Operationalizing neighborhood accessibility for land use-travel behavior research and regional modeling. J Plann Educ Res 22:270–287
Lynch K (1960) The image of the city. MIT Press, Cambridge
Mont O, Neuvonen A, Lähteenoja S (2014) Sustainable lifestyles 2050: stakeholder visions, emerging practices and future research. J Clean Prod 63:24–32
Porta S, Renne JL (2005) Linking urban design to sustainability: formal indicators of social urban sustainability field in Perth, Western Australia. Urban Des Int 10(1):51–64
Pretty JN (1995) Participatory learning for sustainable agriculture. World Dev 23(8):1247–1263
Rawls J (1971) A theory of justice, Claredon Press, Oxford
Rocha EM (1997) A ladder of empowerment. J Plann Educ Res 17(1):31–41
Roy B, Bouyssou D (1993) Aide multicritère à la décision: méthodes et cas. Economica, Paris
Saelens BE, Handy SL (2008) Built environment correlates of walking: a review. Med Sci Sports Exerc 40(7):550–566
Tsoukias A, Montibeller G, Lucertini G, Belton V (2013) Policy analytics: an agenda for research and practice. EURO J Decis Process 1:115–134
Wilson C, Crane L, Chryssochoidis G (2015) Why do homeowners renovate energy efficiently? Contrasting perspectives and implications for policy. Energy Res Soc Sci 7:12–22
Woodward DG (1997) Life cycle costing—theory, information acquisition and application. Int J Project Manage 15(6):335–344
Wouters M, Anderson JC, Wynstra F (2005) The adoption of total cost of ownership for sourcing decisions—a structural equations analysis. Acc Organ Soc 30(2):167–191

FSC
www.fsc.org
MIX
Papier | Fördert gute Waldnutzung
FSC® C083411

Kolibri 360 GmbH
Ferdinand-Jühlke-Straße 7
99095 Erfurt, Deutschland
produktsicherheit@kolibri360.de